ENVIRONMENTAL SOCIOLOGY

From Analysis to Action

FOURTH EDITION

Edited by

LESLIE KING

Smith College

DEBORAH McCARTHY AURIFFEILLE

College of Charleston

ROWMAN & LITTLEFIELD

Lanham • Boulder • New York • London

Executive Editor: Nancy Roberts
Assistant Editor: Megan Manzano
Senior Marketing Manager: Amy Whitaker
Interior Designer: Ilze Lemesis

Credits and acknowledgments for material borrowed from other sources, and reproduced with permission, appear on the appropriate page within the text.

Published by Rowman & Littlefield
An imprint of The Rowman & Littlefield Publishing Group, Inc.
4501 Forbes Boulevard, Suite 200, Lanham, Maryland 20706
www.rowman.com

6 Tinworth Street, London SE11 5AL, United Kingdom

British Library Cataloguing in Publication Information Available

Library of Congress Cataloging-in-Publication Data
Names: King, Leslie, editor. | Auriffeille, Deborah McCarthy. 1966– editor.
Environmental sociology: from analysis to action / [edited by] Leslie King, Smith College,
    Deborah McCarthy Auriffeille, College of Charleston. — 4th Edition. | Lanham :
    ROWMAN & LITTLEFIELD, [2019] | Includes bibliographical references and index.
Identifiers: LCCN 2018044291 (print) | LCCN 2018055545 (ebook) | ISBN 9781538116791
    (electronic) | ISBN 9781538116777 (cloth : alk. paper) | ISBN 9781538116784
    (paper : alk. paper)
Subjects: LCSH: Environmental sociology. | Environmental justice. | Environmentalism—
    North America.
Classification: LCC GE195 (ebook) | LCC GE195. E588 2019 (print) | DDC 333.72—dc23
LC record available at https://lccn.loc.gov/2018044291

♾™ The paper used in this publication meets the minimum requirements of American National Standard for Information Sciences—Permanence of Paper for Printed Library Materials, ANSI/ NISO Z39.48–1992.

Printed in the United States of America

Contents

Acknowledgments

For help, assistance, and great ideas on this fourth edition, we would like to thank first and foremost our student assistants: Zoe Tilden, Samantha Peikes, Kathryn Maurer, and Shea Leibow. Their insightful comments and questions on dozens upon dozens of chapters and articles helped us narrow down our choices. Zoe also helped us proofread and edit the pieces we included. In addition, we thank our editor at Rowman & Littlefield, Nancy Roberts, and her assistant Megan Manzano.

Preface

We both strongly believe that humans have come to a turning point in terms of our destruction of ecological resources and the endangerment of human health. A daily look at the major newspapers points, without fail, to worsening environmental problems and sometimes (but not often enough) a hopeful solution. Humans created these problems, and we have the power to resolve them. Naturally, the longer we wait, the more devastating the problems will become; the more we ignore the sociological dimensions of environmental decline, the more our proposed solutions will fail.

Out of our concern for and dedication to bringing about a more sustainable future, we have worked hard to develop environmental sociology courses that not only educate students about environmental issues but also show them their potential role as facilitators of well-informed change. This reader results in large part from our commitment to the idea that sociology can be a starting point for social change, and we have sought to include in it work that reflects our vision. Sociology, however, can be good at critiquing social arrangements and not as good at highlighting positive change and explaining how that change has come about. We tried to include a few selections that show how groups of people have been able to effect positive changes, but be warned that some of the selections in this reader, reflecting the discipline of sociology, reveal problems in which solutions may seem elusive.

Anthologies seek to accomplish different things. One of our goals has been to provide students and their instructors with shortened versions of fairly recent academic research. Mostly, the articles and chapters included here were originally intended for an academic audience; to make them more accessible to students new to the discipline, we have shortened most of the selections and tried to provide a bit of context for each one.

We actively looked for readings that interest, motivate, and make sense to an undergraduate audience. Choosing which selections to include has been exciting and thought provoking, but it was not without a few dilemmas. For example, a good deal of research in the subdiscipline of environmental sociology is quantitative. Some undergraduate students have the skills to read and interpret this type of work, but many do not. Thus, we have leaned toward qualitative work that is often more accessible to generalist audiences. Our selection process has evolved over the four editions, and this edition is most pointedly focused on pieces that would provoke productive discussion, whether for students in small seminars or for students in larger classes. We do not include "classical" or foundational works; instead, we provide an overview of more recent work in the field to give students a sense of what types of research environmental sociologists are currently engaging in. This field is relatively new and it's evolving quickly—there are many exciting new directions to discover.

In addition, several other good edited volumes and readers include the "classics," so we did not see a need to reinvent the wheel. One of our most difficult decisions was to leave out many "big name" researchers who have profoundly influenced the field. Some of this work represents a dialogue with a long and intertwined body of thought and research. Understanding such a dialogue would require reading the lineage of research leading up to it. In addition, much of the theoretical work in environmental sociology (as in most of our subdisciplines) engages important, but very specialized, issues.

As a way of providing students with a beginning understanding of this lineage, our introductory chapter presents a brief overview of the field for students wishing to explore specific theoretical perspectives in greater depth. The works in the book itself balance this introductory chapter—recent articles and book chapters illustrate a wide variety of ways that sociologists might address environmental questions. The field of environmental sociology has changed dramatically since the first edition of this reader, published in 2004. Our main challenge, as editors, in putting together the first edition was to find enough contemporary pieces. By contrast, the main challenge that we faced in putting together the fourth edition was choosing which excerpts to include given the abundance of great work. This, as they say, is a good problem to have. We have watched the field grow from a fringe sociology subdiscipline to a major force in research on interdisciplinary environmental issues. It's an exciting, and hopeful, time to be an environmental sociologist. With this in mind, the introductory chapter of the fourth edition, and the pieces republished in this volume, reflect changes and growth that have occurred in the discipline over the last several years.

We also wanted this reader to be accessible to a maximum number of instructors, whether or not they are specialists in environmental sociology. Most sociologists and social scientists we know speak the language of inequalities, political economy, and social constructionism; we tend not to be as fluent in the biological and mechanical details of energy production, watershed management, or climate change. Thus, we organized our reader not by environmental issue but by sociological perspective. The reader frames the issues in terms of sociological concepts and seeks to show students how sociologists go about examining environment-related issues. We do want to emphasize, though, that in developing the reader's conceptual blocks, we were careful to cover a broad range of topics—from coal mining to overfishing to climate change.

Ultimately, we think the most important feature of the reader is not the topics we chose or how we decided to organize the different chapters into categories; rather, it is the connection between power and environmental decision making that is woven throughout the collection in the choice of material. Most of the chapters address systems of power (e.g., inequalities in the distribution of toxic waste or who gets blamed for environmental problems, among others). We believe that good environmental decision making must incorporate sociological perspectives, and we hope that activists, policy makers, and academics will benefit from exposure to these frameworks.

Environmental Problems Require Social Solutions

Deborah McCarthy Auriffeille and Leslie King

What Is Environmental Sociology?

What is environmental sociology? The answer, of course, involves exploring two ideas: sociology and environment. Sociology is, above all else, a way of viewing and understanding the social world. It allows us to better understand social organization, inequalities, and all sorts of human interaction. Sociology is a multifaceted discipline that researchers use in diverse ways and, along with many others (e.g., Feagin and Vera 2001), we think it has the potential to help us create a more just world. Like sociology, *environment* can be an elusive term. Is the environment somewhere outside, "in nature," untouched by humans? Or are humans part of the environment? Does it include places where you live and work and what you eat and breathe? Or is it more remote: the rolling valleys of the Blue Ridge Mountains, the pristine waters of Lake Tahoe, the lush rain forests of Brazil?

For environmental sociologists, the answer is that the "environment" encompasses the most remote regions of the earth as well as all the bits and pieces of our daily lives—from the cleaners we use to wash our carpets to the air we breathe on our way to work each day. Most environmental sociologists assume, first and foremost, that humans and nature are part of each other and are part of the environment and that environment and society can only be fully understood in relation to each other. We build on this understanding to point to fissures that are developing in the relationship between humans and nature. These are problems that humans both have contributed to and are feverishly attempting to solve. Our lack of understanding about the human-nature relationship has led to some of our worst environmental problems—climate change, toxic waste, deforestation, and so on—and has limited our ability to solve those problems.

In fact, some attempts to solve environmental problems have actually made them worse. The Green Revolution, initiated in the mid-twentieth century, is an example. Promoted by U.S.-based organizations, including the Rockefeller Foundation, the Green Revolution entailed the transfer of large-scale agricultural technologies and practices first to Mexico and later to several countries in Asia

(Dawson, Martin, and Sikor 2016). The stated goal of the Green Revolution was to increase world food yields through the transfer of Western agricultural techniques, fertilizers, pesticides, knowledge, and equipment to lower-income countries. The resulting shift massively reorganized agricultural production on a global scale. Although the global rates of food production increased, the revolution, with its focus on export crop production, for-profit rather than sustainable agriculture, mechanization, and heavy pesticide and fertilizer use, contributed to the destabilization of social, political, and ecological systems in many regions of the world (Dowie 2001: 106–40). For example, farmers must now engage in a money economy in order to pay for pesticides and herbicides, and, as a result, many small-scale farmers have lost their land (Bell 1998). What is more, as Peter Rosset and colleagues (2000) point out, an increase in food production does not necessarily lead to a decrease in hunger. In their words, "Narrowly focusing on increasing production—as the Green Revolution does—cannot alleviate hunger because it fails to alter the tightly concentrated distribution of economic power, especially access to land and purchasing power. . . . In a nutshell—if the poor don't have the money to buy food, increased production is not going to help them." Importantly, the Green Revolution is not something that happened only in the past. The Green Revolution continues to operate, and its impacts continue to be felt (Sekhon 2017). One recent article documents that some of the negative impacts of the Green Revolution in India include the loss of soil fertility, erosion of soil, soil toxicity, diminishing water resources, pollution of underground water, and increased incidence of human and livestock diseases (Rahman 2015). Recently, Green Revolution strategies are being promoted in African countries where, because it still primarily benefits large farms that are geared toward production rather than subsistence, it is resulting in some of the same deleterious effects, including poverty exacerbation and impairment of local systems of knowledge (Dawson, Martin, and Sikor 2016).

With a better understanding of how political, cultural, and economic structures shape decision making, perhaps we can prevent such problems from occurring again and build a more socially and environmentally sustainable future. The collection of readings in this book represents a broad sample of work by many writers and researchers who are attempting to illuminate social structures and practices with a view toward creating a more socially just and ecologically sustainable world.

Environmental Problems Are Social Problems

Sociologists, by focusing their research on questions of inequality, culture, economy, power, and politics, bring a perspective to environmental questions and problems that is quite different from that of most natural and physical scientists. Take the following examples: the devastating impact of the 2005 Hurricane Katrina on communities in Louisiana and Mississippi; the 2010 Deepwater Horizon oil spill in the Gulf of Mexico; and the 2011 Fukushima Daiichi nuclear disaster in Japan. Why are those not uniquely "natural science" or engineering issues? While scientists and engineers train their lenses on weather patterns, wetland loss, or technological and engineering questions, sociologists look at how social organization—a series of identifiable managerial steps, a collection of beliefs, a set of regulations, or other social structures—contributes to or causes what are often labeled as "natural disasters" or "accidents." In the following summaries, we consider several of these disasters in detail.

- *Hurricane Katrina:* Hurricane Katrina hit the Gulf Coast of Louisiana and Mississippi on August 29, 2005. With a storm surge of 20–32 feet, Katrina did enormous damage; over 1,800 people were killed, and between 700,000 and 1.2 million people were displaced (Gabe et al. 2005; Picou and Marshall 2007). New Orleans was devastated as its levees were breeched and much of the city was flooded. While hurricanes are typically considered "natural disasters," Katrina's extreme consequences must be considered the result of social and political failures. Prior to Katrina, it was known that New Orleans was at risk for flooding in the event of a powerful storm. According to Jenni Bergal (2007: 4), "Numerous studies before Katrina cautioned that storm protection plans weren't moving fast enough, that the levees might not hold in a strong hurricane, that the U.S. Army Corps of Engineers had used out-dated data in its engineering plans to build the levees and floodwalls and that the wetlands buffering the area from storms were disappearing." Coastal land that protected New Orleans had been lost due to human activities including settlement, the building of canals to promote shipping, and the digging of channels for oil and gas pipelines (Hiles 2007). The levee system that provided additional protection was vulnerable due, among other things, to design errors, and the emergency response that should have assisted residents in the aftermath of the storm proved grossly inadequate (Kroll-Smith, Baxter, Jenkins 2015; McQuaid 2007).

- *Deepwater Horizon Oil Spill:* The Deepwater Horizon oil spill of 2010 happened while British Petroleum (BP) was drilling for oil 5,000 feet under the sea floor in the Gulf of Mexico (Krauss 2012). An explosion on the Deepwater Horizon rig destroyed the drilling platform, caused the death of eleven people and resulted in a breached wellhead; oil spilled for three months, and by the time the well was capped, an estimated 4.9 million barrels (or about 200 million gallons) of oil had flowed into the Gulf (Freudenburg and Gramling 2011). In addition, 1.8 million gallons of dispersants to dissolve the oil were applied, and there are serious safety concerns about these dispersants (Foster 2011). The U.S. Department of Interior (2012) reported that the impacts of the spill were extensive and affected "important species and their habitats across a wide swath of the coastal areas of Alabama, Florida, Louisiana, Mississippi, and Texas, and a huge area of open water in the Gulf of Mexico. When injuries to migratory species such as birds, whales, tuna and turtles are considered, the impacts of the Spill could be felt across the United States and around the globe."

   The stage was set for this disaster by a series of human decisions guided by BP's desire to cut costs and the failure of the U.S. government to strictly regulate and monitor the actions of the companies drilling in the Gulf of

Mexico (Freudenburg and Gramling 2011). According to a report by Public Broadcasting Service's (PBS) *Frontline* and ProPublica (PBS 2010) BP, in order to cut costs and increase profits, had a history of failing to prioritize for the safety of workers and the environment. Prior to the Deepwater Horizon disaster, there had been an explosion in one of BP's Texas refineries in 2005 that killed twenty-six people, and there had been a major oil spill in Alaska in 2006 that resulted from a ruptured pipeline. A reporter for the New Orleans *Times Picayune* (Hammer 2010) wrote, "The rig's malfunctioning blowout preventer ultimately failed, but it was needed only because of human errors. . . . The engineers repeatedly chose to take quicker, cheaper and ultimately more dangerous actions, compared with available options. Even when they acknowledged limited risks, they seemed to consider each danger in a vacuum, never thinking the combination of bad choices would add up to a total well blowout." Thus, to understand an event such as the Deepwater Horizon blowout, it is important to consider not only "technological failures" but also the social organization that allows for a series of risky decisions by powerful actors.

- *Fukushima Nuclear Disaster:* In March 2011, a powerful earthquake struck off the coast of Japan in the Pacific Ocean, generating an enormous tsunami and ultimately causing catastrophic meltdowns at the Fukushima Daiichi Nuclear Plant. The tsunami disabled the backup generators that would have been used to cool reactors that were shut down as a result of the earthquake. The resulting disaster was the worst nuclear accident since the meltdown at the Chernobyl plant in Ukraine in 1986. Radiation releases at the Fukushima plant necessitated the evacuation of 90,000 people (Fackler 2012) and caused extensive damage to Japan's food and water supplies.

  The Japanese government and the power company that ran the plant, Tokyo Electric Power Company (TEPCO), were subsequently criticized for failing to institute adequate safety measures, given that the plant was built in an earthquake- and tsunami-prone area (Fackler 2012; Nöggerath, Geller, and Gusiakov 2011). The plant was designed in the 1960s, when knowledge and understanding of earthquakes and tsunamis was somewhat limited; however, since that time, scientists have compiled a substantial amount of data on these phenomena. A group of scientists writing in the *Bulletin of Atomic Sciences* stated, "The knowledge, generally available by about 1980, that magnitude 9 mega-quakes existed as a class should probably have triggered a re-examination of the earthquake and tsunami counter-measures at the Fukushima power station, but it did not" (Nöggerath, Geller, and Gusiakov 2011: 40).

  New York Times journalist Martin Fackler (2012) explained that a cozy relationship between government officials and industry leaders led to a lax regulatory climate. For example, a number of years before the disaster, a seismologist serving on a high-level committee on offshore earthquakes in northeastern Japan "warned that Fukushima's coast was vulnerable to tsunamis more than twice as tall as the forecasts of up to 17 feet put forth by regulators and TEPCO." The seismologist was completely ignored, and to the *New York Times*, he stated, "They completely ignored me in order to save TEPCO money." Similar to the aforementioned examples, the accident at the Fukushima plant was not attributable to "natural" causes (i.e., the earthquake and subsequent tsunami) alone. And while one could argue that design flaws led to the accident, it is also clear that money and power—social relations—played an important role in the decisions that were made by the Japanese government and TEPCO.

Hurricane Katrina, Deepwater Horizon, and Fukushima represent crises that occurred in a flash and left immediately visible human and ecological tragedies in their wakes. Environmental disasters also occur in slow motion and inflict damages that are harder to detect but are no less severe. Such "slow-motion" disasters, including global air pollution and climate warming, among numerous environment-related problems, also have their roots in human decision making and social structure.

- *Air Pollution in the United States:* Indoor and outdoor air pollution is a major environmental problem, with serious health implications for millions of people around the world. According to the World Health Organization (WHO 2014), 7 million premature deaths worldwide could be attributed to air pollution in 2014.

  In the United States, as in other parts of the world, the ongoing danger presented by a host of air pollution problems has roots in social processes, one being the inability of governmental policies and laws to regulate industry amid the rise of neoliberal agendas, which emphasize deregulation and corporate rights. The 1970 Clean Air Act (and its 1977 and 1990 amendments) sets standards for air pollution levels, regulates emissions from stationary sources, calls for state implementation plans for achieving federal standards, and sets emissions standards for motor vehicles (Rosenbaum 2011). While U.S. air quality has improved since 1970 (EPA 2012), the nation's air remains unhealthy with one in four Americans living in counties with unhealthful levels of ozone or particle pollution (American Lung Association 2018). This is especially true of two kinds of air pollution—ground-level ozone (which forms when sunlight reacts with dirty air) and fine particulates (particles smaller than the diameter of a human hair); both of these are created by fuel combustion. Both ground-level ozone and fine particulate matter can cause pulmonary inflammation, decreased lung function, exacerbation of asthma, and other pulmonary diseases. Fine particulate matter is also associated with cardiovascular morbidity and mortality (Laumbach 2010). The U.S. Environmental Protection Agency (EPA 2016) documents that most hazardous air pollutants come from human-made sources, including mobile sources (cars, buses, planes, trucks, and trains) and stationary sources (factories, refineries, and power plants). Hazardous air pollution is not just an engineering problem in need of a solution but is related to how our economy and culture have become increasingly dependent on and intertwined with fossil fuel.

- *Global Climate Change:* While air pollution, as well as other forms of pollution, poses significant challenges to human and ecosystem health, solutions do exist. Laws could be enacted that would have almost immediate positive effects on the quality of air or water. The threat of global climate change from the release of greenhouse gases, on the other hand, is increasing at an alarming rate, also much of the warming is irreversible. Global warming, which is caused primarily by the increasing release of human produced greenhouse gases, like carbon dioxide, which is created in large part by the burning of fossil fuels, cannot easily be turned back (IPCC 2014). Once carbon dioxide is released, it accumulates and will remain in the atmosphere for thousands of years. Even if we stopped human-caused carbon emissions today, it would take forty years for the climate to stabilize at a temperature that is higher than what we now experience (Rood 2014). This means that even with reductions in greenhouse gas emissions, significant changes in climate patterns worldwide are inevitable. The question is not "whether" climate change is happening, the question is how much we can reduce our greenhouse gas emissions so that we can slow the rate of climate change and limit the total overall temperature increase. We must also ask: how can we adapt to the climactic changes that have occurred already and those that are inevitably coming in the future?

The warming of the planet is problematic because it comes with an increased rate of species loss, extreme weather events, coastal flooding, and water scarcity in addition to lowered crop yields, among other ecological problems. These ecological changes pose myriad challenges and risks to human communities including, but not limited to, loss of homes, compromised food supplies, lost work capacity, and increased food- and water-borne diseases (IPCC 2014). While some media pundits would have us question that climate change is problematic or even that it is taking place at all, the U.S. Department of Defense (DoD) is certain. According to a 2018 Pentagon report, the DoD documented and expressed concern that many of their military installations are highly vulnerable to a variety of climate change–related weather events including increases in heat waves, flooding, drought, and wildfires (Office of the Under Secretary of Defense for Acquisition, Technology, and Logistics 2018). Significant warming has already caused "far more damage than most scientists expected," including the loss of summer sea ice in the Arctic, more acidity in the oceans, and a wetter atmosphere above the oceans, increasing the likelihood of floods (McKibben 2012).

When we first published this reader in 2005, we wrote one paragraph detailing global climate change projections, and we discussed some of the likely problems it would bring in the *future*. Fourteen years later, the future is here. Many people around the globe already feel the effects of rising sea levels, flooding, extreme temperatures, and so on. According to the Union of Concerned Scientists (UCS), U.S. coastal communities are experiencing increased climate-related flooding. By 2017, ninety U.S. communities faced "chronic inundation" from flooding, and that number is expected to double in the next twenty years (UCS 2017a). One of the editors of this reader lives in one of these chronically flooded communities, Charleston, South Carolina, where between 2015 and 2017 alone high-tide floods, including sunny day floods, increased from thirty-eight to fifty per year (UCS 2017b). Many communities are also, already, suffering the effects of increased heat waves (UCS 2014a). Witness the 2015 heat wave in Pakistan that killed 1,300. Increasing temperatures have also caused the wildfire season in California to get longer and hotter (UCS 2014a). Look too at climate exacerbated insect infestations that are on the rise; the recent bark beetle infestation in North America killed 46 million acres of trees (more acreage than any other known infestation in North America) (UCS 2014b).

In spite of these terrifying examples, the world's leaders have so far failed to collaborate in any meaningful way. The United States remains the only signatory that has failed to ratify the Kyoto Protocol (an international treaty that commits nations to reduce greenhouse gas emissions); and even those countries that have signed on have found it difficult to agree on specific action plans (Parks and Roberts 2010).

While climate scientists and other specialists can do the work of tracking and predicting weather patterns and describing current and future results of climate change, effective solutions have eluded us because, in the end, climate change is a social, political, and economic problem as much as it is a scientific and technological one. Proposed solutions to climate change and other environmental problems need to be better grounded in the type of sociological research, much of which is already available, that reveals, for example, how national and international political practices and regulatory structures place the short-term interests of corporations over the long-term interests of citizens and, more generally, life on the planet.

Scientists have given the name "Anthropocene"—or the age of humans—to the geologic era dating from the beginning of profound human impact on the ecosphere and biosphere (there is debate over when exactly that began, but certainly human impact has increased dramatically since the beginning of the Industrial Revolution in the early nineteenth century) and into our contemporary time. However, sociologists note that such a label implies that all humans have an equal impact on carbon emissions and climate change; this elides the role, for example, of powerful corporate actors. Thus, historical sociologist Jason Moore (2017) suggests we call the current era the "Capitalocene" to emphasize that the beginning of the "Anthropocene" was enabled by global capitalism. Through the unequal and often destructive workings of global capitalism, human systems have changed the "nature" of earth. What is significant to us about this new epoch is that it is also a moment for global-wide ecological and cultural transformation.

A number of the essays in this fourth edition document activities, both technological and communal, that communities around the world are engaging in to make a difference. In all cases, it is not enough. But they represent a growing global awareness of our special point in history—of the connections between inequality, environmental problems, and economic structures; and the positive potential of our connections to each other, through new communicative forms made available by the Internet, social media, etc. On the other hand, we are at risk of missing this moment—of getting stuck inside our own limited understanding of our contemporary, biased experience of this particular time period. As Ulrich Beck and other "risk society" theorists have pointed out, modern people have come to expect environmental and other risks and the cultural anxiety that comes with living with that insecurity (e.g., Beck 1992). It has become normal. That is the real danger. But true tension does exist between two current ways of being: the way of "getting fed up and making change, any change" and the way of just "accepting that risk comes with modernization."

While the problems outlined in this introduction are profoundly serious, we believe there is still hope. Sociology, as a body of knowledge and as a way of seeing the world, has to be a part of this problem solving. Sociologists have the tools to follow the crumb trail of social facts left behind by climate change and other ecological crises to envision a more just and sustainable future. Why should we assume that human ingenuity begins and ends with the invention of a fossil fuel-based economy? We have put our minds together to invent automobiles (which contribute to sprawl, asthma, and greenhouse gas production), polyvinyl chlorides (which are carcinogenic), and nuclear energy (which has led to the nuclear waste problem). Can we not apply our ingenuity to invent new alternative and equitable forms of energy production and regulate or ban harmful carcinogens like polyvinyl chlorides? We invite you to use the readings in this collection as an opportunity to consider these questions.

Environmental sociologists teach us that environmental problems are inextricably linked to societal issues (such as inequality, governance, and economic practices). Endocrine-disrupting hormones, bioengineered foods, ocean dumping, deforestation, asthma, and so on are each interwoven with economics, politics, culture, television, religious worldviews, advertising, philosophy, and a whole complex tapestry of societal institutions, beliefs, and practices. Environmental sociologists tend not to ask whether something—such as the methyl isocyanate used to produce pesticides in Bhopal, India—is inherently good or bad. Rather, the social organization of the pesticide industry is problematic. Blaming methyl isocyanate for the massive death toll in Bhopal's 1982 factory leak is like cursing the chair after you stub your toe on it. Humans invent pesticides, and humans decide how to manage those chemicals once they have been produced—or they can decide to halt production. Who decides whether to ban or regulate a toxin? How do we decide this? Do some citizens have more say than others? How do the press, corporate advertising, and other forms of media shape our understanding of a dangerous chemical or pollutant? These are just a whisper of the chorus of questions sociologists are asking about environmental problems and their solutions.

Sociology can also help us see that environmental concerns are not merely about individual choices. Some people disregard environment-related problems, explaining that "life itself is a risky business and the issue is ultimately about choice and free will." Along this line, the argument is that just as we choose to engage in any number of risky activities (like downhill skiing without a helmet or eating deep-fried Twinkies), we also choose to risk the increased cancer rates that are associated with dry-cleaning solvents in order to have crisp, well-pressed suits and dress shirts. A sociologist, however, would begin by pointing out that not everyone is involved in that choice and not every community experiences the same level of exposure to that toxin. The person who lives downstream from a solvent factory may not be the same person who chooses the convenience (and the risk) of dry-cleaned clothing (Steingraber 2000).

Finally, sociologists can likewise help us see past the limitations of individualist-based approaches. "Mainstream" environmentalists tend to promote individual-level, consumer-based solutions, such as green purchasing and recycling (Maniates 2002). Perhaps because we feel most comfortable when acting as consumers and because many of us don't know how to engage politically, environmental organizations often propose some variation of "20 Things I Can Do to Save the Planet." Often the suggestions involve "green" purchases, such as hybrid cars or organic food. Consumer-based environmentalism is problematic in that it rarely creates system-wide, structural change. In addition, only an elite few can afford most "green" solutions. Another problematic example of individual-based environmentalism (promoted nearly everywhere as an important environmental action) is recycling. Samantha MacBride (2012) has shown that the growth of the recycling industry allowed corporate interests to define waste as an individual, consumer problem, rather than the responsibility of the corporations that produce all of those bottles, cans, and plastic containers. Recycling has allowed companies to produce "one-way" containers and disposable packaging while not having to bear the cost of its post-use transport and disposal. What is more, recycling is mostly an ineffective solution; once the costs—financial and in carbon emissions—of transporting recyclables are taken into consideration, it is not at all clear that recycling is a net environmental gain. Green purchasing and recycling may help a bit—but they may also be harmful if they draw attention away from the fact that larger structures, like the automobile industry, are encouraging people to, for instance, buy hybrid cars. The development of a hybrid vehicle market, then, potentially, deflects public support and dollars away from investments in alternative transit (Maniates 2002).

A Brief History of Environmental Sociology

In the earlier sections, we defined environmental sociology and showed how sociological understandings of "environment" or "ecology" typically differ from those of the natural and physical sciences and also from mainstream environmentalism. In this section, we provide a broad and, by necessity, partial overview of the subdiscipline of environmental sociology. Auguste Comte first coined the term *sociology* in 1838. However, it wasn't until the late 1960s and the 1970s that a significant number of sociologists began studying the impacts of natural processes on humans and the reverse—the social practices that organize our development and the use of products and technologies that impact nonhuman ecological *and* human communities. Certainly many people in the early part of the twentieth century were interested in a wide range of environmental issues—from the conservation of vast expanses of wilderness to the improvement of urban spaces (Taylor 1997). So why the long wait for an environment-focused sociology?

Part of the answer lies in the efforts of early sociologists to establish the discipline of sociology as separate from other areas of study, especially the natural sciences (Hannigan 1995). Sociology provides an important counterpoint to the natural sciences by showing how social interaction, institutions, and beliefs shape human behavior—not just genetics, physiology, and the "natural" environment. In addition, the sociological perspective has been a crucial tool for dismantling attempts to use the natural sciences to justify ethnocentrism, racism, sexism, and homophobism. Sociologists have traditionally been reluctant, therefore, to venture outside the study of how various social processes (e.g., politics, culture, and economy) interact to look at human-nature interactions.[1] Writing about this trend in the discipline, Riley Dunlap and William Catton argued in the 1970s that sociologists should claim the study of the environment and not leave the "natural" world to natural and physical scientists (Catton and Dunlap 1978; Dunlap 1997). Catton and Dunlap thought environmental sociology ought to examine how humans alter their environments and also how environments affect humans. They developed a "new ecological paradigm," which represented an initial attempt to explore society-environment relations.

During the first decades of environmental sociology (the 1970s and 1980s), researchers, most of them in the United States and Western Europe (Lidskog, Mol, and Oosterveer 2015), focused primarily on the same issues that the emerging environmental movement highlighted, including air and water pollution, solid and hazardous waste dumping, litter, urban decay, the preservation of wild areas and wildlife, and fossil fuel dependence. These problems were easy to measure and see: think of polluted rivers catching on fire, visible and smelly urban smog, ocean dumping of solid and hazardous wastes, and the appearance of refuse along the side of the road. Most early sociological studies focused on people's attitudes toward problems and the impacts of those problems on demographic trends (for instance, trends in health and mortality).[2]

During these early years of the subdiscipline, some environmental sociologists drew a distinction between the "realists," who preferred not to question "the material truth of environmental problems" (Bell 1998: 3), and the "constructionists," who emphasized the creation of meaning—including the meaning of "environment" and "environmental problems"—as a social phenomenon (Bell 1998; Lidskog 2001). Social constructionism—which emphasizes the process through which concepts and beliefs about the world are formed (and reformed) and through which meanings are attached to things and events—has always been a big part of sociology, and a number of environmental sociologists have used this framework to understand environmental issues (e.g., Greider and Garkovich 1994; Hannigan 1995; Burningham and Cooper 1999; Scarce 2000; Yearley 1992). Realists worried that a focus on how meanings are contingent would detract attention from what they saw as real and worrisome environmental degradation. The debate between "realists" and "constructionists" has, however, largely disappeared from the work of environmental sociologists. All sociology is, to some extent, constructionist (Burningham and Cooper 1999). All along, the difference was more in the extent to which the authors *emphasized* the process through which meanings are created.

Potential and currently existing hazards that are socially, politically, and technologically complex, difficult to detect, potentially catastrophic, sometimes long-range in impact, and attributable to multiple causes (e.g., environmental racism, rain forest destruction, loss of biodiversity, technological accidents, and climate change) began to attract the attention of the nascent environmental sociology community beginning in the 1980s.[3] Environmental sociologists of the late twentieth and early twenty-first centuries are studying a broad range of issues—from environmental racism to the international trade in electronic waste to lead poisoning.[4]

Social scientists have developed a number of specialist lenses to explore the increasingly complex relationships between environments and societies. Environmental sociologists frequently focus on power and inequalities. They tend to ask who (which groups or individual actors) have the power not only to make policy decisions but also to create knowledge and set the terms of debate. In addition, most sociologists see all things—from material items to institutions to "nature"—as imbued with meaning by humans.

Environmental sociologists interested in power dynamics often explore intersections between political and economic practices and structures; we often use the term *political economy* to characterize this work. According to Rudel, Roberts, and Carmin (2011: 222), "The political economy of the environment refers to how people control and, periodically, struggle for control over the institutions and organizations that produce and regulate the flows of materials that sustain people (corporations and the state)." Often, researchers using political economy frameworks see environmental devastation and resource exhaustion as inevitable consequences of capital accumulation (the process of increasing the monetary value of an investment).[5]

There are numerous ways to approach a political economy of the environment (see Rudel, Roberts, and Carmin 2011 for a review). Sociologist Brett Clark and colleagues (2018) discuss three theoretical perspectives taken up by researchers concerned with understanding the social and environmental effects of a growth-based economic system.

The "treadmill of production," developed by Allan Schnaiberg (1980), emphasizes the tendency of capitalist production to constantly seek to expand. According to Schnaiberg and his colleagues, this emphasis on growth leads to increased resource consumption as well as the increased generation of wastes and pollutants (both from the by-products of production and from consumption). Thus, according to this perspective, capitalist production, by its very nature, is at odds with efforts to clean up or improve the environment. The exploitation of people and the destruction of resources, however, continue and are often legitimated by institutions like the media (Bonds 2016) because, though everyone suffers in the end, many also benefit in the short term, even if the benefits are unequal.

A second perspective emphasizes metabolic processes, examining the interchange of matter and energy between humans and the larger ecosystem (Clark, Auerbach, and Zhang 2018; Foster 2010). John Bellamy Foster coined the term *metabolic rift* to extend Marx's views on ecological crisis. For Foster, the metabolic rift occurs when through capitalist processes, especially capitalist agriculture, energy is transferred out of an ecological resource and is not replaced. For instance, the overuse of fertilizers can deplete the soil (Clark and Foster 2009; Foster 2010). Researchers using this framework have also examined the rift between other exploitative processes, such as fossil fuel production, and natural ecological processes (Clark and York 2005; Foster and Clark 2012). In all cases, it is assumed that capitalist processes inevitably exceed their own human labor and natural resource needs. The idea of the metabolic rift has more recently been elaborated on to include less "binary" thinking—emphasizing "capitalism *in* nature" rather than "capitalism *and* nature" (Moore 2011). As Jason Moore explains, "Capitalism does not develop upon global nature so much as it emerges through the messy and contingent relations of humans with the rest of nature" (110).

A third perspective has been labeled "ecologically unequal exchange" (Clark, Auerbach, and Zhang 2018) and focuses on inequalities between countries. Historically and currently, the labor and natural resources of countries in the "Global South" (less affluent counties, most of them former colonies) have been exploited by elites in wealthy countries. In addition, Global South countries are increasingly the recipients of unwanted products of industrial production and consumption. For example, electronic waste, which is often hazardous, is routinely exported from wealthy countries to Global South countries (Little and Lucier 2017).

Not all sociologists, however, believe that capitalism is, by its very nature, environmentally exploitative. "Ecological modernization," developed mainly by European social scientists, is less critical of current capitalist political economic systems (e.g., Mol 1997; Spaargarten and Mol 1992). Ecological modernization calls our attention to the ways in which environmental degradation may be reduced or even reversed within our current system of institutions. Theorists working within this framework believe that our institutions may be capable of transforming themselves through the use of increasingly sophisticated technologies and that production processes in the future will have fewer negative environmental consequences. Maurie Cohen (2006) has argued that the inability of the U.S. environmental movement to embrace the principles of ecological modernization to produce change has deprived the United States of some of the sustainability successes witnessed in northern European countries, especially in regard to energy efficiency and the development of clean production technologies.

Another important theoretical perspective developed in large part by European scholars focuses on risk and science. Ulrich Beck (1999), for example, has argued that people in modern times feel increasingly at risk, due in large part to environmental degradation. Beck developed the concept of the *risk society*. According to Beck (1999: 72), we are now at "a phase of development in modern society in which the social, political, ecological and individual risks created by the momentum of innovation increasingly elude the control and protective institutions of industrial society." In other words, even powerful actors in society may be subject to harm from dangerous technologies they themselves have created and/or profited from. Beck examines how risks, and especially the social stresses associated with our perceptions of risks, are fostering deterioration in our quality of life.

The study of risks and the study of health are necessarily interconnected, and many environmental sociologists focus their research on the health consequences of environmental risks and other types of degradation. For example, Kari Marie Norgaard (2007) documents how in a rural region of Northern California controversy arose over the U.S. Forest Service's proposed use of herbicides on spotted knapweed. The Karuk tribe and other community members, who perceived that they would suffer disproportionate health risks, successfully fought the Forest Service's plans for herbicide application. Norgaard shows that the question of who faces risks from modern technology, as well as perceptions about the existence of risks, is influenced by sociological factors such as gender, race, and power. In the case study by Norgaard, conflict arose partly because citizens had been left out of policy deliberations over how to deal with the weed problem. Phil Brown and his colleagues (e.g., Brown and Mikkelsen 1990; Brown 2007), meanwhile, have documented and promoted the idea of "popular epidemiology"—a process whereby nonscientist citizens become active producers and users of scientific data.[6]

Theoretical perspectives such as the treadmill of production, metabolic rift, ecologically unequal exchange, ecological modernization, the risk society, and popular epidemiology have been developed specifically to help us better understand the human-environment nexus. In addition, many environmental sociologies call our attention to environmental inequalities. The study of environmental justice is a major development that derives from a combination of social scientists' long-standing interest in inequalities and social movement activism that has sought to remedy environmental inequities. Finally, sociologists have long been interested in the cause and consequences of social movement activism, and researchers working in this area of sociology have increasingly examined environmental social movement organizations—such as Greenpeace or the Sierra Club—and environmental movement objectives—such as reducing toxins (see Pellow and Brehm 2013 for a review of research in environmental sociology as well as several disciplines that speak to—or have influenced—it).

Lidskog and colleagues detailed in a 2015 article that while environmental sociology research was at first centered in the United States and Europe, in more recent decades, regions across the globe have been contributing to the discipline—especially East Asia and Latin America. Work conducted by researchers in these regions are bringing new perspectives, new issues, and a more global focus to environmental sociology. In Japan, for instance, environmental sociologists focus much of their research on region-specific case studies of victims of environmental pollution, most recently, victims of the Fukushima nuclear disaster. The Fukushima research is especially important as very little sociological research on nuclear power and/or disasters has appeared in the twenty-first century. Brazilian research, on the other hand, has taken a broader focus on socioenvironmental problems in developing countries in general (Lidskog, Mol, and Oosterveer 2015). The Brazilian work on food and agriculture has applicability to the many agrarian regions of the developing world. Lidskog and colleagues express hope that as environmental sociology takes on a more global presence and focus, it will "prevent the development of poorly connected/integrated place-based environmental sociology communities and . . . allow for much more cross fertilization between different traditions, frames and approaches" (Lidskog, Mol, and Oosterveer 2015: 356).

A Brief Look at What's Included in the Reader

This reader provides a sampling of excerpts from recently published sociological research on environment-related issues. The first selection, by Hillary Angelo and Colin Jerolmack, shows how social forces affect how we see and understand "nature" (chapter 1). Next, we include four works that use a political economy perspective, examining, for example, how economic growth often leads to environmental destruction. Some of these authors use ideas developed by Karl Marx. Many of the authors in this section of the reader argue that there is a fundamental conflict between capitalism's need for constant expansion and the protection of the environment, and they are wary of technological solutions that, they argue, often fail to produce safer, healthier environmental conditions.

For example, in a provocative essay, John Bellamy Foster argues that we have on our hands an ecological crisis so dire as to warrant an "ecological revolution" that would also be a social revolution because, he contends (as do many environmental sociologists), without addressing fundamental inequalities, we cannot tackle climate change and other environmental issues (chapter 2).[7] Daniel Faber, in chapter 3, explains how environmental inequalities have deepened as a result of globalization. Stefano B. Longo and Rebecca Clausen use a Marxist framework to examine overfishing of tuna in the Mediterranean (chapter 4). In an article very different from the previous three, Benjamin Vail (chapter 5) describes ecological modernization, the idea that technological advances, within the current capitalist system, will result in necessary changes to bring about sustainability. To do this, he conducts a case study of Sweden's broad-reaching environmental policy and asks to what extent it has been guided by the precepts of ecological modernization. Next, Richard York and his colleagues (chapter 6) use a political economy lens to examine the ecological footprint of China, India, Japan, and the United States from 1962 to 2003. The authors show that, contrary to the ideas of ecological modernization, increased efficiency does not lead to smaller ecological footprints.

In the third section of the reader, we present readings that examine how race, class, and gender intersect with environment. Brett Clark and colleagues use an intersectional lens to study the historic case of guano production in nineteenth-century Peru (chapter 7). Next, Karida Brown, Michael Murphy, and Appollonya Porcelli investigate how African American coal miners created meaning out of their Appalachian landscapes (chapter 8). In chapter 9, Valerie Stull, Michael Bell, and Mpumelelo Ncwadi use the concept of "environmental apartheid" to show how environmental injustices are not just enacted in space but space itself can also be used as a mechanism for marginalization of vulnerable groups. Finally, the piece by Lois Bryson, Kathleen McPhillips, and Kathryn

Robinson (chapter 10) examines the gender and class dimensions of Australian lead pollution policies and public health projects.

Because much of the scientific information that reaches the public does so through the news media, the readings in the next section emphasize the newspaper coverage of environmental issues. In chapter 11, Norah MacKendrick analyzes how the Canadian news media has covered "body burdens" (internal contaminant load that most of us—especially those of us in industrialized nations—carry in our bodies), and in chapter 12, Eric Bonds investigates how the U.S. news media covered (and failed to cover) civilian harm due to the U.S. military's use of open burn pits in Afghanistan and Iraq.

"Disaster" has become an important topic of research for sociologists. In this section, the piece by Liesel Ritchie, Duane Gill, and Steven Picou (chapter 13) reflects on the social psychological impacts of the *Exxon Valdez* and BP oil spills on the people living along affected coastlines. The excerpt by Thomas Beamish (chapter 14) uses organizational theories to illuminate an ongoing, slow-motion "accident" (an oil spill that continued for many years) and how workers and managers in the corporation responsible failed to respond. And Steve Kroll-Smith, Vern Baxter, and Pam Jenkins show how residents of two New Orleans neighborhoods contended with and made sense out of evacuation after Hurricane Katrina (chapter 15).

We find that, as students become aware of environmental and social inequalities and injustices, many naturally want to use their sociological knowledge to change things. Some find it frustrating that while sociology reveals many grave societal problems, it does not tell them what to do about those problems. Indeed, there is no one grand sociological recipe for change. To some extent, just learning more about specific issues is the beginning of change. We have to know about and understand issues such as body burdens in order to even think about doing something about that problem. The last two sections of this reader contain research on social movements and other types of change efforts. Chapter 16, an excerpt from work by Kari Norgaard, is a case study on perceptions about global warming in Norway; the author examines what prevents some people from engaging in environmental change activities. Students reading this excerpt may very well ask themselves what environmental realities they are "choosing" to ignore. In the next chapter, Thomas Shriver, Alison Adams, and Rachel Einwohner investigate citizen action against environmentally destructive state policies under the authoritarian, communist regime of Czechoslovakia (chapter 17). Then, David Pellow (chapter 18) describes how a small group of activists in Mozambique succeeded, with the help of transnational social movement actors, in halting an effort (funded by a Danish development organization) to burn obsolete fertilizer and pesticide stocks.

In the final section of the reader, we provide more examples and ideas for how change might occur. Karen Liftin asks what lessons can be learned from ecovillages—intentional communities in which participants agree to live "lightly" on the earth (chapter 19). Next, Amy Lubitow, Bryan Zinschlag and Nathan Rochester examine bike path planning and show how, when community concerns are not taken into account, even the most popular green ideas will fail. Likewise, they also show how small, community groups, who truly listen to the concerns of residents, can make real change, in this case, increased participation in alternative forms of transit (chapter 20). Focusing a lens on food purchasing by colleges and universities, Peggy Barlett shows how institutions of higher learning are at the forefront imagining new forms of sustainable food production (chapter 21). The last chapter by David Pellow and Hollie Brehm shows how social movements' conceptualization of environment and ecology has changed and developed over the years and how a new frame, which the authors call "total liberation," is emerging. Among other things, this frame merges social justice with ecological concern. The authors conclude, and we agree, that this is the "way to go."

A Final Word

There are two main reasons we wanted to edit this book. First, we know that people are increasingly interested in ecological change. Environmental interest and support is especially evident in academic institutions, and new courses and programs relevant to environmental studies appear each year (Galbraith 2009; Johnston 2012; Jones, Selby, and Sterling 2010; Schmit 2009). Universities and colleges not only offer environmental courses but also are becoming more active as positive models of environmental change. Around the United States, institutions of higher education are increasingly engaging in environmentally sustainable practices (Barlett 2011; Barlett and Chase 2004). Some colleges and universities are working to "green" their dining services by including, for example, more organic and locally produced foods or offering the possibility to compost food waste (Barlett 2011). College campuses are also increasingly stressing energy and water conservation, green building, recycling, and other environment-oriented actions (Barlett 2011; Elefante 2011). As part of our work as professors, we, as well as many of our students, are a part of this ongoing dialogue and activity on our own campuses. We see this book as another way we can contribute to the work we find so important.

Our second reason for compiling this book is to make it easier for nonspecialist sociologists to teach environmental sociology courses. We have spent a decade and a half on our own and many hours in collaboration with each other to design environmental sociology courses that reflect our values and teaching philosophies. We believe that this is a tremendously important field of research and one that students should have the option to explore as part of their college experience. As sociologists continue to learn about the interconnectedness of our ecological problems with other social problems (e.g., toxic waste and racism), it becomes ever clearer that our ecological health is dependent on our understanding of broader social issues (e.g., democracy, inequality, economics, and mass media) and vice versa. The degree to which we progress in one area (or regress) will echo into every other area of social thought and change. Environmental problems involve power relations; they derive from cultural and institutional practices, and they are unequally distributed among populations. Environmental problems require social solutions.

Notes

[1] It is important to note that sociological thought has always, to some degree, included the consideration of ecological factors. For example, some scholars at the Institute for Social Research in Frankfurt worked on environment-related issues, such as the question of domination of nature, in the mid-twentieth century (see Merchant 1999 for a review). Brief summaries of the history of sociological thought on nature and environment can be found in Hannigan (1995) and Buttel and Humphrey (2002). In addition, more recently, environmental sociologists have begun to draw on previously overlooked ideas in classical sociological theory and have revealed that the classical theorists were more ecologically oriented than many subsequent sociologists realized (see Pellow and Nyseth 2013).

[2] Sample titles of articles from sociological journals during this time include "Support for Resource Conservation: A Prediction Model" (Honnold and Nelson 1979); "The Costs of Air Quality Deterioration and Benefits of Air Pollution Control: Estimates of Mortality Costs for Two Pollutants in 40 U.S. Metropolitan Areas" (Liu 1979); "The Public Value for Air Pollution Control: A Needed Change of Emphasis in Opinion Studies" (Dillman and Christenson 1975); and "The Impact of Political Orientation on Environmental Attitudes and Actions" (Dunlap 1975).

[3] Sample titles of articles from sociological journals during the 1980s include "The Social Ecology of Soil Erosion in a Colombian Farming System" (Ashby 1985); "Manufacturing Danger: Fear and Pollution in Industrial Society" (Kaprow 1985); "Cultural Aspects of Environmental Problems: Individualism and Chemical Contamination of Groundwater" (Fitchen 1987); "Blacks and the Environment" (Bullard and Wright 1987); and "Exxon Minerals in Wisconsin: New Patterns of Rural Environmental Conflict" (Gedicks 1988).

[4] For a summary of the history of environmental sociology and the change in focus of environmental studies over time, see Dunlap, Michelson, and Stalker (2002).

[5] For an early overview of this body of theory, see O'Connor (1988: 11–38). For a collection of O'Connor's work, see the 1998 book *Natural Causes: Essays in Ecological Marxism*. More recent work in this area is being conducted by scholars such as John Bellamy Foster and others (see, e.g., the excerpts in this reader).

[6] For example, Brown and Mikkelson's (1990) book, *No Safe Place*, describes how residents of Woburn, Massachusetts, identified and sought remediation for a cancer cluster caused by toxic chemicals in the water supply in certain parts of the town.

[7] With a few notable exceptions (e.g., Spaargarten and Mol 1992), sociologists have strong reservations about the long-term sustainability of capitalism.

References

American Lung Association. 2018. *State of the Air 2018*. Washington, DC: American Lung Association. http://www.lung.org/local-content/california/our-initiatives/state-of-the-air/2018/state-of-the-air-2018.html (accessed June 14, 2018).

Ashby, Jacqueline A. 1985. "The Social Ecology of Soil Erosion in a Colombian Farming System." *Rural Sociology* 50 (3) (Fall): 377–96.

Barlett, Peggy. 2011. "Campus Sustainable Food Projects: Critique and Engagement." *American Anthropologist* 113 (1): 101–15.

Barlett, Peggy F., and Geoffrey W. Chase (eds.). 2004. *Sustainability on Campus: Stories and Strategies for Change*. Cambridge, MA: The MIT Press.

Beck, Ulrich. 1992. *Risk Society: Towards a New Modernity*. London: Sage.

——. 1999. *World Risk Society*. Malden, MA: Polity Press.

Bell, Michael M. 1998. *An Invitation to Environmental Sociology*. Thousand Oaks, CA: Pine Forge Press.

Bergal, Jenni. 2007. "The Storm." In *City Adrift: New Orleans before and after Katrina*, edited by Jenni Bergal, Sara Shipley Hiles, Frank Koughan, John McQuaid, Jim

Morris, Katy Reckdahl, and Curtis Wilkie, 1–6. Baton Rouge: Louisiana State University Press.

Bonds, Eric. 2016. "Legitimating the Environmental Injustices of War: Toxic Exposures and Media Silence in Iraq and Afghanistan." *Environmental Politics* 25 (3): 395–413.

Brown, Phil. 2007. *Toxic Exposures: Contested Illnesses and the Environmental Health Movement*. New York: Columbia University Press.

Brown, Phil, and Edwin J. Mikkelsen. 1990. *No Safe Place: Toxic Waste, Leukemia, and Community Action*. Berkeley: University of California Press.

Bullard, Robert D., and Beverly Hendrix Wright. 1987. "Blacks and the Environment." *Humboldt Journal of Social Relations* 14 (1–2) (Fall–Summer): 165–84.

Burningham, Kate, and Geoff Cooper. 1999. "Being Constructive: Social Constructionism and the Environment." *Sociology* 33 (2): 297–316.

Buttel, Frederick H., and Craig R. Humphrey. 2002. "Sociological Theory and the Natural Environment." In *Handbook of Environmental Sociology*, edited by Riley E. Dunlap and William Michelson, 32–69. Westport, CT: Greenwood Press.

Catton, William R., and Riley E. Dunlap. 1978. "Environmental Sociology: A New Paradigm." *American Sociologist* 13: 41–49.

Clark, Brett, Daniel Auerbach, and Karen Xuan Zhang. 2018. "The Du Bois Nexus: Intersectionality, Political Economy, and Environmental Injustice in the Peruvian Guano Trade in the 1800s." *Environmental Sociology* 4 (1): 54–66.

Clark, Brett, and John Bellamy Foster. 2009. "Ecological Imperialism and the Global Metabolic Rift: Unequal Exchange and the Guano/Nitrates Trade." *International Journal of Comparative Sociology* 50 (3–4): 311–34.

Clark, Brett, and Robert York. 2005. "Carbon Metabolism." *Theory and Society* 34: 391–428.

Cohen, Maurie J. 2006. "Ecological Modernization and Its Discontents: The American Environmental Movement's Resistance to an Innovation-Driven Future." *Futures* 38: 528–47.

Dawson, Neil, Adrian Martin, and Thomas Sikor. 2016. "Green Revolution in Sub-Saharan Africa: Implications of Imposed Innovation for the Wellbeing of Rural Smallholders." *World Development* 78: 204–18.

Dillman, Don A., and James A. Christenson. 1975. "The Public Value for Air Pollution Control: A Needed Change of Emphasis in Opinion Studies." *Cornell Journal of Social Relations* 10 (1) (Spring): 73–95.

Dowie, Mark. 2001. "Chapter 6: Food." In *American Foundations: An Investigative History*, 106–40. Cambridge, MA: The MIT Press.

Dunlap, Riley E. 1975. "The Impact of Political Orientation on Environmental Attitudes and Actions." *Environment and Behavior* 7 (4): 428–54.

———. 1997. "The Evolution of Environmental Sociology: A Brief History of the American Experience." In *The International Handbook of Environmental Sociology*, edited by Michael Redclift and Graham Woodgate, 21–39. Northampton, MA: Edward Elgar.

Dunlap, Riley E., William Michelson, and Glenn Stalker. 2002. "Environmental Sociology: An Introduction." In *Handbook of Environmental Sociology*, edited by Riley E. Dunlap and William Michelson, 1–32. Westport, CT: Greenwood Press.

Elefante, Carl. 2011. "The Full and True Value of Campus Heritage." *Planning for Higher Education* 39 (3): 190–200.

Fackler, Martin. 2012. "Nuclear Disaster in Japan Was Avoidable, Critics Contend." *New York Times*, March 9.

Feagin, Joe, and Hernán Vera. 2001. *Liberation Sociology*. Boulder, CO: Westview Press.

Fitchen, Janet M. 1987. "Cultural Aspects of Environmental Problems: Individualism and Chemical Contamination of Groundwater." *Science, Technology, and Human Values* 12 (2): 1–12.

Foster, Joanna. 2011. "Impact of Gulf Spill's Underwater Dispersants Is Examined." *New York Times*, August 26.

Foster, John Bellamy. 2010. "Why Ecological Revolution?" *Monthly Review* 61 (8): 1–18.

Foster, John Bellamy, and Brett Clark. 2012. "The Planetary Emergency." *Monthly Review* 64 (7): 1–25.

Freudenburg, William R., and Robert Gramling. 2011. *Blowout in the Gulf: The BP Oil Spill Disaster and the Future of Energy in America*. Cambridge, MA: The MIT Press.

Gabe, Thomas, Gene Falk, Maggie McCarty, and Virginia Mason. 2005. *Hurricane Katrina: Social-Demographic Characteristics of Impacted Areas*. Washington, DC: Congressional Research Service, Library of Congress.

Galbraith, Kate. 2009. "Sustainability Field Booms on Campus." *New York Times*, August 19.

Gedicks, Al. 1988. "Exxon Minerals in Wisconsin: New Patterns of Rural Environmental Conflict." *Wisconsin Sociologist* 25 (2–3): 88–103.

Greider, Thomas, and Lorraine Garkovich. 1994. "Landscapes: The Social Construction of Nature and the Environment." *Rural Sociology* 59 (1): 1–24.

Hammer, David. 2010. "5 Key Human Errors, Colossal Mechanical Failure Led to Fatal Gulf Oil Rig Blowout." *Times Picayune*, September 5.

Hannigan, John. 1995. *Environmental Sociology: A Social Constructionist Perspective*. New York: Routledge.

Hiles, Sara Shipley. 2007. "The Environment." In *City Adrift: New Orleans before and after Katrina*, edited by Jenni Bergal, Sara Shipley Hiles, Frank Koughan, John McQuaid, Jim Morris, Katy Reckdahl, and Curtis Wilkie, 7–19. Baton Rouge: Louisiana State University Press.

Honnold, Julie A., and Lynn D. Nelson. 1979. "Support for Resource Conservation: A Prediction Model." *Social Problems* 27 (2): 220–34.

Intergovernmental Panel on Climate Change (IPCC). 2014. *Summary for Policymakers. Climate Change 2014: Impacts, Adaptation, and Vulnerability. Part A: Global and Sectoral Aspects. Contribution of Working Group II to the Fifth Assessment Report of the Intergovernmental Panel on Climate Change*. https://www.ipcc.ch/pdf/assessment-report/ar5/wg2/ar5_wgII_spm_en.pdf (accessed June 17, 2018).

Johnston, Lucas F. (ed.). 2012. *Higher Education for Sustainability: Cases, Challenges, and Opportunities from across the Curriculum*. New York: Routledge.

Jones, Paula, David Selby, and Stephen Sterling (eds.). 2010. *Sustainability Education: Perspectives and Practice across Higher Education*. New York: Routledge.

Kaprow, Miriam Lee. 1985. "Manufacturing Danger: Fear and Pollution in Industrial Society." *American Anthropologist* 87 (2): 342–56.

Krauss, Clifford. 2012. "In BP Indictments, U.S. Shifts to Hold Individuals Accountable." *New York Times*, November 15.

Kroll-Smith, Steve, Baxter, Vern, and Pam Jenkins. 2015. *Left to Chance: Hurricane Katrina and the Story of Two New Orleans Neighborhoods*. Austin: University of Texas Press.

Laumbach, Robert J. 2010. "Outdoor Air Pollutants and Patient Health." *American Family Physician* 81 (2): 175–80.

Lidskog, Rolf. 2001. "The Re-Naturalization of Society? Environmental Challenges for Sociology." *Current Sociology* 49 (1): 113–36.

Lidskog, Rolf, Arthur P. J. Mol, and Peter Oosterveer. 2015. "Towards a Global Environmental Sociology? Legacies, Trends, and Future Directions." *Current Sociology* 63 (3): 339–68.

Little, Peter C., and Christina Lucier. 2017. "Global Electronic Waste, Third Party Certification Standards, and Resisting the Undoing of Environmental Justice Politics." *Human Organization* 76 (3): 204–14.

Liu, Ben-Chieh. 1979. "The Costs of Air Quality Deterioration and Benefits of Air Pollution Control: Estimates of Mortality Costs for Two Pollutants in 40 U.S. Metropolitan Areas." *American Journal of Economics and Sociology* 38 (2): 187–95.

Maniates, Michael. 2002. "Individualization: Plant a Tree, Buy a Bike, Save the World?" in *Confronting Consumption*, edited by T. Princen, M. Maniates, and K. Conca, 43–66. Cambridge, MA: The MIT Press.

MacBride, Samantha. 2012. *Recycling Reconsidered.* Cambridge: The MIT Press.

McKibben, Bill. 2012. "Global Warming's Terrifying New Math." *Rolling Stone*, July 19.

McQuaid, John. 2007. "The Levees." In *City Adrift: New Orleans before and after Katrina*, edited by Jenni Bergal, Sara Shipley Hiles, Frank Koughan, John McQuaid, Jim Morris, Katy Reckdahl, and Curtis Wilkie, 20–41. Baton Rouge: Louisiana State University Press.

Merchant, Carolyn (ed.). 1999. *Key Concepts in Critical Theory: Ecology.* Amherst, NY: Humanity Books.

Mol, Arthur P. 1997. "Ecological Modernization: Industrial Transformations and Environmental Reform." In *The International Handbook of Environmental Sociology*, edited by Michael Redclift and Graham Woodgate, 138–49. Northampton, MA: Edward Elgar.

Moore, Jason. 2011. "Ecology, Capital, and the Nature of Our Times: Accumulation & Crisis in the Capitalist World Ecology." *Journal of World-Systems Research* XVII (1): 107–146.

Moore, Jason. 2017. "The Capitalocene, Part I: On the Nature and Origins of Our Ecological Crisis." *Journal of Peasant Studies* 44 (3): 594–630.

National Research Council. 2010. *Advancing the Science of Climate Change.* Washington, DC: The National Academies Press, National Research Council. http://nas-sites.org/americasclimatechoices/sample-page/panel-reports/87-2 (accessed November 2018).

Nöggerath, Johannis, Robert J. Geller, and Viacheslav K. Gusiakov. 2011. "Fukushima: The Myth of Safety, the Reality of Geoscience." *Bulletin of the Atomic Scientists* 67 (5): 37–46.

Norgaard, Kari Marie. 2007. "The Politics of Invasive Weed Management: Gender, Race, and Risk Perception in Rural California." *Rural Sociology* 72 (3): 450–77.

O'Connor, James. 1988. "Capitalism, Nature, Socialism: A Theoretical Introduction." *Capitalism, Nature, Socialism* 1 (Fall): 11–38.

———. 1998. *Natural Causes: Essays in Ecological Marxism.* New York: Guilford Press.

Office of the Under Secretary of Defense for Acquisition, Technology, Logistics. 2018. Department of Defense Climate-Related Risk to DoD Infrastructure Initial Vulnerability Assessment Survey (SLVAS) Report. January. https://climateandsecurity.files.wordpress.com/2018/01/tab-b-slvas-report-1-24-2018.pdf (accessed June 18, 2018).

Parks, Bradley C., and J. Timmons Roberts. 2010. "Climate Change, Social Theory and Justice." *Theory, Culture and Society* 27 (2–3): 134–66.

Public Broadcasting Service (PBS). 2010. "The Spill." PBS *Frontline* and ProPublica. http://www.pbs.org/wgbh/pages/frontline/the-spill/ (accessed December 2012).

Pellow, David, and Hollie Nyseth Brehm. 2013. "An Environmental Sociology for the 21st Century." *Annual Review of Sociology* 39: 185–212.

Picou, J. Steven, and Brent Marshall. 2007. "Katrina as Paradigm Shift: Reflections on Disaster Research in the Twenty-First Century." In *The Sociology of Katrina: Perspectives on a Modern Catastrophe*, edited by David Brunsma, David Overfelt, and J. Steven Picou, 1–20. New York: Rowman & Littlefield.

Rahman, Saidhur. 2015. "Green Revolution in India: Environmental Degradation and Impact on Livestock." *Asian Journal of Water, Environment and Pollution* 12 (1): 75–80.

Rood, Richard. 2014. "What Would Happen to the Climate if We Stopped Emitting Greenhouse Gases Today?" *The Conversation*, December 11. http://theconversation.com/what-would-happen-to-the-climate-if-we-stopped-emitting-greenhouse-gases-today-35011 (accessed June 18, 2018).

Rosenbaum, Walter A. 2011. *Environmental Politics and Policy*. Washington, DC: CQ Press.

Rosset, Peter, Joseph Collins, and Frances Moore Lappe. 2000. "Lessons from the Green Revolution." *Tikkun Magazine*, March 1.

Rudel, Thomas K., J. Timmons Roberts, and JoAnne Carmin. 2011. "Political Economy of the Environment." *Annual Review of Sociology* 37: 221–38.

Scarce, Rik. 2000. *Fishy Business: Salmon, Biology, and the Social Construction of Nature*. Philadelphia: Temple University Press.

Schmit, Julie. 2009. "As Colleges Add Green Majors and Minors, Classes Fill Up." *USA Today*, December 28.

Schnaiberg, Allan. 1980. *The Environment: From Surplus to Scarcity*. New York: Oxford University Press.

Sekhon, Navreet Kaur. 2017. "An Assessment of Mutual Impact of the Green Revolution, Environmental Crisis and Media in Punjab." *International Journal of Contemporary Sociology* 54 (1): 47–57.

Spaargarten, Gert, and Arthur P. Mol. 1992. "Sociology, Environment, and Modernity: Ecological Modernization as a Theory of Social Change." *Society and Natural Resources* 54 (4): 323–444.

Steingraber, Sandra. 2000. "The Social Production of Cancer: A Walk Upstream." In *Reclaiming the Environmental Debate: The Politics of Health in a Toxic Culture*, edited by Richard Hofrichter, 19–38. Cambridge, MA: The MIT Press.

Taylor, Dorceta E. 1997. "American Environmentalism: The Role of Race, Class and Gender in Shaping Activism 1820–1995." *Race, Gender & Class* 5 (1): 16–62.

Union of Concerned Scientists (UCS). 2014a. "Playing with Fire: The Soaring Costs of Western Wildfires." https://www.ucsusa.org/global_warming/science_and_impacts/impacts/climate-change-development-patterns-wildfire-costs.html#.Wyk60KdKgps (accessed June 19, 2018).

———. 2014b. "Rocky Mountain Forests at Risk." https://www.ucsusa.org/global-warming/science-and-impacts/impacts/climate-change-impacts-rocky-mountain-forests.html#.Wyk44adKgps (accessed June 19, 2018).

———. 2017a. "When Rising Seas Hit Home: Hard Choices Ahead for Hundreds of US Coastal Communities." https://www.ucsusa.org/global-warming/global-warming-impacts/when-rising-seas-hit-home-chronic-inundation-from-sea-level-rise#.Wyk2-6dKgps (accessed June 19, 2018).

———. 2017b. "Historic Communities Face New Challenges as Sea Levels Rise along the South Carolina Coast." https://www.ucsusa.org/global_warming/science_and_impacts/impacts/historic-communities-sea-level-rise-south-carolina-coast#.Wyk-E6dKgps (accessed June 19, 2018)

U.S. Department of Interior. 2012. "Deepwater Horizon Oil Spill Phase I Early Restoration Plan and Environmental Assessment." http://www.doi.gov/deepwaterhorizon/upload/Final-ERP-EA-ES-041812.pdf (accessed December 2012).

U.S. Environmental Protection Agency (EPA). 2012. "Our Nation's Air—Status and Trends through 2010." www.epa.gov/airtrends/2011 (accessed November 9, 2018).

———. 2016. "Technology Transfer Network—Air Toxics Web Site." https://www3.epa.gov/airtoxics/pollsour.html (accessed June 15, 2018).

World Health Organization (WHO). 2011. "Air Quality and Health." Fact Sheet No. 313. http://www.who.int/mediacentre/factsheets/fs313/en/index.html (accessed February 1, 2013).

———. 2014. "Media Centre: 7 Million Premature Deaths Annually Linked to Air Pollution." Media Centre. http://www.who.int/mediacentre/news/releases/2014/air-pollution/en/ (accessed June 14, 2018).

Yearley, Steven. 1992. *The Green Case*. London: Routledge.

PART I IMAGINING NATURE

Nature's Looking Glass 1

Hillary Angelo and Colin Jerolmack

The authors of this first piece use a case study of New York City hawks, Violet and Bobby, and their human webcam followers to illustrate how people hold very different views on the meaning of nature. Some mystify nature and view "pure" nature as separate from people. In this "asocial nature" view, just the presence of humans might despoil nature. Others see a "social nature" in which animals are adaptable rather than mystic. According to this viewpoint, "social nature" is neither pure nor impure; the mix of human society with other animal societies and ecosystems is taken as a given. These differing perspectives can lead to conflicts over how, and whether, humans should intervene in phenomena such as a hawk pair nesting on a New York City ledge. The authors suggest that both views of nature are important. Belief in a "pure nature" may motivate us to protect nature. On the other hand, the understanding that nature is always, at least on some level, social encourages us to appreciate the potential and real biodiversity surrounding us.

After a month of courtship and nest-building, two red-tailed hawks began patiently tending to three speckled eggs in April 2011. Given that red-tailed hawks are a common American species, the event would seem no more than a footnote in the rites of spring; but this nest happened to be on a 12th floor window ledge of NYU's Bobst Library, overlooking Washington Square Park in Manhattan.

Within days, the birds, christened "Bobby" and "Violet," had their own Facebook page and Twitter account, and were stars of a streaming reality television show (thanks to a web camera installed by the *New York Times*). Building on New Yorkers' recent infatuation with the famous 5th Avenue hawks known as Pale Male and Lola, Violet and Bobby soon gathered a devoted following who watched anxiously as the pair raised their hatchling ("Pip") on squirrels and rats captured from the park. The public participated with "tweets," chatroom conversations, blog posts, and face-to-face meet-ups. As articulated by the *Times* and reflected in statements by NYU's president, witnessing the hawks' efforts to start a family—up-close—enabled a rare moment of human encounter with "transcendent" nature in the city.

The environmental historian William Cronon writes that people turn to nature because it is seen as untainted by the "social ills" of civilization. He calls this the "wilderness fantasy." Yet paradoxically, rather than facilitating a socially unspoiled nature experience in the comfort of urban homes, the "hawk cam" instead revealed its impossibility: the band around Violet's swollen foot marked her as a raptor that had already encountered human hands; the Department of Environmental Protection opened up a case file on the hawks, declaring the state's authority over them; hawk devotees debated online whether humans should intervene to remove Violet's band, fortify the nest, or even help break open the eggs; and the hawks' choices of inorganic nesting materials—a thick sheet of white plastic, a soiled hand towel, and artificial Easter grass—frustrated viewers, who criticized the hawks' use of these "unnatural" items.

Although some viewers embraced the hawks as "city birds" whose resourcefulness reflected an admirable adaptation to the urban environment they inhabited, fans seeking an asocial nature experience protested when signs of the social world interrupted their image of the hawks as a kind of miniature wilderness. Much as Frederick Olmsted's design for Central Park had aimed to create an experience of nature that hid the city surrounding it, many hawk watchers had a strong desire to maintain the myth of wild, unsullied nature.

The hawk cam, just one of hundreds of popular web cameras streaming the trials and tribulations of wild animals to a global audience, exemplifies a central paradox of our urbanized society's relationship to the environment: on the one hand, we believe that encounters with nature allow us to transcend social life, and yet, on the other, our experience of nature is profoundly shaped by social forces. To emphasize the fact that our relationships with the environment are always socially mediated, sociologists often place quotes around "nature" or refer to *socio-nature*. While the idea of nature as a realm free of social interests may indeed be a fantasy, as a cultural ideal it nonetheless organizes the ways people experience the environment. The conception of asocial nature, in interaction with the reality of "social nature," impacts how humans interpret and respond to the nonhuman world. Perhaps the incursion of the social world spoils the imagery of Bobby and Violet's nest as a microcosm of "pure" wilderness, but the collective human response to this alleged interruption is itself a microcosm of our social struggles over how to live with nature.

Asocial Nature

Classical sociological concerns regarding modernization have centered on topics like people's migration from rural villages to large cities, the rise of capitalism, and the reorganization of community life. More recently, though, social scientists have begun looking at how society's relationship to the physical environment has been transformed. After all, the Industrial Revolution, Karl Marx and Friedrich Engels observed, was founded on the "subjection of nature's forces to man."

Much as the rationalization of work and social life was said to produce a sort of collective malaise (what social theorist Max Weber called "disenchantment"), sociologist James Gibson tells a parallel story of contemporary society's increasing estrangement from nature. Gibson reports that many pre-modern societies revered nature and believed it was animated by spirits, but that the rise of science promoted a mechanistic view of nature "as inert matter." In turn, capitalism reduced nature to a commodity. Gibson and other environmental scholars argue that the physical defilement of the natural world and people's increasing geographic separation from nature through urbanization have led to nature's spiritual profanation. The art and literary critic John Berger looks to our relationships with animals as clear evidence of this estrangement: the transformation of the wolf to working farm dog and then to handbag Chihuahua simultaneously marks the retreat of wild, asocial nature and the disappearance of our respect and awe for it. Scholars like Gibson and Berger see our desecration and humanization of nature as harmful to both the environment (which we are socialized to exploit as a means rather than an end) and society (which is spiritually impoverished because the plants and animals around us are no longer endowed with other-worldly significance).

If the entrance of social interests into our relationship with nature despoils it conceptually and physically, then solving the ecological crisis requires that we supersede instrumentalist orientations and reinvest nature with sacred meaning. Gibson believes it's already happening. He finds a growing number of people "who long to rediscover and embrace nature's mystery and grandeur" and "who look to nature for psychic regeneration and renewal." Gibson interprets New Yorkers' fascination with the red-tailed hawks Pale Male and Lola—and, presumably, Bobby and Violet—as an expression of their desire to seek momentary sanctuary from the concrete jungle in the experience of "kinship" with nature. In his view, the hawks are ambassadors of the wild and our celebration of them is a consecration of nature, a transcendence of the social. Strands of this "culture of enchantment" are evident in the narratives of the environmental movement, which often focuses its attention on the majesty of rugged mountain peaks or charismatic fauna (rather than, say, a suburban park or snails). The implied ideal is preserving, or even reconstructing, "pristine" nature.

The enchanted nature thesis guides many preservationists' efforts and compellingly captures the asocial wilderness ideal that many use to frame their interpretations of the nonhuman world. Yet, as Cronon has said, by reproducing "the dualism that sets humanity and nature at opposite poles," this quixotic standard does not offer a blueprint of "what an ethical, sustainable, *honorable* human place in nature might look like." As Gibson concedes, "unrealistic expectations of purity" can impede appreciation and respect for the hybrid landscapes we actually inhabit. Stories such as *Grizzly Man* and *Into the Wild* show how efforts to realize the wilderness fantasy can in fact be harmful or tragic. By threatening to pigeonhole socially mediated relations with nature—which, in reality, includes *all* relations with nature—as somehow less "pure" or "meaningful" than nature encounters that appear to be asocial, this perspective misses an opportunity to examine how the explicit comingling of the natural and the social can be beneficial for both the environment and society.

Social Nature

A second sociological tradition, dating back to classical sociologist Emile Durkheim, highlights how the social and the natural are inextricably intertwined.

In studying pre-modern aboriginal clans that practiced totemism, a belief system in which a clan adopts a particular plant or animal as its symbol and considers it sacred, Durkheim rejected the assumption that aborigines actually worshipped nature. The totem species was sacred, rather, because it stood for the clan; worshipping it was a way of expressing social solidarity. Though perhaps in different ways, our relations to the nonhuman world remain inherently social. Every landscape, sociologists Thomas Greider and Lorraine Garkovich write, is a "symbolic environment created by the human act of conferring meaning." The meanings we ascribe to the environment reflect our self-definitions and are grounded in particular social contexts. For example, in studying an English exurban village, sociologist Michael Bell found that residents identified as "country people" who, by living close to nature, believed that they led more authentic and wholesome lives than city dwellers. Still, Bell found important differences in villagers' relationships with nature that were patterned by their class position. Wealthier residents, for instance, prized open vistas of the countryside and sculpted their backyards to create clear sight lines, but working-class residents had a more hands-on relationship with the land and paid less attention to "the view," even planting shrubs that impeded it. Though all of the villagers believed nature offered them an escape from the ills of society, in reality their experience of the environment still mapped onto familiar social categories.

"Social nature" analyses undermine the wilderness fantasy by exposing how the social and natural worlds are conjointly constituted. They also help us understand the social motivations behind idealizing nature. Bell found that salient class tensions threatened to undermine group solidarity in the village he studied, and that grounding their sense of selves in nature was a way for villagers to overcome social divisions and rally around a shared identity—a sort of secular totemism. Historically, Cronon argues that urban elites in the U.S.—who themselves never lived close to the land—cultivated the wilderness fantasy as a national project because untamed nature was a monument to America's heroic frontier myth and "the last bastion of rugged individualism." Culture and context always frame our understandings of nature. The idea of asocial nature is itself a social frame.

The principle that nature is just as social as race or gender challenged assumptions about environment-society relations and helped open up a domain of inquiry long considered the realm of the natural sciences. It forces us to see the encroachment of the social world into Bobby and Violet's nest as an inherent part of our experience of nature, not contamination. The chatroom discussions, "expert" opinions rendered in the news, and the face-to-face meetups under the nest aren't ancillary to the "first hand" experience of the hawks—they are the means by which the hawks become meaningful and knowable to people.

While the pristine, asocial nature ideal spurred important environmental legislation like the creation of treasured national parks, scholars like Cronon point to the troubling social and material consequences of the wilderness fantasy: it has too often meant evicting those who inhabited the land (like Native Americans and farmers) and erasing any trace of their existence; and it can foster apathy toward the "impure" environments that most humans actually inhabit. Further, the "not in my backyard" environmental activism of the middle and upper classes may lead to the siting of toxic facilities in poor and minority communities. "Social nature" analyses can help us see environmental justice and social justice as inherently linked ideals.

Scholarship on the social experience of nature falls short of its potential if it simply debunks the wilderness fantasy or subjects it to an ideological critique. For, while the notion that nature is always social may now be sociological commonsense, the experience of nature that people strive to create and maintain is an asocial one. Bell acknowledges that we must continue to account for the imaginative pull of this paradigmatic frame in the urbanized Western world. For most people, there is something fundamentally distinct about encountering the "wild." We can warn hawk-cam viewers that their quest for asocial nature is foolhardy, yet they will still fix their eyes on the screen hoping to capture the moment when Violet feeds her baby beak-to-beak. Interactions with superficially asocial nature in the city—a hike on Central Park's wooded trails or a glimpse of soaring red-tailed hawks—give us an experiential analog that we reflexively connect to a grander image of the kind of nature we wish to preserve and immerse ourselves in, the rugged mountain peak or lion on the savannah we may never know. Few consider the animal-filled circus or weeds in a vacant lot "nature experiences," and most of Violet and Bobby's fans remain unmoved by the pigeons picking at the remains of a bagel below the hawks' nest.

Frame Breaks

If asocial nature is an interpretive frame, then what sociologist Erving Goffman called "frame breaks"—moments when the taken-for-granted way we see the world is disrupted—are telling. Because these interruptions contain *both* the enchanted ideal and the social reality, they provide analytically rich data and moments for education and reflection.

Environmental educators are fluent in putting these contrasts to work. A visit to a natural history museum is unlike wilderness immersion in a wooded park in part because, in order to educate people about "nature," the exhibits quite deliberately play with the tension between social and asocial nature frames. Like going to a movie, habitat dioramas (or, more recently, IMAX films) are self-consciously short-term immersive experiences explicitly designed to mimic an at-risk "natural" environment in order to show us what we would be missing if it disappeared. They are effective in part because the frame is *visible*: stepping back, you see the edges of the glass, hear the echoing cries of the children around you, and feel the wilderness recede. There is no question that this window is an illustration, that the dioramas use a clearly fabricated image of asocial nature to provoke critical reflection about humanity's alteration of the environment. Although proponents of animal welfare decry the collision of caged animals and entertainment, zoos increasingly rely on similar visual techniques to try to educate about conservation through what might be called "living dioramas."

When it comes to the camera that frames Bobby and Violet, however, sublime nature seekers treat it not as a manmade window that illustrates some aspect of the natural world, but as a window into "real," asocial nature. That is why viewers are apt to complain when human hands intervene in Violet's nest, yet watch in fascination as a zookeeper feeds peanuts to an elephant or a museum employee exposes a diorama's artificiality by entering to repair a perfectly rendered plant or animal form.

Yet the frame breaks that hawk-cam viewers continually confront need not be a tragedy. Like a diorama, these moments can foster awareness and even appreciation for the ways that the natural and the social are co-constituted. The presence of a plastic bag in Violet's nest can be a platform for discussing the adaptability of animals (and humans) to built environments and for thinking about how we can alter the cityscape to create a more sustainable habitat for "wild" species. And, rather than deploring the Department of Environmental Protection's intervention in Bobby and Violet's affairs as unnatural or simply holding it up as evidence that nature is always social, we can use the moment to illuminate the constellation of institutions that influence our relationship with other species in the city. Understanding this organizational ecology is more than an academic exercise; it can help environmental advocates find their bureaucratic allies and apply pressure to their opponents. It can also reveal the social structures that govern how access to "natural goods" like clean air, safe drinking water, green spaces, and farmer's markets is distributed in the city.

Spared the toil of the farm or the awesome but terrifying experience of taming the wilderness frontier, contemporary urbanites idealize nature as an environment uncorrupted by people. And though that asocial nature is a chimera, our *belief* in it often serves as a prime motivator for helping the environment. It infuses our encounters with the "wild" with a sense of mystery. By the same token, seeing the social in nature need not threaten our desire to associate with and protect it. The hawk chatroom and in-person meet-ups continued long after Pip fled the nest. The community that sprang up around the hawks abetted people's sense of connection to Bobby, Violet, their baby, and, in some cases, the urban ecology around them. And catching a glimpse of Pip eating her first pigeon becomes meaningful not only because hawk devotees bear witness to the cycle of life, but also because they can blog about it or talk about it in person with each other.

Our own research highlights how the social experience of nature can enhance our appreciation of it and how asocial and social experiences of nature are interconnected. In studying a group of Turkish male immigrants in Berlin who kept domestic pigeons, for instance, Jerolmack was able to show how the birds enabled the men to experience a connection to nature in the midst of the city. Yet the men also said they appreciated keeping pigeons in Berlin because it enabled them to experience a connection to their homeland and express their ethnicity. This social tie to *nation* augmented, rather than mitigated, the immigrants' enjoyment of *nature*. And, in a study of two very different types of bird enthusiasts, Angelo demonstrated the interdependence of asocial and social nature experiences. While ornithologists draw the ire of birdwatchers by shooting and skinning birds to create specimens, their social practices produce the knowledge—in the form of field guides—that birdwatchers rely upon to identify avian species while imaginatively immersing themselves in "wildernesses" big and small. Socially mediated "nature," in the form of bird specimens in a scientific collection, enables the asocial nature experience that birdwatchers enjoy through binoculars. Both practices—though very different—can be understood as driven by, and resulting in, admiration for the nonhuman world.

The journalist Robert Sullivan writes that nature is prospering in cities like New York (Queens contains more species of birds than Yellowstone and Yosemite Parks combined), but that many residents haven't noticed because this isn't the nature they are looking for—it is less precious, less "pure." Yet these species, many of which have learned to adapt their behaviors to the habits of people, powerfully demonstrate the extent to which the social and the natural are intertwined. Pigeons and squirrels "beg" humans for food in Washington Square, and crows crack open nuts by dropping them into car traffic. Given the likelihood of continued ecological disruption, ecologists predict a future in which more and more species will fashion their survival strategies around human societies. Learning to appreciate and promote the biodiversity hiding in our cities expands the terrain of environmental conservation and may in fact offer clues for how to model sustainable human and nonhuman cohabitation.

William Cronon writes that, while humans must be conscious of our place in the natural world in order to live mindfully within it, we must also "recognize and honor nonhuman nature as a world we did not create." The magical experience of asocial nature is not something to blindly embrace, nor is it something that we should—or could—rid ourselves of. Rather than conceptualizing "asocial nature" and "social nature" as moral or experiential opposites, we would do better to think of them as overlapping modes of experience that can both foster greater appreciation of, and concern for, the environment *and* society.

Note

Hillary Angelo and Colin Jerolmack. 2012. "Nature's Looking Glass." *Contexts* 11 (1): 24–29.

References

Bell, Michael. 1994. *Childerley: Nature and Morality in a Country Village*. Chicago: University of Chicago Press. Describes how a rural community uses the idea of asocial nature to construct an image of themselves as moral people.

Cronon, William. 1996. "The Trouble with Wilderness." *Environmental History* 1 (1): 7–28. Examines the origins, and negative environmental consequences, of the wilderness ideal.

Freudenburg, William R., Scott Frickel, Robert Gramling. 1995. "Beyond the Society/Nature Divide: Learning to Think about a Mountain." *Sociological Forum* 10 (3): 361–92. Surveys the ways in which the social and the natural are conjointly constituted.

Gibson, James. 2009. *A Reenchanted World: The Quest for a New Kinship with Nature*. New York: Holt. Frames recent environmentalist efforts to protect animals and open spaces as reflecting a "culture of enchantment" that attempts to make nature sacred once again.

PART **II** POLITICAL ECONOMY

Why Ecological Revolution? 2

John Bellamy Foster

In the next four selections, the authors examine the intersections between politics, economics, and the environment. The sociological study of politics and economics challenges our tendency to take current systems, like capitalism, for granted by critically examining how power and wealth intersect with ecology and environment. Many environmental sociologists draw on the ideas of classical political/economic theorists, like Karl Marx and Friedrich Engels, to argue that capitalism's need for constant expansion contributes to resource depletion and environmental degradation. The articles in this section encourage us to question a modern global, capitalist economic system that must constantly expand and grow in order to thrive. Although the messages of these articles are often cautionary, they can also be read as positive. For, while the current levels and future predictions of environmental degradation are frightening, they are human made. And if we can make a system of production and consumption that harms the environment then we can unmake it.

John Bellamy Foster's chapter sounds the alarm bells. He argues that the earth is experiencing not just an urgent ecological crisis, but potentially the final crisis. Drawing on research indicating that environmental problems, like global climate change, are accelerating and in danger of reaching irreversible "tipping points," Foster contends that technological fixes are not enough and only an ecological revolution—a major change in our social relations and our relationship to the environment—will save our planet, and ourselves, from collapse.

It is now universally recognized within science that humanity is confronting the prospect—if we do not soon change course—of a planetary ecological collapse. Not only is the global ecological crisis becoming more and more severe, with the time in which to address it fast running out, but the dominant environmental strategies are also forms of denial, demonstrably doomed to fail, judging by their own limited objectives. This tragic failure, I will argue, can be attributed to the refusal of the powers that be to address the roots of the ecological problem in capitalist production and the resulting necessity of ecological and social revolution.

The term "crisis," attached to the global ecological problem, although unavoidable, is somewhat misleading, given its dominant economic associations. Since 2008, we have been living through a world economic crisis—the worst economic downturn since the 1930s. This has been a source of untold suffering for hundreds of millions, indeed billions, of people. But insofar as it is related to the business cycle and not to long-term factors, expectations are that it is temporary and will end, to be followed by a period of economic recovery and growth—until the advent of the next crisis. Capitalism is, in this sense, a crisis-ridden, cyclical economic system. Even if we were to go further, to conclude that the present crisis of accumulation is part of a long-term economic stagnation of the system—that is, a slowdown of the trend-rate of growth beyond the mere business cycle—we would still see this as a partial, historically limited calamity, raising, at most, the question of the future of the present system of production.[1]

When we speak today of the world ecological crisis, however, we are referring to something that could turn out to be *final*, i.e., there is a high probability, if we do not quickly change course, of a *terminal crisis*—a death of the whole Anthropocene, the period of human dominance of the planet. Human actions are generating environmental changes that threaten the extermination of most species on the planet, along with civilization, and conceivably our own species as well.

What makes the current ecological situation so serious is that climate change, arising from human-generated increases in greenhouse gas emissions, is not occurring gradually and in a linear process, but is undergoing a dangerous acceleration, pointing to sudden shifts in the state of the earth system. We can therefore speak, to quote James Hansen, director of NASA's Goddard Institute of Space Studies, and the world's most famous climate scientist, of "tipping points . . . fed by amplifying feedbacks."[2] Four amplifying feedbacks are significant at present: (1) rapid melting of arctic sea ice, with the resulting reduction of the earth's albedo (reflection of solar radiation) due to the replacement of bright, reflective ice with darker blue sea water, leading to greater absorption of solar energy and increasing global average temperatures; (2) melting of the frozen tundra in northern regions, releasing methane (a much more potent greenhouse gas than carbon dioxide) trapped beneath the surface, causing accelerated warming; (3) recent indications that there has been a drop in the efficiency of the carbon absorption of the world's oceans since the 1980s, and particularly since 2000, due to growing ocean acidification (from past carbon absorption), resulting in faster carbon build-up in the atmosphere and enhanced warming; (4) extinction of species due to changing climate zones, leading to the collapse of ecosystems dependent on these species, and the death of still more species.[3]

Due to this acceleration of climate change, the time line in which to act before calamities hit, and before climate change increasingly escapes our control, is extremely short. . . .

Many of the planetary dangers associated with current global warming trends are by now well-known: rising sea levels engulfing islands and lowlying coastal regions throughout the globe; loss of tropical forests; destruction of coral reefs; a "sixth extinction" rivaling the great die-downs in the history of the planet; massive crop losses; extreme weather events; spreading hunger and disease. But these dangers are heightened by the fact that climate change is not the entirety of the world ecological crisis. For example, independently of climate change, tropical forests are being cleared as a direct result of the search for profits. Soil destruction is occurring, due to current agribusiness practices. Toxic wastes are being diffused throughout the environment. Nitrogen run-off from the overuse of fertilizer is affecting lakes, rivers, and ocean regions, contributing to oxygen-poor "dead zones."

Since the whole earth is affected by the vast scale of human impact on the environment in complex and unpredictable ways, even more serious catastrophes could conceivably be set in motion. One growing area of concern is ocean acidification due to rising carbon dioxide emissions. As carbon dioxide dissolves, it turns into carbonic acid, making the oceans more acidic. Because carbon dioxide dissolves more readily in cold than in warm water, the cold waters of the arctic are becoming acidic at an unprecedented rate. Within a decade, the waters near the North Pole could become so corrosive as to dissolve the living shells of shellfish, affecting the entire ocean food chain. At the same time, ocean acidification appears to be reducing the carbon uptake of the oceans, speeding up global warming.[4]

There are endless predictive uncertainties in all of this. Nevertheless, evidence is mounting that the continuation of current trends is unsustainable, even in the short-term. The only rational answer, then, is a radical change of course. Moreover, given certain imminent tipping points, there is no time to be lost. Catastrophic changes in the earth system could be set irreversibly in motion within a few decades, at most.

The IPCC [United Nations Intergovernmental Panel on Climate Change], in its 2007 report, indicated that an atmospheric carbon dioxide level of 450 parts per million (ppm) should not be exceeded, and implied that this was the fail-safe point for carbon stabilization. But these findings are already out of date. "What science has revealed in the past few years," Hansen states, "is that the safe level of carbon dioxide in the long run is no more than 350 ppm," as compared with 387 ppm today. That means that carbon emissions have to be reduced faster and more drastically than originally thought, to bring the overall carbon concentration in the atmosphere down. The reality is that, "if we burn all the fossil fuels, or even half of remaining reserves, we will send the planet toward the ice-free state with sea level about 250 feet higher than today. It would take time for complete ice sheet disintegration to occur, but a chaotic situation would be created with changes occurring out of control of future generations." More than eighty of the world's poorest and most climate-vulnerable countries have now declared that carbon dioxide atmospheric concentration levels must be reduced below 350 ppm, and that the rise in global average temperature by century's end must not exceed 1.5°C.[5]

Strategies of Denial

The central issue that we have to confront, therefore, is devising social strategies to address the world ecological crisis. Not only do the solutions have to be large enough to deal with the problem, but also all of this must take place on a world scale in a generation or so. The speed and scale of change necessary means that what is required is an ecological revolution that would also need to be a social revolution. However, rather than addressing the real roots of the crisis and drawing the appropriate conclusions, the dominant response is to avoid all questions about the nature of our society, and to turn to technological fixes or market mechanisms of one sort or another. In this respect, there is a certain continuity of thought between those who deny the climate change problem altogether, and those who, while acknowledging the severity of the problem at one level, nevertheless deny that it requires a revolution in our social system.

We are increasingly led to believe that the answers to climate change are primarily to be found in new energy technology, specifically increased energy and carbon efficiencies in both production and consumption. Technology in this sense, however, is often viewed abstractly as a *deus ex machina*, separated from both the laws of physics (i.e., entropy or the second law of thermodynamics) and from the way technology is embedded in historically specific conditions. With respect to the latter, it is worth noting that, under the present economic system, increases in energy efficiency normally lead to increases in the scale of economic output, effectively negating any gains from the standpoint of resource use or carbon efficiency—a problem known as the "Jevons Paradox.". . .

Technological fetishism with regard to environmental issues is usually coupled with a form of market fetishism. So widespread has this become that even a militant ecologist like Bill McKibben, author of *The End of Nature*, recently stated: "There is only one lever even possibly big enough to make our system move as fast as it needs to, and that's the force of markets."[6]

Green-market fetishism is most evident in what is called "cap and trade"—a catch phrase for the creation, via governments, of artificial markets in carbon trading and so-called "offsets." The important thing to know about cap and trade is that it is a proven failure. Although enacted in Europe as part of the implementation of the Kyoto Protocol, it has failed where it was supposed to count: in reducing emissions. Carbon-trading schemes have been shown to be full of holes. Offsets allow all sorts of dubious forms of trading that have no effect on emissions. Indeed, the only area in which carbon trading schemes have actually been effective is in promoting profits for speculators and corporations, which are therefore frequently supportive of them. Recently, Friends of the Earth released a report entitled *Subprime Carbon?* which pointed to the emergence, under cap and trade agreements, of what could turn out to be the world's largest financial derivatives market in the form of carbon trading. All of this has caused Hansen to refer to cap and trade as "the temple of doom," locking in "disasters for our children and grandchildren."[7]

. . .

Recognizing that world powers are playing the role of Nero as Rome burns, James Lovelock, the earth system scientist famous for his Gaia hypothesis, argues that massive climate change and the destruction of human civilization as we know it may now be irreversible. Nevertheless, he proposes as "solutions" either a massive building of nuclear power plants all over the world (closing his eyes to the enormous dangers accompanying such a course)—or geoengineering our way out of the problem, by using the world's fleet of aircraft to inject huge quantities of sulfur dioxide into the stratosphere to block a portion of the incoming sunlight, reducing the solar energy reaching the earth. Another common geoengineering proposal includes dumping iron filings throughout the ocean to increase its carbon-absorbing properties.

Rational scientists recognize that interventions in the earth system on the scale envisioned by geoengineering schemes (for example, blocking sunlight) have their own massive, unforeseen consequences. Nor could such schemes solve the crisis. The dumping of massive quantities of sulfur dioxide into the stratosphere would, even if effective, have to be done again and again, on an increasing scale, if the underlying problem of cutting greenhouse gas emissions were not dealt with. Moreover, it could not possibly solve other problems associated with massive carbon dioxide emissions, such as the acidification of the oceans.[8]

The dominant approach to the world ecological crisis, focusing on technological fixes and market mechanisms, is thus a kind of denial; one that serves the vested interests of those who have the most to lose from a change in economic arrangements. Al Gore exemplifies the dominant form of denial in his new book, *Our Choice: A Plan to Solve the Climate Crisis*. For Gore, the answer is the creation of a "sustainable capitalism." He is not, however, altogether blind to the faults of the present system. He describes climate change as the "greatest market failure in history" and decries the "short-term" perspective of present-day capitalism, its "market triumphalism," and the "fundamental flaws" in its relation to the environment. Yet, in defiance of all this, he assures his readers that the "strengths of capitalism" can be harnessed to a new system of "sustainable development."[9]

The System of Unsustainable Development

In reality, capitalism can be defined as *a system of unsustainable development*. In order to understand why this is so, it is useful to turn to Karl Marx, the core of whose entire intellectual corpus might be interpreted as a critique of the political economy of unsustainable development and its human and natural consequences.

Capitalism, Marx explains, is a system of generalized commodity production. There were other societies prior to capitalism in which commodity markets played important roles, but it is only in capitalism that a system emerges that is centered entirely on the production of commodities. A "commodity" is a good produced to be sold and exchanged for profit in the market. We call it a "good" because it is has a use value, i.e., it normally satisfies some use, otherwise there would be no need for it. But it is the exchange value, i.e., the money income and the profit that it generates, that is the exclusive concern of the capitalist.

What Marx called "simple commodity production" is an idealized economic formation—often assumed to describe the society wherein we live—in which the structure of exchange is such that a commodity embodying a certain use value is exchanged for money (acting as a mere means of exchange), which is, in turn, exchanged for another commodity (use value) at the end. Here, the whole exchange process from beginning to end can be designated by the shorthand C-M-C. In such a process, exchange is simply a modified form of barter, with money merely facilitating exchange. The goal of exchange is concrete use values, embodying qualitative properties. Such use values are normally consumed—thereby bringing a given exchange process to an end.

Marx, however, insisted that a capitalist economy, in reality, works altogether differently, with exchange taking the form of M-C-M'. Here money capital (M) is used to purchase commodities (labor power and means of production) to produce a commodity that can be sold for more money, M' (i.e., M + Δm or surplus value) at the end. This process, once set in motion, never stops of its own accord, since it has no natural end. Rather, the surplus value (profit) is reinvested in the next round, with the object of generating M''; and, in the following round, the returns are again reinvested with the goal of obtaining M''', and so on, *ad infinitum.*[10]

For Marx, therefore, capital is self-expanding value, driven incessantly to ever larger levels of accumulation, knowing no bounds. "Capital," he wrote, "is the endless and limitless drive to go beyond its limiting barrier. Every boundary is and has to be a [mere] barrier for it [and thus capable of being surmounted]. Else it would cease to be capital—money as self-reproductive." It thus converts all of nature and nature's laws as well as all that is distinctly human into a mere means of its own self-expansion. The result is a system, fixated on the exponential growth of profits and accumulation. "Accumulate, accumulate! That is Moses and the prophets!"[11]

Any attempt to explain where surplus value (or profits) comes from must penetrate beneath the exchange process and enter the realm of labor and production. Here, Marx argues that value added in the working day can be divided into two parts: (1) the part that reproduces the value of labor power (i.e., the wages of the workers) and thus constitutes necessary labor; and (2) the labor expended in the remaining part of the working day, which can be regarded as surplus labor, and which generates surplus value (or gross profits) for the capitalist. Profits are thus to be regarded as residual, consisting of what is left over after wages are paid out—something that every businessperson instinctively understands. The ratio of surplus (i.e., unpaid) labor to necessary (paid) labor in the working day is, for Marx, the rate of exploitation.

The logic of this process is that the increase in surplus value appropriated depends on the effective exploitation of human labor power. This can be achieved in two ways: (1) either workers are compelled to work longer hours for the same pay, thereby increasing the surplus portion of the working day simply by adding to the total working time (Marx calls this "absolute surplus value"); or (2) the value of labor power, i.e., the value equivalent of workers' wages, is generated in less time (as a result of increased productivity, etc.), thereby augmenting the surplus portion of the working day to that extent (Marx calls this "relative surplus value").

In its unrelenting search for greater (relative) surplus value, capitalism is thus dependent on the revolutionization of the means of production with the aim of increasing productivity and reducing the paid portion of the working day. This leads inexorably to additional revolutions in production, additional increases in productivity, in what constitutes an endless treadmill of production/accumulation. The logic of accumulation concentrates more and more of the wealth and power of society in fewer and fewer hands, and generates an enormous industrial reserve army of the unemployed.

This is all accompanied by the further alienation of labor, robbing human beings of their creative potential, and often of the environmental conditions essential for their physical reproduction. "The factory system," Marx wrote, "is turned into systematic robbery of what is necessary for the life of the worker while he is at work, i.e., space, light, air and protection against the dangerous or the unhealthy contaminants of the production process."[12]

For classical political economists, beginning with the physiocrats and Adam Smith, nature was explicitly designated as a "free gift" to capital. It thus did not directly enter into the determination of exchange value (value), which constituted the basis of the accumulation of private capital. Nevertheless, classical political economists did see nature as constituting public wealth, since this was identified with use values, and included not only what was scarce, as in the case of exchange values, but also what was naturally abundant, e.g., air, water, etc.

Out of these distinctions arose what came to be known as the Lauderdale Paradox, associated with the ideas of James Maitland, the eighth Earl of Lauderdale, who observed in 1804 that private riches (exchange values) could be expanded by destroying public wealth (use values)—that is, by generating scarcity in what was formerly abundant. This meant that individual riches could be augmented by landowners monopolizing the water of wells and charging a price for what had previously been free—or by burning crops (the produce of the earth) to generate scarcity and thus exchange value. Even the air itself, if it became scarce enough, could expand private riches, once it was possible to put a price on it. Lauderdale saw such artificial creation of scarcity as a way in which those with private monopolies of land and resources robbed society of its real wealth.[13]

Marx (following Ricardo) strongly embraced the Lauderdale Paradox, and its criticism of the inverse relation between private riches and public wealth. Nature, under the system of generalized commodity production, was, Marx insisted, reduced to being merely a *free gift to capital* and was thus robbed. Indeed, the fact that part of the working day was unpaid and went to the surplus of the capitalist meant that an analogous situation pertained to human labor power, itself a "natural force." The worker was allowed to "work for his own life, *to live*, only in so far as he works for a certain time gratis for the capitalist . . . [so that] the whole capitalist system of production turns on the prolongation of this gratis labour by extending the working day or by developing the productivity, i.e., the greater intensity of labour power, etc." Both nature and the unpaid labor of the worker were then to be conceived in analogous ways as free gifts to capital.[14]

Given the nature of this classical critique, developed to its furthest extent by Marx, it is hardly surprising that later neoclassical economists, exercising their primary role as apologists for the system, were to reject both the classical theory of value and the Lauderdale Paradox. The new marginalist economic orthodoxy that emerged in the late nineteenth century erased all formal distinctions within economics between use value and exchange value, between wealth and value. Nature's contribution to wealth was simply defined out of existence within the prevailing economic view. However, a minority of heterodox economists, including such figures as Henry George, Thorstein Veblen, and Frederick Soddy, were to insist that this rejection of nature's contribution to wealth only served to encourage the squandering of common resources characteristic of the system. "In a sort of parody of an accountant's nightmare," John Maynard Keynes was to write of the financially driven capitalist system, "we are capable of shutting off the sun and the stars because they do not pay a dividend."[15]

For Marx, capitalism's robbing of nature could be seen concretely in its creation of a rift in the human-earth metabolism, whereby the reproduction of natural conditions was undermined. He defined the labor process in ecological terms as the "metabolic interaction" between human beings and nature. With the development of industrial agriculture under capitalism, a rift was generated in the nature-given metabolism between human beings and the earth. The shipment of food and fiber hundreds, and sometimes thousands, of miles to the cities meant the removal of soil nutrients, such as nitrogen, phosphorus, and potassium, which ended up contributing to the pollution of the cities, while the soil itself was robbed of its "constituent elements." This created a rupture in "the eternal natural condition for the lasting fertility of the soil," requiring the "systematic restoration" of this metabolism. Yet, even though this had been demonstrated with the full force of natural science (for example, in Justus von Liebig's chemistry), the rational application of scientific principles in this area was impossible for capitalism. Consequently, capitalist production simultaneously undermined "the original sources of all wealth—the soil and the worker."[16]

Marx's critique of capitalism as an unstainable system of production was ultimately rooted in its "preconditions," i.e., the historical bases under which capitalism as a mode of production became possible. These were to be found in "primitive accumulation," or the expropriation of the commons (of all customary rights to the land), and hence the expropriation of the workers themselves— of their means of subsistence. It was this expropriation that was to help lay the grounds for industrial capitalism in particular. The turning of the land into private property, a mere means of accumulation, was at the same time the basis for the destruction of the metabolism between human beings and the earth.[17]

. . .

Marx's whole critique thus pointed to the reality of capitalism as a system of unsustainable development, rooted in the unceasing exploitation and pillage of human and natural agents. As he put it: "*Après moi le déluge!* is the watchword of every capitalist and of every capitalist nation. Capital therefore takes no account of the health and the length of life of the worker [or the human nature metabolism], unless society forces it to do so. . . ."[18]

Toward Ecological Revolution

If the foregoing argument is correct, humanity is facing an unprecedented challenge. On the one hand, we are confronting the question of a terminal crisis, threatening most life on the planet, civilization, and the very existence of future generations. On the other hand, attempts to solve this through technological fixes, market magic, and the idea of a "sustainable capitalism" are mere forms of ecological denial: since they ignore the inherent destructiveness of the current system of unsustainable development—capitalism. This suggests that the only rational answer lies in an ecological revolution, which would also have to be a social revolution, aimed at the creation of a just and sustainable society.

In addressing the question of an ecological revolution in the present dire situation, both short-term and long-term strategies are necessary, and should complement each other. One short-term strategy, directed mainly at the industrialized world, has been presented by Hansen. He starts with what he calls a "geophysical fact": most of the remaining fossil fuel, particularly coal, must stay in the ground, and carbon emissions have to be reduced as quickly as possible to near zero. He proposes three measures: (1) coal burning (except where carbon is sequestered—right now not technologically feasible) must cease; (2) the price of fossil fuel consumption should be steadily increased by imposing a progressively rising tax at the point of production: well head, mine shaft, or point of entry—redistributing 100 percent of the revenue, on a monthly basis, directly to the population as dividends; (3) a massive, global campaign to end deforestation and initiate large-scale reforestation needs to be introduced. A carbon tax, he argues, if it were to benefit the people directly—the majority of whom have below average per-capita carbon footprints, and would experience net gains from the carbon dividends once their added energy costs were subtracted—would create massive support for change. It would help to mobilize the population, particularly those at the bottom of society, in favor of a climate revolution. Hansen's "fee and dividend" proposal is explicitly designed not to feed the profits of vested interests. Any revenue from the carbon tax, in this plan, has to be democratically structured so as to redistribute income and wealth to those with smaller carbon footprints (the poor), and away from those with the larger carbon footprints (the rich).[19]

Hansen has emerged as a leading figure in the climate struggle, not only as a result of his scientific contributions, but also due to his recognition that at the root of the problem is a system of economic power, and his increasingly radical defiance of the powers that be. Thus, he declares: "the trains carrying coal to power plants are death trains. Coal-fired plants are factories of death." He criticizes those such as Gore, who have given in to cap and trade, locking in failure. Arguing that the unwillingness and inability of the authorities to act means that desperate measures are necessary, he is calling for mass "civil resistance." In June 2009, he was arrested, along with thirty-one others, in the exercise of civil resistance against mountain top removal coal mining.[20]

In strategizing an immediate response to the climate problem, it is crucial to recognize that the state, through government regulation and spending programs, could intervene directly in the climate crisis. Carbon dioxide could be considered an air pollutant to be regulated by law. Electrical utilities could be mandated to obtain their energy increasingly from renewable sources. Solar panels could be included as a mandatory part of the building code. The state could put its resources behind major investments in public environmental infrastructure and planning, including reducing dependence on the automobile through massive funding of public transportation, e.g., intercity trains and light rail, and the necessary accompanying changes in urban development and infrastructure.

Globally, the struggle, of course, has to take into account the reality of economic and ecological imperialism. The allowable carbon-concentration limits of the atmosphere have already been taken up as a result of the accumulation of the rich states at the center of the world system. The economic and social development of poor countries is, therefore, now being further limited by the pressing need to impose restrictions on carbon emissions for the sake of the planet as a whole—despite the fact that underdeveloped economies had no role in the creation of the problem. The global South is likely to experience the effects of climate change much earlier and more severely than the North, and has fewer economic resources with which to adapt. All of this means that a non-imperialistic, and more sustainable, world solution depends initially on what is called "contraction and convergence"—a drastic *contraction* in greenhouse gas emissions overall (especially in the rich countries), coupled with the *convergence* of per-capita emissions in all countries at levels that are sustainable for the planet.[21] Since, however, science suggests that even low greenhouse gas emissions may be unsustainable over the long run, strategies have to be developed to make it economically feasible for countries in the periphery to introduce solar and renewable technologies—reinforcing those necessary radical changes in social relations that will allow them to stabilize and reduce their emissions.

For the anti-imperialist movement, a major task should be creating stepped-up opposition to military spending (amounting to a trillion dollars in the United States in 2007) and ending government subsidies to global agribusiness—with the goal of shifting those monies into environmental defense and the meeting of the social needs of the poorest countries, as suggested by the *Bamako Appeal*.[22] It must be firmly established as a principle of world justice that the wealthy countries owe an enormous ecological debt to poorer countries, due to the robbing by the imperial powers of the global commons and the pillage of the periphery at every stage of world capitalist development.

Already, the main force for ecological revolution stems from movements in the global South, marked by the growth of the Vía Campesina movement, socialist organizations like Brazil's MST, and ongoing revolutions in Latin America (the ALBA countries) and Asia (Nepal). Cuba has been applying permaculture design techniques that mimic energy-maximizing natural systems to its agriculture since the 1990s, generating a revolution in food production. Venezuela, although, for historic reasons, an oil power economically dependent on the sale of petroleum, has made extraordinary achievements in recent years by moving toward a society directed at collective needs, including dramatic achievements in food sovereignty.[23]

Reaching back into history, it is worth recalling that the proletariat in Marxian theory was the revolutionary agent because it had nothing to lose, and thus came to represent the universal interest in abolishing, not only its own oppression, but oppression itself. As Marx put it, "the living conditions of the proletariat represent the focal point of all inhuman conditions in contemporary society. . . . However, it [the proletariat] cannot emancipate itself without abolishing the conditions which give it life, and it cannot abolish these conditions without abolishing all those inhuman conditions of social life which are summed up in its own situation."[24]

Later Marxist theorists were to argue that, with the growth of monopoly capitalism and imperialism, the "focal point of inhuman conditions" had shifted from the center to the periphery of the world system. Paul Sweezy contended that, although the objective conditions that Marx associated with the proletariat did not match those of better-off workers in the United States and Europe in the 1960s, they did correspond to the harsh, inhuman conditions imposed on "the masses of the much more numerous and populous underdeveloped dependencies of the global capitalist system." This helped explain the pattern of socialist revolutions following the Second World War, as exemplified by Vietnam, China, and Cuba.[25]

Looking at this today, I think it is conceivable that the main historic agent and initiator of a new epoch of ecological revolution is to be found in the third world masses most directly in line to be hit first by the impending disasters. Today the ecological frontline is arguably to be found in the inhabitants of the Ganges-Brahmaputra Delta and of the low-lying fertile coast area of the Indian Ocean and China Seas—the state of Kerala in India, Thailand, Vietnam, Indonesia. They, too, as in the case of Marx's proletariat, have nothing to lose from the radical changes necessary to avert (or adapt to) disaster. In fact, with the universal spread of capitalist social relations and the commodity form, the world proletariat and the masses most exposed to sea level rise—for example, the low-lying delta of the Pearl River and the Guangdong industrial region from Shenzhen to Guangzhou—sometimes overlap. This, then, potentially constitutes the global epicenter of a new environmental proletariat.[26]

The truly planetary crisis we are now caught up in, however, requires a world uprising transcending all geographical boundaries. This means that ecological and social revolutions in the third world have to be accompanied by, or inspire, universal revolts against imperialism, the destruction of the planet, and the treadmill of accumulation. The recognition that the weight of environmental disaster is such that it would cross all class lines and all nations and positions, abolishing time itself by breaking what Marx called "the chain of successive generations," could lead to a radical rejection of the engine of destruction in which we live, and put into motion a new conception of global humanity and earth metabolism. As always, however, real change will have to come from those most alienated from the existing systems of power and wealth. The most hopeful development within the advanced capitalist world at present is the meteoric rise of the youth-based climate justice movement, which is emerging as a considerable force in direct action mobilization and in challenging the current climate negotiations.[27]

What is clear is that the long-term strategy for ecological revolution throughout the globe involves the building of a society of substantive equality, i.e., the struggle for socialism. Not only are the two inseparable, but they also provide essential content for each other. There can be no true ecological revolution that is not socialist; no true socialist revolution that is not ecological. This means recapturing Marx's own vision of socialism/communism, which he defined as a society where "the associated producers govern the human metabolism with nature in a rational way, bringing it under their collective control . . . accomplishing it with the least expenditure of energy and in conditions most worthy and appropriate for their human nature."[28]

One way to understand this interdependent relation between ecology and socialism is in terms of what Hugo Chávez has called "the elementary triangle of socialism" (derived from Marx) consisting of: (1) social ownership; (2) social production organized by workers; and (3) satisfaction of communal needs. All three components of the elementary triangle of socialism are necessary if socialism is to be sustained. Complementing and deepening this is what could be called "the elementary triangle of ecology" (derived even more directly from Marx): (1) social use, not ownership, of nature; (2) rational regulation by the associated producers of the metabolic relation between humanity and nature; and (3) satisfaction of communal needs—not only of present but also future generations (and life itself).[29]

As Lewis Mumford explained in 1944, in his *Condition of Man*, the needed ecological transformation required the promotion of "basic communism," applying "to the whole community the standards of the household," distributing benefits "according to need, not ability or productive contribution." This meant focusing first and foremost on "education, recreation, hospital services, public hygiene, art," food production, the rural and urban environments, and, in general, "collective needs." The idea of "basic communism" drew on Marx's principle of substantive equality in the *Critique of the Gotha Programme*: "from each according to his ability, to each according to his needs!" But Mumford also associated this idea with John Stuart Mill's vision, in his most socialist phase, of a "stationary state"—viewed, in this case, as a system of economic production no longer geared to the accumulation of capital, in which the emphasis of society would be on collective development and the quality of life.[30] For Mumford, this demanded a new "organic person"—to emerge from the struggle itself.

An essential element of such an ecological and socialist revolution for the twenty-first century is a truly radical conception of sustainability, as articulated by Marx:

> From the standpoint of a higher socio-economic formation, the private property of particular individuals in the earth will appear just as absurd as the private property of one man in other men [i.e., slavery]. Even an entire society, a nation, or all simultaneously existing societies taken together, are not the owners of the earth. They are simply its possessors, its beneficiaries, and have to bequeath it in an improved state to succeeding generations as *boni patres familias* [good heads of the household].[31]

Such a vision of a sustainable, egalitarian society must define the present social struggle; not only because it is ecologically necessary for human survival, but also because it is historically necessary for the development of human freedom. Today we face the challenge of forging a new organic revolution in which the struggles for human equality and for the earth are becoming one. There is only one future: that of sustainable human development.[32]

Notes

Foster, John Bellamy. 2010. "Why Ecological Revolution?" *Monthly Review: An Independent Socialist Magazine* January, 61 (8): 1.

[1] On the long-term aspects of the current financial-economic crisis, see John Bellamy Foster and Fred Magdoff, *The Great Financial Crisis* (New York: Monthly Review Press, 2009).

[2] James E. Hansen, "Strategies to Address Global Warming" (July 13, 2009), http//www.columbia.edu.

[3] Ibid.; "Seas Grow Less Effective at Absorbing Emissions," *New York Times*, November 19, 2009; S. Khatiwala. F. Primeau and T. Hall, "Reconstruction of the History of Anthropogenic CO_2 Concentrations in the Ocean," *Nature* 462, no. 9 (November 2009), 346–50.

[4] "Arctic Seas Turn to Acid, Putting Vital Food Chain at Risk," October 4, 2009, http://www.guardian.com.uk.

[5] Hansen, "Strategies to Address Global Warming"; AFP, "Top UN Climate Scientist Backs Ambitious CO2 Cuts," August 25, 2009.

[6] Bill McKibben, "Response," in Tim Flannery, *Now or Never* (New York: Atlantic Monthly Press, 2009), 116; Al Gore, *Our Choice: A Plan to Solve the Climate Crisis* (Emmaus, PA: Rodale, 2009), 327.

[7] Friends of the Earth, "Subprime Carbon?" (March 2009), http://www.foe.org/subprimecarbon, and *A Dangerous Obsession* (November 2009), www.foe.co.uk/resources/reports/dangerous_obsession.pdf; James E. Hansen, "Worshipping the Temple of Doom" (May 5, 2009), http//www.columbia.edu.

[8] James Lovelock, *The Revenge of Gaia* (New York: Basic Books, 2006), and *The Vanishing Face of Gaia* (New York: Basic Books, 2009), 139–58; Gore, *Our Choice*, 314–15. Hansen, it should be noted, also places hope in the development of fourth generation nuclear power as part of the solution. See James Hansen, *Storms of My Grandchildren* (New York: Bloomsbury USA, 2009), 194–204.

[9] Gore, *Our Choice*, 303, 320, 327, 330–32, 346.

[10] Karl Marx, *Capital*, vol. 1 (London: Penguin 1976), 247–80. On how Marx's M-C-M' formula serves to define the "regime of capital," see Robert Heilbroner, *The Nature and Logic of Capitalism* (New York: W.W. Norton, 1985), 33–77.

[11] Karl Marx, *Grundrisse* (London: Penguin, 1973), 334–35, 409–10, and *Capital*, vol. 1, 742; John Bellamy Foster, "Marx's Grundrisse and the Ecological Contradictions of Capitalism," in Marcelo Musto, *Karl Marx's Grundrisse* (New York: Routledge, 2008), 100–02.

[12] Marx, *Capital*, vol. 1, 552–53.

[13] The discussion of the Lauderdale Paradox is based on John Bellamy Foster and Brett Clark, "The Paradox of Wealth," *Monthly Review* 61, no. 6 (November 2009): 1–18.

[14] Karl Marx, *Capital*, vol. 3 (London: Penguin, 1981), 949, *Critique of the Gotha Programme* (New York: International Publishers, 1938), 3, 15.

[15] John Maynard Keynes, "National Self-Sufficiency," in *Collected Writings* (London: Macmillan/Cambridge University Press, 1982), vol. 21, 241–42.

[16] Karl Marx, *Capital*, vol. 1, 636–39, *Capital*, vol. 3, 948–50, and *Capital*, vol. 2 (London: Penguin 1978), 322; Foster, *The Ecological Revolution*, 161–200.

[17] See Foster, "Marx's Grundrisse and the Ecological Contradictions of Capitalism," 98–100.

[18] Marx, *Capital*, vol. 1, 381.

[19] James Hansen, et al., "Target Atmospheric CO2: Where Should Humanity Aim?" *Open Atmospheric Science Journal* 2 (2008): 217–31; James E. Hansen, "Response to Dr. Martin Parkinson, Secretary of the Australian Department of Climate Change" (May 4, 2009), http://www.columbia.edu; Hansen, "Strategies to Address Global Warming" and "Worshipping the Temple of Doom"; Frank Ackerman, et al. "The Economics of 350," October 2009, www.e3network.org, 3–4.

[20] James E. Hansen, "The Sword of Damocles" (February 15, 2009), "Coal River Mountain Action" (June 25, 2009), and "I Just Had a Baby, at Age 68" (November 6, 2009), http://www.columbia.edu; Ken Ward, "The Night I Slept with Jim Hansen" (November 11, 2009), www.grist.org.

[21] Tom Athanasiou and Paul Baer, *Dead Heat* (New York: Seven Stories Press, 2002).

22 John Bellamy Foster, Hannah Holleman, and Robert W. McChesney, "The U.S. Imperial Triangle and Military Spending," *Monthly Review* 60, no. 5 (October 2008), 9–13. The Bamako Appeal can be found in Samir Amin, *The World We Wish to See* (New York: Monthly Review Press, 2008), 107–34.

23 An important source in understanding Cuban developments is the film "The Power of Community: How Cuba Survived Peak Oil," http://www.powerofcommunity. org/cm/index.php. On Venezuela see Christina Schiavoni and William Camacaro, "The Venezuelan Effort to Build a New Food and Agriculture System," *Monthly Review* 61, no. 3 (July–August 2009): 129–41.

24 Karl Marx and Frederick Engels, *The Holy Family* (Moscow: Foreign Languages Publishing House, 1956), 52. Translation follows Paul M. Sweezy, *Modern Capitalism and Other Essays* (New York: Monthly Review Press, 1972), 149.

25 Sweezy, *Modern Capitalism*, 164.

26 John Bellamy Foster, "*The Vulnerable Planet* Fifteen years Later," *Monthly Review* 54, no. 7 (December 2009): 17–19.

27 On the climate justice movement see Tokar, "Toward Climate Justice."

28 Marx, *Capital*, vol. 3, 1959.

29 On the elementary triangles of socialism and ecology see Foster, *The Ecological Revolution*, 32–35. The failure of Soviet-type societies to conform to these elementary triangles goes a long way toward explaining their decline and fall, despite their socialist pretensions. See John Bellamy Foster, *The Vulnerable Planet* (New York: Monthly Review Press, 1999), 96–101.

30 Lewis Mumford, *The Condition of Man* (New York: Harcourt Brace Jovanovich, 1973), 411; Marx, *Critique of the Gotha Programme*, 10: John Stuart Mill, *Principles of Political Economy* (New York: Longmans, Green and Co., 1904), 453–55.

31 Marx, *Capital*, vol. 3, 911, 959.

32 Paul Burkett, "Marx's Vision of Sustainable Human Development," *Monthly Review* 57, no. 5 (October 2005): 34–62.

The Unfair Trade-off
Globalization and the Export of Ecological Hazards

Daniel Faber

Globalization is the process by which technological, communications, and political changes have intensified the worldwide exchange of money, goods, people, and culture. In this chapter, Daniel Faber discusses how globalization has also accelerated the exchange of ecological hazards between nations. Due to imbalances between countries in terms of national-level environmental governance and due to weak global structures of environmental governance, this exchange has led to a worldwide system of environmental "in"justice, in which affluent nations (where environmental laws are relatively strict) export ecological hazards to poor nations (where environmental laws are relatively weak). The poorer countries (often referred to as "the Global South") have limited capacity to adequately evaluate and manage the risks associated with such hazards; transnational corporations are thus endangering people's health and environments in many areas of the world. Faber suggests that only by achieving greater social governance over trade, lending institutions, and regulatory bodies can the process that leads some national and corporate leaders to sacrifice human and environmental health in order to compete in the world economy be overcome.

The Global Ecological Crisis

. . . The creation of modern global communications and transportation systems, and the development of advanced infrastructure in the newly industrializing countries, are granting industrial capital the geographic mobility to take advantage of more favorable business climates abroad. This is especially true in the countries of East Asia and the global South with large supplies of cheap and highly disciplined wage laborers, abundant natural resources and energy supplies, tax advantages, and weaker environmental regulations. The commodities and surplus-profits produced by the factories are then exported back into the United States and other advanced capitalist countries. The pollution, however, remains behind. Even worse, the toxic waste, industrial pollution, discarded consumer goods, and other forms of "anti-wealth" produced in the United States are also becoming increasingly mobile, and end up in the "pollution havens" of the Third

49

World. Prior to the invention of environmental protection laws in the United States and elsewhere, it was not necessary (let alone cost-effective) to export environmental problems to other countries. This is no longer the case. . . .

Dumping on the Third World: The Export of Ecological Hazards and Environmental Justice

. . . The worsening ecological crisis in the global South is directly related to an international system of economic and environmental stratification in which the United States and other advanced capitalist nations are able to shift or impose the environmental burden on weaker states. In fact, one of the primary aims of U.S. economic planners is to cut costs by displacing environmental problems [externalities] onto poorer Southern nations—countries with little power in global environmental policy decision-making institutions. Lawrence Summers, former Undersecretary of the Treasury of International Affairs and key economic policy-maker under the Clinton administration [and former President of Harvard University], is infamous for writing a 1991 memo as a chief economist at the World Bank that argued,

> Just between you and me, shouldn't the World Bank be encouraging more migration of the dirty industries to the LDCs [less developed countries]? . . . I think the economic logic behind dumping a load of toxic waste in the lowest wage country is impeccable and we should face up to that. . . . I've always thought that under-populated countries in Africa are vastly under-polluted.

The Summers memo reflects the "thinking" of many U.S. policy makers aligned with the interests of U.S. multinational corporations: that human life in the Third World is worth much less than in the United States. If the poor and underemployed masses of Africa become sick or die from exposure to pollution exported from the United States, it will have a much smaller impact on the profits of international capital. Aside from the higher costs of pollution-abatement in the United States, if highly-skilled and well-compensated American workers fall prey to environmentally-related health problems, then the expense to capital and the state can be significant. Although morally reprehensible, under the capitalist system it pays business to shift pollution onto the poor in the less developed countries.

Given the willingness of undemocratic governments in the global South to trade-off their environmental protection for economic growth, the growing mobility of capital (in all forms) is facilitating the export of ecological problems from the advanced capitalist countries to the third world and sub-peripheral states. This *export of ecological hazard* from the United States and other Northern countries to the less developed countries takes place:

> (1) . . . in the form of foreign direct investment (FDI) in domestically-owned hazardous industries, as well as destructive investment schemes to gain access to new oil fields, forests, agricultural lands, mining deposits, and other natural resources; (2) . . . with the relocation of polluting and environmentally

hazardous production processes and polluting facilities owned by transnational capital to the South; (3) . . . as witnessed in the marketing of more profitable but also more dangerous foods, drugs, pesticides, technologies, and other consumer/capital goods; and (4) . . . with the dumping of toxic wastes, pollution, discarded consumer products, trash, and other forms of "anti-wealth" produced by Northern industry.

Hence, corporate-led globalization is facilitating the displacement of ecological hazards from richer to poorer countries. Although a few international agreements (such as the Basel Convention) have been put into place, they are for the most part ineffective at stemming the transfer of hazards. Since few peripheral countries have the ability to adequately evaluate and manage the risks associated with such hazards, the export practices of transnational corporations are increasing the health, safety, and environmental problems facing many peripheral countries. . . .

Antigreen Greenbacks: The Export of Ecologically Hazardous Investment Capital

. . . Since the mid-1970s, the U.S. regime of environmental regulation has resulted in stricter laws, increased delays due to permitting, and higher costs related to pollution control technology, liability and insurances cases, and worker health and safety. These costs are especially significant for companies involved in the production of heavy metals, asbestos-containing products, copper and lead smelting, and leather tanning, and has led these industries to relocate overseas. On the other hand, the competition for foreign investment among the developing countries is fierce, and combined with the imposition of structural adjustment policies by the International Monetary Fund (IMF) and World Bank on indebted developing countries, more and more nations are opening themselves up to increased FDI by weakening environmental standards.

. . . As a consequence, the rate of growth in hazardous industries in the developing countries is now greater than the overall industrial growth in those same countries, indicating that the cost advantages stemming from weaker environmental protection is attracting investment. Again, this trend began in the late 1970s, just as environmental regulations became more stringent in the United States and other advanced capitalist countries.

China: The New Economic Superpower, or Ecological Nightmare?

. . . Although China has excelled in providing low-quality consumer goods, the country is now ramping up to create more advanced industries, adding state-of-the-art capacity in cars, specialty steel, petrochemicals, and microchips. So, while American petrochemical makers have invested in little new capacity inside the United States over the past decade, over 12,000 workers are constructing a $2.7 billion petrochemical complex in Nanjing, China. This facility will be among the world's biggest, most modern complexes for making ethylene, the basic ingredient in plastics. Constructing such a plant in China offers sizable cost advantages over rival facilities in the United States, Europe, and Japan due to the lower environmental costs of doing business. The Chinese government allows industry to freely pollute the air, water and ground, which (combined with the low cost of labor) easily allows industry to undercut the prices charged by companies abiding by strict standards elsewhere in the world. However, the economic incentives offered to foreign capital to invest in China, including few controls over pollution and worker health and safety violations, have created an ecological nightmare. As stated by the journalists Joseph Kahn and Jim Yardley,

> Environmental woes that might be considered catastrophic in some countries can seem commonplace in China: industrial cities where people rarely see the sun; children killed or sickened by lead poisoning or other types of local pollution; a coastline so swamped by algal red tides that large sections of the ocean no longer sustain marine life. China is choking on its own success.

The magnitude of this ecological crisis is apparent in a 2007 draft report by the World Bank and China's State Environment Protection Agency. The study finds that 750,000 people die prematurely in China each year, mainly from air pollution in the large cities. . . . Incredibly, only 1 percent of the country's 560 million city dwellers breathe air considered safe by the European Union. And air quality is getting worse. The central government's most recent report put the cost of air pollution at $64 billion in 2004.

Of the twenty most polluted cities in the world, according to the World Bank, sixteen are located in China. About one-third of China's lakes, rivers, and coastal waters are so polluted that they pose a threat to human health, according the Organization for Economic Cooperation and Development. As a result, 300 million Chinese don't have access to clean drinking water, resulting in 60,000-odd premature deaths a year. Acid rain falls over 30 percent of the country. Industrial pollution is so extensive that the country's birth defect rate is triple that of the developed nations. At least a million Chinese babies born each year have birth defects. As acknowledged in the World Bank report, China's poor are disproportionately affected by these environmental health burdens. The World Bank puts the cost of China's pollution at 8 percent of GDP, although some economists say it as high as 10 percent of GDP, which is equal to the country's rate of economic growth. Fed up with the pollution, a number of environmental riots have erupted in China in recent years, and are likely to become more numerous in the future.

Africa's Black Gold: Investing in Repression and Environmental Injustice

. . . Nigeria has seen huge investments of U.S. capital to develop the oil fields, and is now the eighth leading exporter of oil in the world (and the largest oil producer in Africa). More than $300 billion in oil has been exported since 1975. Petroleum companies such as Chevron and Royal Dutch Shell have invaded the oil-rich Niger Delta, home to the Ogoni people, and one of the most populated regions in all of Africa. At the invitation of a brutally repressive Nigerian government, the international oil companies ignore standard environmental protection measures in order to cuts costs and maximize profits. Enjoying a complete lack of government oversight, the oil companies have created what the European Parliament calls "an environmental nightmare" for the Ogoni people. A constant barrage of oil spills—an average of 300 per year—have significantly contaminated waterways and groundwater, killed fish and other wildlife on which the local people are dependent, and decimated the resource base of numerous subsistence economies in the region. Petroleum pollution in Ogoni streams is 680 times greater than European Community permissible levels. Leaking pipes have also caught fire, exploded, and killed hundreds of people. Toxic wastes dumped in unlined pits litter the countryside, while continuous gas flares pollute nearby villages with 35 million tons of CO_2 a year (76 percent of natural gas in the oil producing areas is flared, compared to 0.6 percent in the United States, along with 12 million tons of methane, which is more than any nation on earth). Local crops will not grow, and acid rain pervades the area.

As the ecological crisis emerged full force in the early 1990s, the Ogoni people organized peaceful protests to raise international awareness of their plight. In response to awareness such actions were generating around the world, the [then] military government reacted with extreme repression. In November of 1995, Ken Saro-Wiwa, the leader of the Movement for the Survival of Ogoni People— a highly respected and renowned playwright in the international community— and eight other Ogoni leaders, were arrested on trumped-up treason charges. They were immediately tried by a military tribunal, found guilty, and executed. Despite the military's unfounded allegations, the world knew the "Ogoni 9" were killed for organizing peaceful protests against the country's large oil exporter, Royal/Dutch Shell. As stated in a recent report, "Shell failed to use its substantial influence with the Nigerian government to stop the execution. Indeed, Shell has publicly admitted that it had invited the Nigerian army to Ogoni land, provided them with ammunition and logistical and financial support for a military operation that left scores dead and destroyed many villages." In defense of the company and the military regime following the execution, Naemeka Achebe, the general manager for Shell Nigeria, stated, "For a commercial company trying to make investments, you need a stable environment . . . Dictatorships can give you that. Right now in Nigeria, there is acceptance, peace, and continuity."

In response to the repression, Nigerian villagers have brought suit in U.S. court against Chevron alleging that the company supported military attacks on protesters in the Niger Delta. A Human Rights Watch investigation uncovered Chevron's use of a covert Nigerian security force known as the "kill-and-go" squad against the movement. In the trial, Chevron stated that the incident was "regrettable" but resulted from attempts by protesters to take control of weapons held by security personnel. Although the company was cleared of direct liability in 2004, the judge in the case noted that a reasonable juror could reason that the company had indirect responsibility, and could be liable for reparations due to its "extraordinarily close relationship" with Chevron Nigeria. In August of 2007, another judge allowed claims of wrongful death and other human rights suits to proceed. . . .

Global Pollution Havens: The Export of Ecologically Hazardous Industry

. . . In the age of corporate-led globalization, free trade and neo-liberal economic policies are encouraging countries to lower wages and environmental standards in order to cuts costs and achieve a comparative advantage in the world econ-omy. By pitting various nations against one another in this "race to the bottom" phenomenon, in which countries lower environmental regulations in order to gain a competitive edge, multinational corporations have acquired greater and greater power vis-a-vis the nation-state. With the increased international mobility of industrial capital, various governments [at all levels] are pressured to reduce the financial burden of environmental regulations, taxation policy, labor rules, and consumer product safety requirements upon industry. Otherwise, the man-ufacturer will simply pick up and move to another part of the world where the business climate is more favorable. The state is left with little choice but to grant such concessions if the jobs and other economic benefits are to be preserved.

Over the last three decades, those U.S.-based industries most heavily impacted by environmental regulations, including lead smelting, dye and chemi-cal manufacturing, asbestos-related production, pesticides, textiles, copper smelt-ing, vinyl chloride, etc., have moved to other countries with weaker rules and enforcement. American companies often make no secret of the fact that more stringent environmental regulations are a major factor in relocating facilities abroad. As stated by the U.S. corporation Chemex, "As a result of tougher envi-ronmental regulations . . . many North American [mineral oil] refineries have ceased operations. Recognizing an opportunity, Chemex redirected its focus to the procurement of quality used refineries" for export to developing countries. This process is now accelerating, as double standards in worker and community health protection become more commonplace in the world, especially in the less developed countries of the Caribbean, Africa, Latin America, and especially Asia. According to a United Nations study, over half of the transnational firms sur-veyed in the Asia-Pacific region adopt lower standards in comparison to their country of origin in the North. As a result, the increased mobility of U.S. capital is serving to relocate many of the worst public health risks and environmental injustices associated with "dirty industry" to the global South. . . .

Mexico: Environmental Troubles South of the Border

. . . Since the passage of the North American Free Trade Agreement (NAFTA) in 1994, Mexico's environmental problems have worsened throughout the country.

NAFTA is a free trade agreement that reduces tariffs and other barriers to trade among Mexico, Canada, and the United States. Aided by the agreement, dirty industries are moving out of the United States to Mexico, where environmental standards are lax, unions are weak, and worker health and safety concerns are ignored. Along the 2,100-mile U.S.-Mexico border running from the Pacific Ocean to the Gulf of Mexico, there are more than 2,000 factories, or *maquiladoras*, including U.S. companies, involved in textiles and clothing, chemicals, and electronics. A 1991 U.S. Government Accounting Office study even found that several Los Angeles furniture manufacturers relocated to Mexico after the establishment of stringent air pollution restrictions in California (80 percent of these businesses cited environmental costs in their decision to move).

The explosive growth of the *maquiladoras* is creating an ecological disaster along both sides of the border. Factories big and small generate huge volumes of pollution (some 87 percent of *maquiladoras* use toxic materials in their production processes). Reports show that industrial waste is seldom treated before it is discharged into rivers, arroyos, the Rio Grande, or the ocean. *Maquiladoras* also generate a substantial amount of hazardous waste, including dangerous solvents such as trichloroethylene, acids, heavy metals like lead and nickel, paints, oils, resins, and plastics. Over 65 percent of such waste is unaccounted for in either the United States or Mexico. The situation is growing worse because NAFTA no longer requires TNCs to return waste to the United States for proper disposal. . . .

Buyer Beware: The Export of Ecologically Hazardous Commodities

In the era of corporate-led globalization, dangerous pesticides and other chemicals, biotechnology, drugs, and other consumer products that are highly restricted in the United States are still manufactured here and routinely exported to other nations. American corporations know that there is little government oversight or public pressure to inspect and regulate such products in overseas markets, and that significant profits are to be made from shipping their hazardous products to unsuspecting consumers all around the world. This process has been underway for many decades, but is accelerating with the expansion of world trade.

One of the most hazardous commodities exported by the United States to the rest of the world are pesticides. Roughly a billion pounds of pesticides are exported each year, or *45 tons per hour*. Tragically, American policy makers have done little to stop the export of pesticides forbidden in the United States. Under the Federal Insecticide, Fungicide and Rodenticide Act (FIFRA), EPA does not review the health and environmental impacts of pesticides manufactured for export only, or what are termed "never registered" pesticides. The most recent figures available indicate that nearly 22 million pounds of these exported pesticides are banned or severely restricted for use in the United States, an average of more than 22 tons per day. Furthermore, an average rate of more than 30 tons per day of "extremely hazardous" chemicals, as rated by the World

Health Organization, are also exported. Nearly 1.1 billion pounds of known or suspected carcinogenic pesticides were exported by the United States between 1997 and 2000, an average rate of almost 16 tons per hour. As a result, many pesticides that the Environmental Protection Agency (EPA) has judged too dangerous for domestic use, as well as pesticides never evaluated by the EPA, are regularly shipped from U.S. ports.

. . . The most dangerous U.S. chemical exports are often destined for Third World countries where the prevailing working conditions—a lack of protective equipment, unsafe application and storage practices, inadequate training of pesticide applicators—greatly magnify the health risks for agricultural workers and their families. In fact, about 57 percent of these products are shipped to the developing world, while most of the remaining chemicals are shipped to ports in Belgium and the Netherlands for reshipment to developing countries. As a result, poisonings continue to mount. The World Health Organization (WHO) estimates that three million severe pesticide poisonings occur each year, and, of these, a minimum of 300,000 people die, many of them children. Some 99 percent of these cases occur in developing countries.

The Global Circle of Poison

The people of the global South are not the only ones being poisoned by pesticides exported from the United States. Third World agricultural exports contaminated with pesticides come back to the United States and other Northern countries in a vicious "circle of poison." Although the U.S. environmental movement was successful in legally restricting or prohibiting the use of many hazardous chemicals such as DDT in the 1970s, multinational corporations continue to manufacture and export these same pesticides to the Third World. The circle of poison closes when U.S. citizens consume Third World exports contaminated with the pesticides. For instance, imports of Chilean grapes, Canadian and Mexican carrots, Mexican broccoli and tomatoes, Argentine and Hungarian apple juice, and Brazilian orange juice are found to have worse levels of pesticide contamination than U.S.-grown crops.

Food and Drug Administration (FDA) data shows that food imports from developing countries are often contaminated with pesticides banned or restricted for health reasons in the United States, including a violation rate of 40.8 for all imports of Guatemalan green peas; 18.4 percent for Mexican strawberries; and 15.6 percent for Mexican lettuce. FDA inspections of Chinese imports have also caught dried apples preserved with a cancer-causing chemicals; frozen catfish laden with banned antibiotics; scallops and sardines coated with bacteria, and mushrooms laced with illegal pesticides. These were among the 107 food imports from China that the FDA detained at U.S. ports in April of 2007, along with more than 1,000 shipments of tainted Chinese dietary supplements and other products.

Both U.S. and foreign corporations know that exporting tainted products into the United States poses little risk of being caught by an underfunded and understaffed Food and Drug Administration. FDA testing of food imports (and domestic products) is infrequent and restricted to only a few choice chemicals. Since 1997, FDA officials have examined just 1.5 percent of all food imports, while shipments skyrocketed from more than 4 million entries in 1997 to more than 15 million in 2006. Under assault from the Bush administration and the polluter-industrial complex, the FDA's regulatory affairs staff is getting leaner—it shrank from a high of 4,003 full-time employees in the 2003 fiscal year to 3,488 in 2007. As a result, noncriminal foreign and domestic inspections carried out by FDA's Center for Food Safety and Applied Nutrition staffers amounted to 9,038 in 2005, down from 11,566 just two years earlier.

In the few instances where testing is done, FDA health standards are inadequate. In some cases, consumption of a single food item contaminated with chemicals at levels allowed by FDA, such as DDT in fish, would expose the consumer to more than 50 times the daily intake levels considered "safe" by the EPA. Persistent organic pollutants such as DDT and PCBs are implicated in a breast cancer epidemic that impacts an estimated 2,044,000 women in the United States, and claims about 40,000 lives each year. In fact, a person eating the USDA's recommended five servings of fruits and vegetables per day will eat illegal pesticides at least 75 times per year. In contrast, the average consumer has to eat about 100 pounds of fresh fruits and vegetables in order to eat from a shipment tested for pesticides by the FDA. This means that the average American is at least 15 times more likely to eat an illegal pesticide than to eat from a shipment tested by the FDA. A form of *toxic trespass*, these dangerous chemicals are invading the bodies of U.S. citizens, and are linked to various types of cancers, learning disabilities and autism, immune system suppression, central nervous disorders, damage to reproductive systems, and numerous other disorders. According to the U.S. Centers for Disease Control (CDC), the American people carry a "body burden" of the pesticides chlorpyrifos (Dursban) and methyl parathion that dramatically exceed acceptable thresholds for chronic exposure. . . .

Dumping on the Third World: The Export of Pollution and Hazardous Waste

The United States is the single largest producer of hazardous wastes in the world. Each year the United States produces some 238 million tons. Meanwhile, the costs of hazardous waste disposal in the United States has grown from $15 per ton in 1980 to over $250 per ton, while the costs of incineration has increased over three-fold to between $1,500 and $3,000 per ton. Although capital has looked to reduce expenses by locating hazardous waste dumps and facilities in poor communities of color throughout the United States, there is also a growing incentive to export wastes to developing countries. The disposal costs per metric ton of hazardous waste in Africa, for instance, has historically hovered around $40—$50 per ton (and in the case of an agreement between the Gibraltar-based company and the Benin Republic government, for as low as $2.50 per ton). These costs are so low because regulations governing toxic waste disposal are virtually nonexistent in developing countries.

The incentive to cut disposal costs by exporting toxics to the global South is also strong among other advanced capitalist countries. Hundreds of cases involving hundreds of millions of pounds of hazardous waste being exported from the advanced capitalist countries to the South have been documented over the last two decades. As if emboldened by the words of former World Bank chief economist Lawrence Summers, "I've always thought that underpopulated countries in Africa are vastly *under*-polluted," dump sites of toxic waste from Western nations can be found throughout Africa, from Senegal to Nigeria, to Zimbabwe, Congo, and even South Africa. In some years, West Africa alone has imported up to 300 million tons of toxic waste from some 24 industrialized countries.

A Toxic Terror in the Ivory Coast

The devastating impacts of hazardous waste trade in Africa and the global South is illustrated by case of the *Proba Koala* in the Ivory Coast. Exemplary of the growing integration of capital on a global scale, the *Proba Koala* was a Korean-built, Greek-managed, Panamanian-flagged tanker chartered by the London branch of a Swiss trading corporation whose fiscal headquarters are in the Netherlands—the multibillion dollar Dutch global oil and metals trading company called Trafigura Beheer BV. The ship had been acting as a storage vessel for unrefined gasoline. In the summer of 2006, the Trafigura had explored disposing of the ship's "washings" after a routine cleaning of the storage hull with caustic soda in Amsterdam. However, due to the $300,000 or more cost estimate for disposing of the waste in that city, the company instead elected to take the ship to the Ivory Coast, even though there are no facilities capable of handling high-level toxic wastes. Upon arrival, the captain of the *Proba Koala* contacted a local company called Compagnie Tommy to dispose of the waste for a mere $15,000, representing a huge savings for Trafigura.

On August 19th of 2006, the *Proba Koala* offloaded 528 tons of the washings onto more than a dozen tanker trucks. The washings were a toxic alkaline mix of water, gasoline, and caustic soda, which gave off many poisonous chemicals, including hydrogen sulfide. After loading up, Compagnie Tommy simply waited until after midnight. Under the cover of darkness, the tanker trucks fanned out to dump the waste in 18 public open-air sites around the country's main city of Abidjan. These sites included the city's main garbage dump, a roadside field beside a prison, a sewage canal, and several neighborhoods. In a scene eerily reminiscent of the Bhopal disaster, citizens throughout the city awoke at night to an overpowering stench that burned their eyes and made it hard to breathe. By morning, nausea, vomiting, diarrhea, nose bleeds, stomach aches, chests pains, and breathing difficulties were affecting thousands of people. Tests later showed the sludge contained excessive levels of mercaptans and hydrogen sulfide, a potent poison that can quickly paralyze the nervous system, and cause blackouts, respiratory failure and death. More than 100,000 Abidjan residents sought medical treatment, and 69 were hospitalized as a result of the dumping. Fifteen people died. The spreading illnesses sparked violent demonstrations from a population convinced that government corruption was to blame for the dumping. The political furor ultimately forced the prime minister and his government to resign in September of 2006 (though many were later reinstated). Nevertheless, this mass resignation is unprecedented in the history of the Ivory Coast, and symbolizes the anger among the African people that their home would be used as a dumping ground by the advanced capitalist nations.

Limitations of the Basel Convention for Controlling Global Dumping of Toxic Waste

In 1989, some 118 countries signed onto the Basel Convention on the Transboundary Movement of Hazardous Wastes and Their Disposal. Enacted in 1992, the treaty was designed to better regulate the movement of hazardous waste between nations. Unfortunately, there were problems left unaddressed. For one, the Convention did *not* prohibit waste exports to any location except Antarctica, and instead merely required a notification and consent system known as "prior informed consent" (PIC). As such, if a nation did consent to accept hazardous wastes for disposal, but did not have the capacity to control and monitor such wastes in a safe and environmentally sound manner, then the prior informed consent rule was meaningless. In addition, a number of key nations, including the Ivory Coast and the United States, undermined the agreement by failing to ratify the main amendment to the Basel Convention. The Convention also did not adequately address the dumping of toxic products and materials through industrial recycling programs. Nevertheless, despite these problems, the Basel Convention established a new international norm that views the export of hazardous wastes from the North to the South as an unacceptable act of ecological imperialism.

Immediately after the adoption of the Convention, the international environmental movement and less developed countries went to work on overcoming its limitations. Over the course of the 1990s, their actions proved successful. The Convention has subsequently been strengthened through the adoption of hundreds of decisions, a protocol, an amendment, and the amendment of annexes. Of these agreements, the Basel Ban is the most important, as it puts into place a global ban on the export of hazardous wastes from members of the Organization for Economic Co-operation and Development (OECD) to non-OECD countries. It has, without question, in the words of the Basel Action Network, "transformed the Basel Convention from a control regime, to a no-exceptions, environmentally-justified trade barrier to hazardous waste."

Unfortunately, even as a non-Party, the United States has vociferously opposed improvements to the Basel Convention. In fact, since the very beginning of the Basel negotiations, both Republican and Democratic administrations alike have joined with the polluter-industrial complex to strongly oppose the concept of a no-exceptions waste trade ban. Furthermore, the United States is attempting to redefine what constitutes hazardous wastes, including efforts to avoid the Basel Convention for the management of end-of-life American ships or to de-list certain types of electronic wastes. . . .

Recycling "Trash for Cash"

In the new millennium, a new wave of waste trade is developing in the form of various "trash for cash" or recycling schemes of post-consumer products. Loopholes in the rules allow waste transfers to legally continue under the auspices of recycling. These exported wastes take the form of used car (lead acid) batteries, cell phones, plastics, heavy metals, old ships laden with asbestos, and lead scrap, which is shipped from the United States to southern China, India, Pakistan, the Philippines, Malaysia, and Taiwan for "recycling." In January of 1993 alone, for instance, the United States sent over 1,985 tons of plastic waste to India. These new types of waste products are becoming a far more serious form of toxic dumping in comparison to the export of toxic chemicals.

. . . Electronic waste (or E-waste) is perhaps the most rapidly growing waste problem in the world. According to the United Nations, about 20 million to 50 million tons of E-waste is generated worldwide annually. Such waste contains toxins like lead, mercury, and other chemicals that can poison waterways, the land, or air (if burned). The United States, which uses most of the world's electronic products and generates most of the E-waste, is able to significantly reduce disposal costs by shipping E-wastes to the developing countries, especially Asia. In addition to U.S. efforts to undermine the Basel Convention, the U.S. government has also intentionally exempted E-wastes from the Resource Conservation and Recovery Act (RCRA). In short, the export of E-waste to developing countries serves as an economic escape valve for American industry. Rather than designing products that are less toxic and that can be more easily rebuilt and reused, American business maximizes profits by building products with very hazardous components with a short life span. Some 20 million computers become obsolete each year in the United States, generating some 5 to 7 million tons of E-waste.

In the era of corporate-led globalization, toxic wastes disposal is running "downhill" on the path of least resistance. About 80 percent of the E-waste handled by traders is exported to Asia, and 90 percent of that is destined for China, where environmental regulations are weak and poorly enforced. E-waste today contains a witches' brew of toxic substances such as lead and cadmium in circuit boards; lead oxide and cadmium in monitor cathode ray tubes (CRTs); mercury in switches and flat screen monitors; cadmium in computer batteries; polychlorinated biphenyls (PCBs) in older capacitors and transformers; and brominated flame retardants on printed circuit boards; plastic casings, cables, and polyvinyl chloride (PVC) cable insulation.

The open burning, acid baths, and toxic dumping around recycling centers releases vast quantities of pollution. Lead levels in drinking water are 2,400 times higher than what the World Health Organization considers to be safe in the Guiyu region of China. Similar environmental problems can be found in India and Pakistan recycling operations. As in the United States, poor communities surrounding the plants bear the greatest health impacts from these operations. . . .

A Better World Is Possible?

In the era of neo-liberalism and corporate-led globalization, environmental justice (EJ) movements in the both the United States and the global South have a mutual interest in developing coordinated strategies. The growing ability of multinational corporations and transnational financial institutions to evade environmental safeguards, worker/community health and safety regulations, and dismantle unions and the social safety nets in the United States is being achieved by crossing national boundaries into politically repressive and economically oppressive countries. And in this context, abetted by "free trade" agreements and economic liberalization enforced by the WTO, various nationalities and governments are increasingly being pitted against one another to attract capital investment by dismantling labor and environmental laws seen as damaging to profits. In this respect, corporate-led globalization is weakening the power of the EJ movement to win concessions from the state and American industry.

At the same time, any potential victory by a community of color in the United States against the disposal of toxic incinerator ash in their own locality is quite limited if the result is the transport and disposal of the same waste in a poor West African community. If multinational corporations flee to the Third World to avoid environmental regulations and liability in the North, then the actions of U.S. environmentalists may be indirectly exacerbating environmental injustices elsewhere in the world. Stringent environmental standards must be applied to all nations in order to foster global environmental justice. A reworking of established "free trade" agreements in favor of more positive "fair trade" agreements are an important first step in the struggle to defeat neo-liberal economic policy. Such a "fair trade" agreement would establish minimum standards or "floors" for regulations rather than "ceilings." In other words, rather than a "race to the bottom," whereby the nation with the weakest environmental regulations sets the standard "ceiling" which all trading partners must accept, a transnational EJ movement must work for a series of mandatory strong standards that apply to all nations. Such a regulatory harmonization process would privilege nations with the strictest environmental laws as establishing a standard "floor" to which all other countries must comply if trade is to be conducted between them.

. . . The implementation of new international agreements and treaties to address the environmental injustices fostered by corporate-led globalization cannot be piecemeal in approach. Strong baseline standards around particular issues is not enough. Agreements must be comprehensive in nature, taking into account *all* of the interconnected processes by which ecological hazards are displaced and transferred between countries, and especially between the North and South. For instance, in response to the Basel Convention (and Basel Ban), there is evidence that as dirty industries are deterred from exporting hazardous wastes abroad, many factories are relocating from their home bases in the United States to more permissive investment locations in the poorer countries. Once relocated, industry is able to take advantage of the less stringent environmental regulations to more cheaply dispose of hazardous waste directly inside the new country. As a result, the intent of the Basel Ban will be defeated.

Unless comprehensive international rules are also put into place to govern foreign direct investment in "toxic" industries, hazardous wastes may still wind up in other countries via this alternative route. The migration of dirty industries to maquiladora zones in Mexico are a strong example of the migration process. . . . [T]here are signals that a new transnational EJ movement devoted to tackling the export of ecological hazards to poor communities of color inside and outside of the United States is beginning to take shape. The Southwest Network for Economic and Environmental Justice (SNEEJ), and the Environmental Health Coalition (EHC), for instance, are placing pressure on multinational corporations and government agencies to clean-up pollution along the U.S.-Mexico border. In addition, a coalition of Canadian, U.S., and Mexican organizations have successfully expanded right-to-know legislation in Mexico, including the establishment of a Pollutant Release and Transfer Register that is similar to those in Canada and the United States. Although still in its infancy, the rise of an environmentalism of the poor in the global South and new transnational networks of EJ organizations in the North are among the most promising vehicles for curbing the ecological horror stories brought about by corporate-led globalization.

In the short run, only by achieving greater social governance over trade and lending institutions and regulatory bodies can the process that leads different countries to sacrifice human and environmental health in order to compete in the world economy be overcome. This includes efforts to reestablish popular control over the United Nations as a counterweight to the WTO. . . . [I]nstitutions, including transnational corporations and large banks, the International Monetary Fund and the World Bank, the United Nations, and the General Agreement on Trade and Tariffs (GATT), must be opened up to greater public participation in decision-making. The antiglobalization movement, in the form of the International Forum on Globalization (IFG), has prepared alternative proposals for building a more just and sustainable international system that ends corporate dominance over the world economy. These IFG proposals include a system of unified global economic governance under a restructured United Nations. In the long run, however, even bigger transformations are necessary. . . .

Note

Faber, Daniel. 2008. "Chapter 4: The Unfair Trade-off: Globalization and the Export of Ecological Hazards." In *Capitalizing Environmental Injustice: The Polluter-Industrial Complex in the Age of Globalization.* New York: Rowman & Littlefield.

The Tragedy of the Commodity

4

The Overexploitation of the Mediterranean Bluefin Tuna Fishery

Stefano B. Longo and Rebecca Clausen

In a famous essay published in 1968, ecologist Garrett Hardin argued that a "tragedy of the commons" occurs when land—or any resource—is held in common. Hardin argued that when herders, for example, all used the same land to graze sheep, that land would ultimately be overused and depleted. This notion has since been widely deployed to justify the privatization of land, water, and other resources. Utilizing the overfishing of the Mediterranean bluefin tuna as a case study, Stefano Longo and Rebecca Clausen provide an alternative explanation; they argue that capitalism, which must constantly expand in order to survive, and the privatization associated with capitalism cause environmental degradation and resource depletion. They call this the "tragedy of the commodity."

> Hegel remarks somewhere that all great, world-historic facts and personages occur, as it were, twice. He has forgotten to add: the first time as tragedy, the second as farce.
>
> —Karl Marx, *The Eighteenth Brumaire of Louis, Bonaparte*, 1852 (Tucker 1978, p. 594)

In the late 20th century, the tragedy of the commons emerged as a leading thesis to explain the social origins of environmental resource depletion (Hardin, 1968). This framework has become one of the most cited theories in environmental social sciences, used consistently to explain a variety of ecological "tragedies" (e.g., Corral-Verdugo, Frías-Armenta, Pérez-Urias, OrduñaCabrera, & Espinoza-Gallego, 2002; McHugh, 1977; Nickler, 1999). Using a political economic framework, this article will develop a critique of this well-known theory. In doing so, we offer an alternative framework, the tragedy of the commodity, for examining the social drivers of environmental degradation. We argue that Marx's explanation of capitalist private property, commodity production, and the general formula for capital provide powerful theoretical guides for clarifying the social relations of production that have driven the overexploitation of fisheries in the recent past. We develop a case study of Atlantic bluefin tuna (ABFT) production, with an emphasis on the traditional Sicilian trap fishery, to illustrate the tragedy of the commodity.

The decline of ABFT provides important perspective on the sociological processes that affect marine ecosystems. By examining a historic fishery in the Mediterranean that is undergoing massive depletion as well as rapid technological change, we provide alternative explanations to oft-misunderstood problems in fisheries management. Relying on historical and qualitative data, the case study demonstrates the ways in which the socioeconomic imperative of capitalist private property toward accumulating surplus-value directs production, reorganizes social relations, and transforms nature into an instrumental input that can more easily serve the needs of capital.

. . . Latest research reveals that every square mile of the world's oceans has been affected by anthropogenic forces, of which overfishing is a primary concern (Halpern et al., 2008). More than 90% of the ocean's top predators have been overexploited and are ecologically threatened, including the bluefin tuna (Myers & Worm, 2003). Throughout the world, fisheries are in fast decline and aquaculture has become a major source of marine protein for human consumption (Food and Agriculture Organization of the United Nations, 2007). Indeed, the depletion of fisheries and growth of aquaculture are interrelated phenomena that can be illuminated by sociological analysis.

Our analysis of social and ecological transformations of global marine fisheries centers on three main themes: (a) refuting the dominant discourse of "inevitable" ocean decline due to the tragedy of the commons, (b) exposing the underlying structural factors of capitalist development that drive ocean depletion and disruptions in ocean ecosystems, and (c) exploring how the political economy of fisheries enables a tragedy of the commodity in modern bluefin tuna production . . .

The Tragedy of the Commons

Overexploitation in modern fisheries is often referred to as a manifestation of the "tragedy of the commons." The concept was developed by Garret Hardin (1968) to describe the conditions that shape the degradation of resources held in "common." Hardin's model is based on the notion that land, or other natural resources that are common property, will be degraded due to the competing individual interests of the users. In the absence of control or coercion by private entities or the state, Hardin maintained that the self-serving motivations of individual users inevitably lead to the destruction of commonly held nature.

. . . The classic illustration of the tragedy of the commons involved herders adding additional livestock to common grazing land to increase individual benefits. Hardin (1968) contended that individual "herdsmen" would attempt to acquire the benefits offered by the commons, while socializing costs to all the herders. By adding an extra animal to the pasture the herder reaps all the benefit but pays only a fraction of the environmental costs. As each actor is motivated by individual maximization of benefit, it is assumed that the herdsmen will increasingly introduce grazing animals into a finite ecological system, which leads to the tragic despoliation of the land. Therefore, Hardin (1968) concluded "freedom in commons brings ruin to all" (p. 1244).

The tragedy of the commons theory assumes that individuals are not constrained by conscience or by custom. This approach is essentially built on the notion that humans, in their inherent nature, are self-serving, will make rational cost-benefit decisions, and that resource users, typically, will not be motivated or constrained by altruism or the common good. This leads to a critical examination of the liberal ideology of personal freedom, in that, in order to avert tragedy, individuals must be constrained by either state and/or private actors (including firms). . . .

Hardin was a staunch defender of the liberal notion of private property, which served as a linchpin to his theory. The approach contends that private property, as exclusionary property relations, can prevent society from undermining resources and the destruction of natural systems. Although acknowledging that this property arrangement is not necessarily a just or fair distribution, Hardin claimed that private property allows for control over resource depletion by barring some users access as well as investment in and/or protection of crucial resources to support long-term production. For Hardin, and many others who have adopted this perspective, private property arrangements are offered as a leading policy solution for avoiding ecological tragedies.

Tragedy of the commons theorists also acknowledge the potential for state action and management as alternative arrangements for promoting resource conservation (Dietz, Dolsak, Ostrom, & Stern, 2002). The type of state control promoted by Hardin's thesis can be best characterized as an approach to resource protection that excludes or coerces users through creating the legal structure to sustain resource extraction within a competitive market economy and capitalist private property arrangements. For example, legal codes enforce limited access through licensing or individual quotas, sometimes referred to as "catch shares" (Costello, Gaines, & Linham, 2008). In essence, the tragedy of the commons theory promoted the notion that the role of the state is to enclose the commons, sometimes resulting in individual entry permits and/or transferable quotas, usually bypassing local users, and creating broad management schemes over vast natural resources (Ostrom, 1999). Generally speaking, the tragedy of the commons was built on the premise that only private control or state (top-down) management can prevent ecological devastation and the tragedy of the commons (Hardin, 1998).

. . . [T]he theory has not been without its critics. McCay, Acheson, and others have made key contributions to the literature on the commons (Acheson, 1975, 1988; Feeny, Berkes, McCay, & Acheson, 1990; McCay, 1998; McCay & Acheson, 1987; McEvoy, 1986). These scholars have examined differing examples and definitions of "the commons" and the ways in which local cultural factors, historical conditions, and community settings can play a role in maintaining viable "common property" fisheries. In addition, Dietz, Ostrom, and others (Dietz et al., 2002; Dietz, Ostrom, & Stern, 2003; Ostrom, 1999; Ostrom, Burger, Field, Norgaard, & Policansky, 1999) have developed major works in this area. Their focus has been on examining the role of common ownership as a viable property regime and contextualizing some economic assumptions of the tragedy of the commons. This research has revealed that private control and state regulation can be conflated or that the "drama of the commons" (Dietz et al., 2002) depends on the composition of state policies and governance, together with political economic and historical conditions. Furthermore, governments can, potentially, promote environmental sustainability or advance environmental tragedies (Ostrom et al., 1999).

. . . We extend the critique of the tragedy of the commons by drawing on Marxist political economy to highlight the fundamental contradictions of capitalist agri-food production that are driving the oceanic crisis. By making use of the theoretical frame provided by the tragedy of the commodity, we aim to clarify the root dynamics that shape the historical context for agri-food production and environmental degradation in the modern era. We illustrate this using Atlantic bluefin tuna as a case study.

Capitalist Production and the Tragedy of the Commodity

Much has been made of the tragedy of the commons theory as an explanatory model. Surely, it provides descriptive power in that it can explain the behaviors of individual actors in given social circumstances. Nevertheless, the crude formulation neglects historically specific conditions, taking them as transhistorical constants. The tragedy of the commodity approach, emerging from a historical materialist perspective, contends that an analysis of human agency must include a thorough understanding of the historical social organization of production and consumption.

Marx's (1977) systematic critique of the categories of political economy in *Capital* is at the same time a systematic elaboration of the "capitalist mode of production, and the relations of production and forms of intercourse [*Verkehrsverhältnisse*] that correspond to it" (p. 90) . . .

[Marx's] political economy described social conditions as the outcome of historical processes and its specific social forms, not a reflection of invariant natural laws (Sweezy, 1942). That is to say,

> Nature does not produce on the one hand owners of money or commodities, and on the other hand men possessing nothing but their own labour-power. This relation has no basis in natural history, nor does it have a social basis common to all periods in human history. It is clearly the result of past historical developments, the product of many economic revolutions, of the extinction of a whole series of older formations of social production. . . . *Definite historical conditions are involved in the existence of the product as a commodity.* (Marx, 1977, p. 273, italics added)

This examination of the commodity unveils the fundamental makeup of the capitalist system of production. Here, the commodity forms a microcosm of the social organization and its consequent relations of production; "a system that appears as an 'immense collection of commodities'" (Marx, 1977, p. 125).

Marx explained the underlying characteristics of commodities as use-value and exchange-value. That is, "Every useful thing . . . may be looked at from the two points of view of quality and quantity" (Marx, 1977, p. 125). Extending this simple formulation by the force of a historical materialist analysis, the dual character of the commodity reveals a central contradiction of capitalist production. That is, the quantitative form of value is divorced from its qualitative dimension, and social production is accordingly organized along this historical form. To elucidate this analytical insight, we provide a brief overview of Marx's analysis of capitalist commodities and valorization.

Value, Growth, and the Exploitation of Nature

In the general formula for capital, Marx (1977) distinguished between simple circulation of commodities and the circulation of money as capital. In the former, depicted as C-M-C, commodities are the beginning and end of the process. Although mediated by money, it reaches conclusion with the use of a commodity, "in short use-value, is therefore its final goal" (Marx, 1977, p. 250). The outcome results in a qualitatively different form that serves a particular need, and circulation is complete. Conversely, in capitalist commodity production, money or exchange-value are the beginning and end result. Depicted as M-C-M, money is put into circulation to return money, a quantity for a quantity, "its driving and motivating force is therefore exchange-value" (p. 250). Accordingly, the process is fundamentally transformed from one organized around the exchange of qualities to the exchange of quantities.

In a capitalist exchange of two quantities, beginning with M and ending with M, it would be nonsensical unless the quantity at the end of the exchange was different (larger) than that existing at the outset. Marx explained that the complete depiction of the general formula for capital consists of MC-M', where M' is equal to M +ΔM. Thus, capitalist circulation requires the realization of surplus-value within ΔM, or what forms the basis of profit and capital accumulation. Providing the impetus for reinvestment, it becomes the central aim of production.[1] Marx unambiguously described how capitalist production prioritizes surplus-value when he stated, "The absolute value of a commodity is, in itself, of no interest to the capitalist who produces it. All that interests him is the surplus-value present in it, which can be realized by sale" (Marx, 1977, p. 437).

Under capitalist social relations of production, labor itself becomes a commodity for sale. The capitalist buys labor power, and "the worker receives a means of subsistence for his labour power" (Tucker, 1978, p. 209). Although labor is often misinterpreted or misunderstood as the single domain of value production, commodified labor-power *and* nature come together to produce values. In the "Critique of the Gotha Program," Marx made clear that both human labor and nature are fundamental to the production process (Tucker, 1978). "Marx does *not* see abstract labour time as an adequate representative or measure of wealth—including the wealth of nature" (Burkett, 2006, p. 28). Human nature acts along with and on extra-human nature, generating a useful form.

As the capitalist valorization process requires a quantitative difference in value when capital is circulated, and labor and nature are providing the material means for producing value, the disparity in value must be found in the materiality of the commodity production process. Profit does not appear through "a mystical source of self-valorization" that results from circulation "independent of its production process," as presented by many modern economists (Marx, 1992, p. 204). The capitalist valorization process requires the exploitation of labor and nature, which forms the locus of surplus-value. Therefore, capital does not pay the entire costs of production. Referring to the "free gifts" of nature, Marx clarified that capitalist commodity production appropriates nature and its use-value, which permits the production and realization of surplus-value. Although nature is not a product of wage labor, it is expropriated *gratis* by capital through commodity production (Burkett, 2006).

Because exchange-value, and its embodied surplus-value, is the ultimate goal of capitalist commodity production, profit realization becomes an endless process. The process begins and ends with money (quantities), and the result of one cycle is the beginning of the next. "The movement of capital is therefore limitless" (Marx, 1977, p. 253). At each stage of circulation there is a new starting point, relatively larger than the last, resulting in an ever-increasing scale of profit accumulation as the quantitative nature of commodity production knows no restriction or limits (Foster, Clark, & York, 2010). To be sure, in order to realize surplus-value, the commodity must be sold, thus fueling the sales effort toward increasing effective demand (Baran & Sweezy, 1966).

. . . [U]nbounded accumulation of capital becomes the motor force of capitalist production (Sweezy, 1942). Under the compulsion of competition, capitalist producers expropriate value to expand and accumulate capital, staying ahead of their competitors. As such,

> The development of capitalist production makes it necessary constantly to increase the amount of capital laid out in a given industrial undertaking, and competition subordinates every individual capitalist to the immanent laws of capitalist production, as external and coercive laws. It compels him to keep extending his capital, so as to preserve it, and he can only extend it by means of progressive accumulation. (Marx, 1977, p. 739)

The capitalist production process is geared toward producing surplus value, and generally this is accomplished by extending the working day (absolute surplus-value) or by increasing productivity (relative surplus-value). Intensifying technological investments so as to expand the productivity of capital allows it to stay ahead of real or potential competitors. As there are defined limits on absolute surplus-value, increasing relative surplus-value becomes an essential method of capital accumulation and growth. Consequently, capitalist production must be continually transformed or revolutionized (Braverman, 1998).

In what Schumpeter (1962) called "creative destruction," capital relentlessly seeks new ways to develop the productive forces of labor and nature and expropriate value. This "creative destruction" is often implemented through the means of technological change, which expands capital efficiency and must, first and foremost, meet the needs of capital. As Braverman (1998) made clear, "Technology instead of simply *producing* social relations is *produced* by the social relations," and furthermore, "The social form of capital, driven to incessant accumulation as the condition for its own existence, *completely transforms technology*" in its own image (p. 14).

The historical tendency of capitalist accumulation was described by Marx (1977) as a process of expropriation, whereby individual private property is "supplanted by capitalist private property, which rests on the exploitation of alien, but formally free labor" (p. 928). This formulation is an important conceptual distinction that aids our understanding of "the great transformation" (Polanyi, 1957) and is particularly relevant to this case study. The need to transform land into capitalist private property is the driving force of the enclosure of the commons. That is, capitalist private property is the social form that promotes enclosures, and the process of primitive accumulation, so as to conform to the needs of commodity production described above. This occurred during the prehistory of capitalist production, but also continues into the present via the power of capital to sweep away traditional production systems, including the individual private property of small-scale industry, that do not easily conform to the essential characteristics of capitalist development, particularly the expropriation of surplus-value. Thus, "It has to be annihilated; it is annihilated" (Marx, 1977, p. 928).

Our case study of the Atlantic bluefin tuna in the Mediterranean illustrates how the tragedy of the commodity leads to both the catastrophe of ABFT depletion as well as the chimera of technological solutions to address the bluefin tuna's recent demise. We suggest that the first consequence of the historically specific form of commodity production is the *tragedy* of resource degradation (i.e., overfishing). As Marx's addendum to Hegel's remark alludes, the consequences of commodification do not end there. They manifest in a second occurrence as the *farce* of technological "solutions" to the original problem (i.e., tuna ranching and aquaculture).

Commodification and Overexploitation

The Mediterranean bluefin tuna fishery is one of the oldest and, at one time, likely one of the most productive fisheries in the world. Human societies have been capturing bluefin tuna in this region for about 10,000 years. Circa 1000 CE, a trapping system developed that fixed this fishery as a Mediterranean icon (Sarà, 1998). Traditional trap fishing allowed for an ecologically sustainable harvest of bluefin tuna that was viable for centuries.

The history of the Sicilian trap fishery, *la tonnara*,[2] can help illuminate the present circumstances in the Mediterranean and the emergence of the tragedy of the commodity. Historical research on the *tonnara* displays a pattern of attention by the managers, owners, and workers regarding the impacts that particular technologies might have on the bluefin populations. Many actors associated with the fishery maintained that practices promoting intensified capture methods would have damaging consequences for the long run of the fishery.

La Mantia (1901) described contestations between active *tonnare* over the amount, location, and technology used in fishing for bluefin tuna as well as other species during the bluefin spawning season. He noted that disputes occurred as early as the 16th century. For example, in the Sicilian city of Palermo, there were concerns among managers and fishers that the location of the *tonnara* in Arinella was too close to that in Mondello, which began proceedings to enact minimum distances between active bluefin fishing. Also, bans were put in place regarding the types of fishing nets that could be used in the vicinity of the *tonnara*, and the capture of smaller tuna species was prohibited by a number of decrees in 1784, 1785, and 1794 in Sicily as a result of concerns among those representing established trap fishing (La Mantia, 1901). Furthermore, D'Amico (1816) discussed the potential damage to the bluefin fishery by practices that resulted in overfishing and the capture of juvenile fish, arguing that these practices "sterilized" the fishery.[3]

Interviews with *tonnaroti* (tuna fishers) revealed a deep understanding of the bluefin reproductive processes and an awareness of issues related to overfishing. On more than one occasion, fishermen clearly expressed that it was necessary to "leave fish for next year" and that this was a traditional fishing ethic. Unlike the herders in Hardin's theory, who disregard the future implication of overgrazing, many participants who took part in this study were unquestionably mindful of the future of the fishery and future generations. For example, this was explained in terms of the *tonnara's* selectivity. Because of its design, methods, and stationary nature, the trap did not deplete the fishery.

Although the traditional Sicilian fishery may be considered a "commons," there were many social arrangements surrounding access, methods of capture, and distribution of resources (Consolo, 1986; Sarà, 1998). Historical research, substantiated by in-depth interviews, suggests that the traditional trap fishery was undoubtedly integrated into community life, and the sustainability of the fishery was associated with the welfare of the community. Thus, legal decrees, fishing practices, and traditions in the trap fishery indicate a relatively sophisticated understanding of resource sustainability. Unfortunately, this was not carried over into the recent phase of bluefin fishing.

Within the last half-century, the Mediterranean ABFT fishery has been dramatically over-fished, first by industrial long-lines and most recently by purseseines (International Commission for the Conservation of Atlantic Tunas [ICCAT], 2010b). As the fishery entered the modern era, new methods of production were sought to ensure increases in production and value. Fishing vessels grew in size, power, and on-board equipment, and the development of capital- and technology-intensive methods significantly reduced the demand for labor (Safina, 2001). Making use of modern technology with lethal efficiency, powerful fishing gear, including radar, sonar, and spotter planes, increased captures in earnest during the 1980s and continued for more than 20 years (ICCAT, 2010a). At the same time, the labor-intensive traditional trap technology was on the decline or had collapsed.

Furthermore, governments and industry groups pushed for expansion of fishing capacity and technology in the name of food security and jobs.[4] This resulted in massive subsidies entering into the fishery offered by regional governments and the European Union. The subsidies have been used to "modernize" the fleet, develop tuna ranching facilities, and research bluefin tuna aquaculture (Bregazzi, 2005; Doumenge, 1999; Tudela, 2002). By the mid-1990s, bluefin tuna "ranching" emerged as the newest technology for providing this high-value commodity for the global market. Soon after, ranches were the leading sources of bluefin tuna originating from the Mediterranean.

The expansion of production is readily observed in the era following the Second World War. The growth of investment in more technologically intensive fishing efforts with the intention of increasing production developed at an unprecedented rate (Bregazzi, 2005). As a result, bluefin tuna captures in this region increased from about 5,000 tons in 1950 to about 40,000 tons in 1994 and close to 60,000 tons in 2007 (ICCAT, 2010b).

During this era, the ABFT became a global commodity that was associated with possibly the highest value of all fish species. With prices that could run as high as $1,000 per kilogram, the potential for large returns drew in transnational capital (Normile, 2009). Fishing technology in the Mediterranean ABFT fishery was shaped by a globalized industry led by transnational firms such as the Mitsubishi Corporation, the largest in the fishery, and fishing industry interests in the region. A thriving global market for sashimi-grade tuna thrust the bluefin tuna into stardom. With skyrocketing prices that reflected its global fame, it became a culinary status symbol, particularly in Japan (Bestor, 2001; Issenberg, 2007).

In the modern era, the global bluefin tuna market grew as a luxury food product or a "boutique species" (Safina, 2001). Thus, its value as a commodity stemmed not from its nutritive capacity but from market exchange. That is, in its commodified form, it has a high exchange-value, but low use-value. While other food products, such as beef, have been historically tied to high status, the sociocultural phenomenon surrounding bluefin tuna is unique. Little physical sustenance is garnered from 10 to 20 grams of sashimi-grade tuna, the largest global market for bluefin tuna. This became a symbol for the global elite, and later the middle class. Socially constructed notions of prestige that display one's position in the social hierarchy and fashionable perceptions of sushi/sashimi shaped global market demand in wealthy nations, which were pushed by marketing and other powerful media sources (Issenberg, 2007).

The political-economic power of the industry is evidenced in the many unsuccessful attempts by nongovernmental organizations to reign in captures through policy and management by ICCAT or the Convention on International Trade in Endangered Species during the first decade of the 21st century (Bregazzi, 2006; Vasquez, 2010). Incidentally, it is indicative that the management practices developed by ICCAT, essentially a private quota system, were very likely influenced by the logic of the tragedy of the commons theory, which maintains that private property arrangements are a practical solution to overfishing.[5]

The traditional bluefin tuna trap fishery was dually affected by the transformations in the Mediterranean. First, the *tonnara* has a limited capacity for expansion and investors slowly abandoned it for more powerful technologies. The industrial fleets increased the size of the total catch and lowered the total costs of labor.[6] Second, over time, fishing by industrial fleets diminished the trap fishery's captures due to the heavy impact on Mediterranean bluefin populations. The capture levels in the traditional fishery showed significant declines in the latter decades of the 20th century (ICCAT, 2010b). Within a short time the Sicilian tonnare disappeared, and the traditional trap fishery throughout the Mediterranean has been decimated. Many practices and relationships that developed in the traditional fishery, such as the *tonnara*, focused on qualitative concerns including community welfare and future generations, which came to stand in contradiction with the concerns of capital.

Ranching Bluefin Tuna

The practice of bluefin tuna ranching involves capturing live ABFT from the open ocean and transferring them to holding pens to increase the fat content of the tuna. As ABFT sushi/sashimi is highly prized for its fatty or oily texture, something that only a short time ago was looked on disapprovingly by consumers, bluefin ranching develops as a method to add value to the commodity (Longo, 2011). Bluefin ranching in the Mediterranean originated in Cueta, Spain, to fatten postspawning tuna (Miyake et al., 2003). Because of the fact that bluefin tuna exert high energy and feed less frequently during spawning, their fat content and weight declines during this phase of the life cycle. Fattening postspawning ABFT could add value when its condition was considered suboptimal for market. However, recent ranching practices began capturing ABFT during the spawning period to further increase fat and weight before they are shed through the reproductive process.

Ranching is a method that can be economically efficient but is ecologically harmful. For example, bluefin tuna food conversion ratio estimates are as high as 30:1, meaning that it can take up to 30 pounds of feed to increase the weight of a bluefin tuna by 1 pound (Relini, 2003; Tudela, 2002). Capturing live ABFT for fattening has no material benefits for human subsistence. That is, there is no increase in the total protein available or calories produced for human development. More energy and protein are consumed in ranching than are gained. Furthermore, tuna ranching operations increase pressure on wild bluefin tuna stocks because they capture live fish for fattening. Capturing ABFT during their spawning migration disrupts their life cycle, a prime example of what has been referred to elsewhere as the metabolic rift (Foster 1999; Longo, 2012). Thus, bluefin tuna ranching has been ecologically damaging to Mediterranean bluefin stocks and ocean ecosystems.

Ranching has played an important role in the impending development of ABFT aquaculture. Aquaculture entails closed captivity of the species during its entire life cycle. These facilities have provided a testing ground for confined production and have been used as resources for research and development. Recent European Union—funded efforts have conducted multimillion dollar research programs to "domesticate" ABFT for commercial production (SELFDOTT, 2010). This has been regarded as the sustainable alternative, as it seeks to avoid the overexploitation of bluefin and maintain industry growth. However, tuna aquaculture is still susceptible to many of the ecological and social contradictions associated with ranching, most notably the challenges associated with bluefin tuna's food conversion ratio.

The Tragedy of the Commodity, Capitalist Valorization, and Bluefin Tuna

The inherent dynamics and logic of the social conditions of "capitalist private property" (Marx, 1977) create the historical context for environmental tragedies to unfold. In the marine environment, the drive to accumulate has spurred the capitalist commodification processes, leading to an oceanic crisis (Clausen & Clark, 2005). This crisis exemplifies both resource overexploitation as well as the unintended ecological and social consequences resulting from technological developments . . .

The traditional trap fishery was not subject to this tragedy until very recently. This fishery operated for close to a millennium and has been devastated within a few decades. Its demise occurred at precisely the time that the species underwent an unprecedented market metamorphosis. The collapse of this traditional fishery has been due, in no small part, to the processes of capitalist valorization, illustrating a tragedy of the commodity. As capital reaches ecological and social boundaries to realizing surplus-value, it seeks to overcome them in various ways (Foster et al., 2010).

To overcome these boundaries, capital searches for new geographic areas for exploitation along with the creation of new commodities and markets for investment and consumption. Bluefin tuna consumption expanded worldwide, particularly in Japan, with the increasing popularity of sashimi-grade bluefin tuna, which became a highlight on the menus of elite eating establishments (Bestor, 2001; Issenberg, 2007). As the Mediterranean became one of the last areas of abundant bluefin tuna stocks, global capital rushed in. In the process, the *tonnara* was "creatively" destroyed, or as Marx stated, it was "annihilated."

Capital slowly took a dominant role in the Mediterranean bluefin fishery, and there was a period of coexistence between traditional trap fishing and modern fishing. Yet even as individual private property, the *tonnara* was limited as a result of social and technical boundaries, and therefore inadequate to the dictates of capitalist private property. By examining this closely, it becomes clear that the "tragedy" of the *tonnara* emerged within the transformations in the social relations of production.

As noted, the capacity for expansion is constrained in the traditional trap fishery. First, it is a passive (stationary) trap and subject to variations in natural systems and contingency. Second, it is a labor- and knowledge-intensive system, which relies little on capital or high-technology intensive methods. The method of production is fundamentally linked to the natural metabolism of the bluefin and its reproductive cycles (Longo, 2012). Thus, in 18th or 19th century Sicily, for example, any potential for growth was constrained to increasing absolute surplus-value by reducing the number of laborers and expanding the working day, or simply good fortune (Calleri, 2006). Because of the social and technical character of the capture system, expansion had clear limits.

The *tonnara* required a work force that had comprehensive knowledge of the local oceanic conditions. This is most clearly evidenced by the senior tuna fisherman or *tonnaroti* (Longo, 2012). Many members of the crew were not easily replaceable. The social, technological, and knowledge barriers to expanding surplus-value could only be superseded by revolutionary changes in the methods of bluefin production, increasing relative surplus-value, and social abandonment of the *tonnara*. The expansion of capitalist private property and social power of capitalist commodity production created a dynamic that undermined sustainable harvesting methods, such as the *tonnara*.

Today, the Mediterranean fishery is at the brink of economic and ecological collapse (ICCAT, 2007; MacKenzie, Mosegaard, & Rosenberg, 2009). ABFT population levels have been depleted to historically low levels. Indeed, captures in the broader Mediterranean have peaked, even with increasing effort[7] (ICCAT, 2010a). In addition, the onset of tuna ranching has only exacerbated the problem by adding other environmental concerns to the mix, including overfishing of feed-fish species. Very recent developments in bluefin tuna domestication and aquaculture are regarded as the technical fix toward achieving sustainability. However, this will be plagued with its own set ecological problems, many of which have been displayed in other modern aquaculture systems based on capitalist private property (Naylor et al., 1998).

These phenomena can be regarded as the latest wave of technologies driven by capitalist valorization. The logic that prevails is bound by capitalist private property, which seeks to revolutionize the means of production in the process of finding new avenues for realizing surplus-value. Capturing and ranching a high-trophic-level species such as ABFT is ecologically inefficient and socially detrimental (Longo, 2012). Nevertheless, it converts low-value "feed" species (e.g., sardines) into high-value fatty bluefin sushi/sashimi (*magurotoro*).

Tuna ranching serves as a clear example of a technological development that expropriates value from nature to maximize revenues, a fact that is unlikely to be disputed by almost anyone in the industry. Yet this erodes ocean resources and externalizes the costs of production. This process will be further complicated by a complete transition to bluefin tuna domestication. ABFT aquaculture in the Mediterranean is not yet commercially viable, but there are ongoing efforts to advance it as a lucrative enterprise. A species such as ABFT, with its high metabolic rate and physical size, is an ecologically irrational choice for domestication and aquaculture. The (ir)rationality of these decisions lies purely in the market realm, where exchange-value determines production outcomes.

It is erroneous to deem the recent technological transformations as driven by the individual fishing boats and self-interested bureaucrats, or a self-driven process of "pure" scientific advancement. The increase in captures was fundamentally tied to the expansion of productive capacity and the effectiveness of technological adaptations that brought high returns for the fishing industry and transnational capital. The tragedy of the commodity approach highlights that this is characteristic of the commodification process, constantly pushing the bounds of productivity seeking relative surplus-value (Braverman, 1998). Thus, commodification drives the ecological tragedy.

. . . Transformations in the ABFT fishery were based on expanding the global bluefin market, led by the social construction of a luxury food item. Nevertheless, for capitalist commodity production, "the nature of the needs, whether they arise, for example, from the stomach, or the imagination, makes no difference" (Marx, 1977, p. 125). Such are the consequences of the capitalist valorization process. The high market value of this global food commodity and the needs of capitalist private property propelled increased fishing effort and caused dramatic alterations in the production process and the ecology of the fishery.[8]

Conclusion

The logic of economic growth at all costs, ignoring ecological boundaries, is essential to capitalist production in part due to a real, or perceived, competitive structure and the drive for capital accumulation. When analyzing complex resource concerns such as those occurring in the Mediterranean bluefin tuna fishery, the seemingly logical tragedy of the commons explanatory model of individual selfishness and negligence is oversimplified. This perspective ignores the broader sociostructural and historical realities that are centrally important when analyzing these issues.

According to the tragedy of the commons theory, problems are formulated largely as relations between humans and nature, neglecting how this relationship is mediated through social relations of production (Foster, 2002). Historically, specific circumstances and social relations affect the ways in which communities access common property resources such as the oceans (Berkes, Feeny, McCay, & Acheson, 1989). The dominant discourse on ABFT, focusing on individual culpability, universal human characteristics, and population, overlooks the systemic structure of capitalist private property and commodity production and its inherent expansionary and exploitative character.

Contrary to Hardin's (1968) theory, the forces of overexploitation in the Mediterranean are not individual selfishness or resource users that are rational utility maximizers. Rather, the drivers of bluefin tuna depletion are fundamentally based in the historically specific relations of capitalist private property and commodity production. The tragedy of the commodity emerged as the quantitative form of value, and the inherent surplus-value embodied in it became divorced from its qualitative dimension. Capitalist private property seeks every avenue to expand into all aspects of social life and extinguish methods of production that can encumber the endless accumulation of surplus-value (Braverman, 1998).

In the case of ABFT, we do *not* find that the tragedy of the commons is causing the demise of the species. Rather, Atlantic bluefin tuna suffers from the tragedy of the commodity as capital creates opportunities for investors to increase the rate of returns, driving production decisions and technological developments. As Burkett (2006) makes clear, "The real tragedy of the commons has been the depletion and despoliation of communal resources by private, market-driven economic activity, that is, the inadequate recognition and enforcement of communal property in the form of strict user rights and responsibilities" (p. 82).

Historically, the traditional trap fishery in Sicily, the *tonnara*, was a system of production that allowed resource users access, yet social restrictions were imposed and material circumstances were limiting. This common resource was organized around social conditions that were not yet fully penetrated by capitalist private property. Modern ABFT production, born of capitalist private property and the unending quest to maximize surplus-value, became the form-determinant of fishing methods, technology, and the labor process in the modern era, resulting in a host of social and ecological contradictions. Thus, ecologically inefficient and environmentally destructive practices are implemented in the name of economic growth.

Applying Hegel's and Marx's aphorism, if overfishing in the Mediterranean is a tragedy, then ABFT ranching and aquaculture are a farce. Overexploitation is one of the manifestations of the tragedy of the commodity and is devastating in and of itself. However, the historical phenomenon of commodification deals a second blow by introducing tuna ranching and aquaculture. The destruction of bluefin tuna populations, as well as other ocean species, can only be addressed by confronting the tragedy of the commodity. In doing so, we begin to challenge the dominant assumptions that have governed modern natural resource management strategies.

Notes

Longo, Stefano B., and Rebecca Clausen. "The Tragedy of the Commodity: The Overexploitation of the Mediterranean Bluefin Tuna Fishery." *Organization and Environment* 24(3): 312–328. Reprinted by permission of Sage Publications.

[1] While seemingly absurd on the surface, the general formula for capital could be said to reach its logical conclusion in M-M', or what is referred to as finance capital, money begetting more money. With this formulation, quality completely gives way to quantity.

[2] The Sicilian traditional trap fishery was considered to be one of the earliest and most productive trapping systems in the region (Mather, 1995; Sará, 1998).

[3] In recent times, some fisheries scientists have maintained that overfishing of juvenile ABFT, particularly those less than 1 year old and weighing as little as 4 kilograms, by modern industrial fishing has negatively affected the fishery (Mather, 1995; Safina, 2001).

[4] The irony that new technologies were decreasing the need for labor in fishery was not lost on local tuna fishers who participated in this research.

[5] ICCAT manages the Mediterranean bluefin fishery by determining the maximum sustainable yield (MSY) and allocating a quota or a total allowable catch (TAC) in tons to each member nation. Member nations are expected to enforce their own TACs. This is done mainly by allocating a quota for the fishing season to each vessel in a nation's fleet.

[6] While there were attempts to lower the costs of labor in the modern history of the *tonnara*, this could not address the ongoing contradictions that the method of production posed for capital.

[7] This phenomenon can be indicative of the precollapse stage of a fishery (Mullon, Fréon, & Cury, 2005).
[8] The rate of expansion on ABFT production far exceeds the rate of population growth. Annual catches of Mediterranean ABFT increased more than 10-fold in the span of about 50 years (ICCAT, 2007).

References

Acheson, J. M. 1975. "The Lobster Fiefs: Economic and Ecological Effects of Territoriality in the Maine Lobster Industry." *Human Ecology* 3: 183–207.

———. 1988. *The Lobster gangs of Maine.* Lebanon, NH: University Press of New England.

Anonymous. 2005, May. "The Tragedy of the Commons, contd." *The Economist.* http://www.economist.com/node/3930586?story_id=3930586.

Baran, P. A., and P. M. Sweezy.1966. *Monopoly Capital: An Essay on the American Economic and Social Order.* New York: Monthly Review Press.

Benjamin, W. 1955. "Theses on the Philosophy of History." In *Illuminations*, edited by W. Benjamin, 255–66. New York: Harcourt, Brace, & World.

Berkes, F., D. Feeny, B. J. McCay, and J. M. Acheson. 1989. "The Benefits of the Commons." *Nature* 340: 91–93.

Bestor, T. C. 2001. "Supply-Side Sushi: Commodity, Market, and the Global City." *American Anthropologist* 103: 76–95.

Braverman, H. 1998. *Labor and Monopoly Capital: The Degradation of Work in the Twentieth Century.* New York: Monthly Review Press.

Bregazzi, R. M. 2005. *The Tuna Ranching Intelligence Unit.* New Orleans, LA: Advanced Tuna Ranching Technologies.

———. 2006. *Thunnus nostrum: Bluefin Tuna Fishing & Ranching in the Mediterranean Sea 2004–2005.* Madrid, Spain: Advanced Tuna Ranching Technologies.

Burkett, P. 2006. *Marx and Ecological Economics: Towards a Red and Green Political Economy.* Boston, MA: Brill.

Calleri, N. 2006. *Un' impresa Mediterranea di pesca: I Pallavicini el le tonnare delle Egadi neisecoli 17–19* [A Mediterranean Fishing Company: The Pallavicini and Traps of the Egadi during the 17th-19th Centuries]. Genoa, Italy: Unioncamere Liguria.

Clausen, R., and B. Clark. 2005. "The Metabolic Rift and Marine Ecology: An Analysis of Ocean Crisis within Capitalist Production." *Organization & Environment* 18: 422–44.

Consolo, V. 1986. *La pesca del tonno in sicilia* [Tuna Fishing in Sicily]. Palermo, Italy: Sellerio Editore.

Corral-Verdugo, P., M. Frías-Armenta, F. Pérez-Urias, V. Orduña-Cabrera, and N. Espinoza-Gallego. 2002. "Residential Water Consumption, Motivation for Conserving Water and the Continuing Tragedy of the Commons." *Environmental Management* 30: 527–35.

Costello, C., S. D. Gaines, and J. Linham. 2008. "Can Catch Shares Prevent Fisheries Collapse?" *Science* 321: 1678–81.

D'Amico, F. C. 1816. *Osservazioni pratiche intorno alla pesca, corso e cammino dei tonni* [Practical Observations around Fishing, Course and Path of the Tuna]. Messina, Italy: Società Tipografica.

Dietz, T., N. Dolsak, E. Ostrom, and P. Stern. 2002. *The Drama of the Commons.* Washington, DC: National Academies Press.

Dietz, T., E. Ostrom, and P. C. Stern. 2003. "The Struggle to Govern the Commons." *Science* 302: 1907–12.

Doumenge, F. 1999. "La storia delle pesche tonniere [The History of Tuna Fishing]." *Biologia Marina Mediterranea* 6: 5–106.

Feeny, D., F. Berkes, B. J. McCay, and J. M. Acheson. 1990. "The Tragedy of the Commons: Twenty-Two Years Later." *Human Ecology* 18: 1–19.

Food and Agriculture Organization of the United Nations. 2007. *The State of the World's Fisheries and Aquaculture 2006*: Rome, Italy: Author.

Foster, J. B. 1999. "Marx's Theory of Metabolic Rift: Classical Foundations for Environmental Sociology." *American Journal of Sociology* 105: 366–405.

———. 2002. *Ecology against Capitalism*. New York: Monthly Review Press.

Foster, J. B., B. Clark, and R. York. 2010. *The Ecological Rift: Capitalism's War on the Earth*. New York: Monthly Review Press.

Fracchia, J. 1991. "Marx's *Aufhebung* of Philosophy and the Foundations of a Materialist Science of History." *History and Theory* 30: 153–79.

———. 2004. "On Transhistorical Abstractions and the Intersection of Social Theory and Critique." *Historical Materialism* 12: 125–46.

Gordon, H. S. 1954. "The Economic Theory of a Common-Property Resource: The Fishery." *Journal of Political Economy* 62: 124–42.

Greenberg, P. 2010, June 22. "Tuna's End." *The New York Times.* http://www.nytimes.com/2010/06/27/magazine/27Tuna-t.html.

Halpern, B. S., S. Walbridge, K. A. Selkoe, C. V. Kappel, F. Micheli, C. D'Agrosa, J. F. Bruno, K. S. Casey, C. Ebert, H. E. Fox, R. Fujita, D. Heinemann, H. S. Lenihan, E. M. Madin, M. T. Perry, E. R. Selig, M. Spalding, R. Steneck, and R. Watson. 2008. "A Global Map of Human Impact on Marine Ecosystems." *Science* 319: 948–52.

Hardin, G. 1968. "The Tragedy of the Commons." *Science* 162: 1243–48.

———. 1974. "Lifeboat Ethics: The Case against Helping the Poor." *Psychology Today* 8: 38–43.

———. 1998. "Extensions of the Tragedy of the Commons." *Science* 280: 682–83.

Heilbroner, R. L. 1986. *The Essential Adam Smith*. New York: W. W. Norton.

International Commission for the Conservation of Atlantic Tunas. 2007. *Report of the Standing Committee on Research and Statistics (SCRS)*. Madrid, Spain: Author.

———. 2010a. *Nominal Catch Information*. Madrid, Spain: Author.

———. 2010b. *Report of the Standing Committee on Research and Statistics (SCRS)*. Madrid, Spain: Author.

International Consortium of Investigative Journalists. 2010. *Looting the Seas: How Overfishing, Fraud, and Negligence Plundered the Majestic Bluefin Population*. http://www.publicintegrity.org/treesaver/tuna/#.

Issenberg, S. 2007. *The Sushi Economy: Globalization and the Making of a Modern Delicacy*. New York: Gotham Books.

La Mantia, V. 1901. *Le tonnare in Sicilia* [The Tuna Traps in Sicily]. Palermo, Italy: Tipografia Giannitrapani.

Longo, S. B. 2012. "Mediterranean Rift: Socio-ecological Transformations in the Sicilian Bluefin Tuna Fishery." *Critical Sociology* 38 (3): 417–36. doi:10.1177/0896920510382930.

———. 2011. "Global Sushi: The Political Economy of the Mediterranean Bluefin Tuna Fishery in the Modern Era." *Journal of World Systems Research* 17: 403–27.

Mather, F. J., J. M. Mason, and A. C. Jones. 1995. *Historical Document: Life History and Fisheries of Atlantic Bluefin Tuna*. Springfield, VA: Southeast Fisheries Science Center.

MacKenzie, B. R., H. Mosegaard, and A. A. Rosenberg. 2009. "Impending Collapse of Bluefin Tuna in the Northeast Atlantic and Mediterranean." *Conservation Letters* 2: 25–34.

Marx, K. 1977. *Capital.* Vol. 1. New York: Vintage Books.

———. 1992. *Capital.* Vol. 2. New York: Penguin Books.

McCay, B. J. 1998. *Oyster Wars and the Public Trust: Property, Law, and Ecology in New Jersey History.* Tucson: University of Arizona Press.

McCay, B. J., and J. M. Acheson. 1987. *The Question of the Commons: The Culture and Ecology of Communal Resources.* Tucson: University of Arizona Press.

McEvoy, A. F. 1986. *The Fisherman's Problem: Ecology and Law in the California Fisheries 1850–1980.* New York: Cambridge University Press.

McGoodwin, J. R. 1990. *Crisis in the World's Fisheries: People, Problems, and Policies.* Stanford: Stanford University Press.

McHugh, J. L. 1977. "Rise and Fall of World Whaling: The Tragedy of the Commons Illustrated." *Journal of International Affairs* 31: 23–33.

McWhinnie, S. F. 2009. "The Tragedy of the Commons in International Fisheries: An Empirical Examination." *Journal of Environmental Economics and Management* 57: 321–33.

Millennium Ecosystem Assessment. 2005. *Ecosystems and Human Well-Being: Synthesis.* Washington, DC: Island Press.

Miyake, P. M., J. M. De la Serna, A. Di Natale, A. Farrugia, I. Katavic, N. Miyabe, and V. Ticina. 2003. *General Review of Bluefin Tuna Farming in the Mediterranean Area.* Madrid, Spain: International Commission for the Conservation of Atlantic Tunas.

Mullon, C., P. Fréon, and P. Cury. 2005. "The Dynamics of Collapse in World Fisheries." *Fish and Fisheries* 6: 111–20.

Myers, R. A., and B. Worm. 2003. "Rapid Worldwide Depletion of Predatory Fishing Communities." *Nature* 423: 280–83.

Naylor, R. L., R. J. Goldburg, H. Mooney, M. C. M. Beveridge, J. Clay, C. Folke, N. Kautsky, J. Lubchenco, J. Primavera, and M. Williams 1998. "Nature's Subsidies to Shrimp and Salmon Farming." *Science* 282: 883–84.

Nickler, P. A. 1999. "A Tragedy of the Commons in Coastal Fisheries: Contending Prescriptions for Conservation, and the Case of the Atlantic Bluefin Tuna." *Environmental Affairs* 26: 549–76.

Normile, D. 2009. "Persevering Researchers Make Splash with Farm-Bred Tuna." *Science* 324: 1260–61.

Ostrom, E. 1999. "Coping with Tragedies of the Commons." *Annual Review of Political Science* 2: 493–535.

Ostrom, E., J. Burger, C. B. Field, R. B. Norgaard, and D. Policansky. 1999. "Revisiting the Commons: Local Lessons, Global Challenges." *Science* 284: 278–82.

Pauly, D., V. Christiensen, J. Dalsgaard, R. Freese, and F. Torres. 1998. "Fishing Down Marine Food Webs." *Science* 279: 860–63.

Polanyi, K. 1957. *The Great Transformation: The Political and Economic Origins of Our Time.* Boston, MA: Beacon Press.

Ponting, C. 2007. *A New Green History of the World: The Environment and Collapse of Great Civilizations.* New York: Penguin.

Relini, G. 2003. "Fishery and Aquaculture Relationship in the Mediterranean: Present and Future." *Mediterranean Marine Science* 4: 125–54.

Revkin, J. 2008. "The (Tuna) Tragedy of the Commons." *The New York Times,* November 26. http://dotearth.blogs.nytimes.com/2008/11/26/the-tuna-tragedy-of-the-commons/.

Safina, C. 2001. "Tuna Conservation." In *Tuna Physiology, Ecology, and Evolution,* edited by B. A. Block and E. D. Stevens, 414–57. New York: Academic Press.

Sarà, R. 1998. *Dal mito all' aliscafo: Storie di tonni e di tonnare* [From Myth to Hydrofoil: Stories of Tuna and Tuna Traps]. Messina, Italy: Editore Raimondo Sará.

Schnaiberg, A., and K. A. Gould. 1994. *Environment and Society: The Enduring Conflict.* New York: St. Martin's Press.

Schumpeter, J. A. 1962. *Capitalism, Socialism, and Democracy.* New York: Harper Perennial.

SELF-DOTT. 2010. *Latest News.* http://sites.google.com/site/self-dottpublic/.

Sweezy, P. M. 1942. *The Theory of Capitalist Development.* New York: Monthly Review Press.

Tucker, R. C. 1978. *The Marx-Engels Reader.* New York: W. W. Norton.

Tudela, S. 2002. "Grab, Cage, Fatten, Sell." *Samudra* 32: 9–16.

United Nations. 2005. *Millennium Ecosystem Assessment Synthesis Report.* New York: Author.

Vasquez, J. C. 2010. *Governments Not Ready for a Trade Ban on Bluefin Tuna.* http://www.cites.org/eng/news/press/2010/20100318_tuna.shtml.

The World Bank. 2004. *Saving Fish and Fishers: Toward Sustainable and Equitable Governance of the Global Fishing Sector.* Report No. 29090-GLB. http://siteresources.worldbank.org/INTARD/Resources/SavingFishandFishers.pdf.

York, R., and M. H. Gossard. 2004. "Cross-national Meat and Fish Consumption: Exploring the Effects of Modernization and Ecological Context." *Ecological Economics* 48: 293–302.

Ecological Modernization at Work?

Environmental Policy Reform in Sweden at the Turn of the Century

Benjamin Vail

Currently, one of the theoretical debates in environmental sociology is whether capitalism can be rendered environmentally sustainable. Proponents of ecological modernization argue that yes, capitalism is incredibly flexible and that businesses will move toward more ecologically sustainable practices, especially with government incentives. Others, such as the authors of the previous two selections, are suspect of this view and tend to see the "treadmill of production"—capitalism's need for constant growth—as antithetical to environmental sustainability.

Benjamin Vail does not engage in this debate but asks whether Sweden's sweeping "sustainable society" policy follows the ideas of ecological modernization. In doing so, he clearly describes ecological modernization and shows how this theory has developed over time.

Sweden is widely viewed as a global leader in positive environmental practices, both in terms of public policy and business-sector activities. Throughout the 1990s and into the 2000s, Swedish society—in the realms of politics, the economy, and civil society—debated the value of sustainable development. The government decided on an ambitious long-term program to achieve a "sustainable society" by 2020. This article seeks to understand the process by which this policy decision was made—as well as its practical implications, which include potentially sweeping impacts on social and economic life in the push for environmental improvements. In pursuit of an explanation, this article asks whether the environmental sociological theory of ecological modernization describes the process of decision-making in Sweden during this time period.

The hypothesis considered here is that Swedish environmental policy reform during this period was largely consistent with the model of social change predicted (and prescribed) by ecological modernization theory (EMT). To test this hypothesis, the Swedish decision-making process is described, ecological modernization theory is introduced, and empirical evidence is presented to compare the Swedish situation with the theory's predictions.

This study gathered data through the use of key informant interviews with decision-makers in government, political parties, and environmental groups. Further information was obtained from government documents, news media accounts, and political party publications. Many sources were available in English; where sources existed only in Swedish all translations were done by the author.

Swedish Environmental Reforms in the 1990s and Early 2000s

After several years of study and deliberation, in the late 1990s the Swedish Parliament voted to approve a new structure for national environmental policy. The purpose of the reforms was to solve environmental problems with the aim of creating a "sustainable society" within one generation. In 1998 existing environmental protection laws were consolidated into a single Environmental Code, the implementation of which was to be guided by fifteen environmental quality objectives—essentially a national "mission statement" for ecological sustainability—to be fulfilled by 2020.

The government officially said its actions were driven by the concept of sustainable development as presented in the 1987 Brundtland Report (World Commission on Environment and Development 1987), which states that development efforts should meet the needs of the present generation without compromising the ability of future generations to meet their own needs (Persson and Lindh 4).

Based on this definition of sustainable development, the government said in a "Statement of Government Policy" in 1996, "Sweden should be a driving force and a model for ecological sustainability" (Swedish Commission 2). The Government elaborated on its understanding of sustainable development in its 1997 Spring Economic Bill and 1998 Budget Bill, stating that ecologically sustainable development involves three objectives: environmental protection, sustainable use of resources, and more efficient use of energy and resources (Persson and Lindh 5 and Sustainable Sweden, "Environmental Quality" 3). Ultimately, "the Government's primary environmental objective is to hand over a society to the next generation in which the major environmental problems have been solved" (Swedish Ministry, 2000c 6).

Under a Social Democrat-led coalition, Parliament approved two different "framework" bills designed to put this vision of a sustainable society into effect, the Environmental Code and fifteen environmental quality objectives.

. . .

In 1997, the government submitted to Parliament the first proposal for the fifteen environmental quality objectives in a bill entitled, "Swedish Environmental Quality Objectives: An Environmental Policy for a Sustainable Sweden" (Swedish

Ministry, 2000c 10 and Swedish Environmental Protection Agency 2001). This bill was passed by Parliament in April 1999.

In September 1999, the government submitted details of how to achieve each objective with interim targets, monitoring strategies, and action plans. A Parliamentary committee, the Committee on Environmental Objectives, was created to study and review the environmental objectives and interim goals. In June 2000, this committee submitted its report, *The Future Environment: Our Common Responsibility.* In April 2001, the government submitted a new environmental objectives bill, *Swedish Environmental Quality Objectives: Interim Goals and Action Strategies,* to Parliament based on the committee's recommendations. This bill was passed by Parliament in November 2001. The new Environmental Code was passed by Parliament in 1998 and took effect on 1 January 1999 (Ministry of the Environment, 2000a 3).

According to Christina Lindbäck, the Director of the Division of Environmental Quality at the Environment Ministry, in the early 1990s the Swedish EPA discovered that there were about 170 environmental goals laid out in a variety of government plans. This fact had led to confusion and missed targets. Stakeholders including members of Parliament, representatives from government ministries, and NGOs and other interest groups sat on the Committee on Environmental Objectives that was appointed to propose targets, measures and strategies to fulfill the fifteen environmental quality objectives. "At the same time, the Government gave tasks to every authority in Sweden having a responsibility under each of the fifteen environmental quality objectives to do an analysis about what are the problems and to come up with proposals for how to solve these problems." Lindbäck said. . . .

The Code was created to consolidate the various laws and rules concerning the environment that were enacted piecemeal over the decades by Parliament and government agencies. The point was to rationalize environmental law to make it more comprehensible to both regulators and the regulated. The Code specified rules about land, water, and air use and protection, and was based on five fundamental principles considered to be the basis for sustainable development: protection of human health, preservation of biological diversity, minimization of use of natural resources, and protection of the natural and cultural environments (Swedish Ministry, 2000b 7).

Effective application of the Environmental Code was predicated on three general action strategies:

1. Improved efficiency in energy use, production, and transportation: energy conservation and public transportation will be promoted largely though [*sic*] education and economic incentives.
2. Cleaner production processes and recycling: the use of hazardous substances must be reduced and phased out and more goods recycled through education and legislation.
3. Improved environmental management practices: agricultural subsidies and more protection of natural areas are key (Committee on Environmental Objectives 11–12).

The Code was intended to facilitate action to achieve improvements in the following fifteen environmental quality objectives (Sustainable Sweden, *Environmental Quality* 4–6).

1. Clean air
2. High-quality groundwater
3. Sustainable lakes and watercourses
4. Flourishing wetlands
5. A balanced marine environment, sustainable coastal areas and archipelagoes
6. No eutrophication
7. Natural acidification only
8. Sustainable forests
9. A varied agricultural landscape
10. A magnificent mountain landscape
11. A good urban environment
12. A non-toxic environment
13. A safe radiation environment
14. A protective ozone layer
15. Limited (influence on) climate change

An Introduction to Ecological Modernization Theory

According to Thomas Nilsson of the Swedish EPA, who worked as a secretary and policy developer for the Committee on Environmental Objectives, these policies put Sweden on the "cutting edge" of international environmental reforms (Interview 5 October 2001). From a sociological perspective, ecological modernization theory (EMT) helps to explain the decision-making process of Swedish environmental reforms.

In brief, the ecological modernization point of view suggests—both as an empirical observation and as a prediction of future trends—that society is moving toward more environmentally friendly economic and social relations. Ecological modernization theorists look at the air and water quality improvements, for example, since the 1970s as proof of social progress in protecting the environment. Mol and Spaargaren state that although economic interests "still play a dominant role in production and consumption, and that they will probably always remain at least as important as—among others—ecological criteria . . . the innovation is that ecological interests and criteria are catching up with economic ones" (42).

While many authors have expounded on EMT and applied ecological modernization thinking to a variety of studies, Buttel argues that it is appropriate to focus on the writings of Mol and Spaargaren "because of all the scholars and researchers in this tradition (at least as far as the literature in English is concerned) they have done the most to articulate a distinctive theoretical argument" (Buttel, "Ecological Modernization" 58).

The origins of ecological modernization (EM) as a system of thought and social inquiry can be traced to the German sociologist Joseph Huber's work in the late 1970s (Mol and Sonnenfeld 4). Seippel says that EMT arose as a response to the radicalization of environmental thought due to the recession of the early 1980s, and also due to changes in the environmental movement, the emergence of new environmental problems, and the expanding availability of alternative discourses (289). Buttel says that EMT emerged in reaction to the ideologies and actions of 1980s radical environmental groups, as a response to the tradition in North American environmental sociology that emphasized the intrinsic nature of environmental degradation in modern society, and as a description of sustainable development efforts in developed nations ("Ecological Modernization" 59–60).

Buttel and Mol describe EMT as having moved through several "stages" of development from the 1980s through the mid-1990s and with contemporary new directions ("Ecological Modernization" 59 and "Global Economy" 2–3). Buttel writes that the "first generation" of EM thinking "was based on the overarching hypotheses that capitalist liberal democracy has the institutional capacity to reform its impact on the natural environment" ("Ecological Modernization" 59) and that the further development of capitalist modes of production—to a kind of "green capitalism"—will result in ecological improvements. This stage tended to emphasize capitalism's ability to adapt to new conditions and was premised largely on the idea of "technical fixes" for environmental problems (see Spaargaren and Mol 334). The second generation of EM literature focused on the specific socio-political processes by which the continued social and economic development of liberal democracies—in other words, further industrialization and modernization—is hypothesized to result in ecological improvements (Buttel "Ecological Modernization" 59).

Mol and Sonnenfeld suggest that in the late 1990s a "third phase" of EMT emerged that focused on issues of private consumption, the spread of ecological modernization to non-European countries, and the influence of globalization on EM processes in individual national-states (5). This phase can be seen as a response to criticism that EMT was preoccupied almost exclusively with production processes, was Eurocentric in its orientation, and did not account for trans-national social and ecological issues (Buttel, "Classical Theory" 31–2).

Mol and Spaargaren offer a version of EMT as a theory of social change and also a political program aimed at achieving a more sustainable society (Spaargaren and Mol; Mol, "Industrial Transformations"; and Seippel 290). They assert that in contrast to other theories of social-environmental change, EMT offers a realistic, achievable vision of a re-ordering of social, economic and political relations to promote environmental improvements (Mol, "Global Economy" 9).

Ecological modernization theory makes five main claims about social-ecological changes taking place in advanced industrialized nations. The primary purpose of EMT is to explain changes in the relationships between social institutions—namely the state, market actors, and civil society—in relation to the environment. According to Mol, EMT proposes the following hypotheses:

1. While science and technology have caused environmental problems, they are also the sources of technical solutions. Preventative technologies can replace command-and-control approaches to remediate environmental problems.

For example, problems such as air and water pollution can be resolved through the application of science and technology to achieve cleaner operating processes that reduce emissions from vehicles and factories.

2. Market forces and economic actors become more important as "social carriers of ecological restructuring, innovation and reform." In other words, the market encourages producers to cut costs by using more efficient technologies and by responding to consumer demand for ecofriendly products.

3. The state's role in environmental protection is transformed. Government adopts decentralized, flexible and collaborative problem-solving methods. Non-governmental organizations (NGOs), such as environmental groups, begin to play a more influential role in policy-making. And international institutions, such as the World Trade Organization and the European Union, begin to undermine the authority and ability of the traditional nation-state to determine and enforce policies by altering the meaning of sovereignty.

4. The roles and ideologies of environmental social movements change as environmental groups increasingly seek to work cooperatively with government and market actors to achieve environmental improvements. Instead of seeking to protect the environment by reducing production and consumption, for example, environmentalists try to work with government and business to make reforms. Although they will abandon their "de-modernization" ideology, these groups will continue to have a "dualistic strategy of cooperation and conflict, and internal debates on the tensions that are a by-product of this duality."

5. Society-wide shifts in discourses lead to the emergence of new ideologies characterized by concern for environmental issues. A heightened awareness of intergenerational solidarity "seems to have emerged as the undisputed core and common principle" guiding environmental reform. ("Global Economy" 5 and "Industrial Transformations" 140–2)

. . . [P]erhaps the most fundamental EMT claim is that economic actors such as manufacturers and investors will begin to take action in support of environmental protection—not in response to government command-and-control dictates or rigid regulations, but because environmental values are permeating corporate culture and because companies find it advantageous to adopt environmentally friendly practices. Mol and Spaagaren and Mol describe this process as "ecologicalization" of the economy and "economization" of ecology. In other words, science and technology make possible new market-driven fixes to environmental problems. . . .

Cleaner technologies, EMT suggests, will lead to "dematerialization" of the economy (Mol and Sonnenfeld 6). This hypothesis suggests "that for each unit of output . . . there will be progressively fewer environmental resources required as inputs into production. . . . At a highly aggregated level, dematerialization of production leads to societies and economies becoming 'decoupled' from resource use" (Buttel, "Environmental Sociology" 21). Efficiency, recycling and the use of renewable resources will reduce the need to exploit virgin natural resources and will reduce wastes, creating product "life cycles" that are environmentally benign.

Ecological modernization offers the optimistic vision of sustainable development without radical changes in the existing social-economic order. This overall view appears consistent with Swedish efforts to achieve a "sustainable society" through political and market reforms that leave the modern industrial consumer society intact.

Putting EMT to the Test in the Swedish Case

To test whether ecological modernization theory describes and explains Swedish environmental reforms, it is useful to start by asking whether evidence in the Swedish case supports Mol's five hypotheses.

The Role of Science and Technology

The Swedish example appears to provide quite strong evidence in support of Mol's hypothesis of a changing role for science and technology. The Committee on Environmental Objectives' report (2000) gave clear indications of how new technology would be necessary—and encouraged by the state—to achieve sustainability.

Improved efficiency in energy use, production, and transportation was the first action strategy declared in the Environmental Code. This policy depends on the use of new technologies such as new fuels and better engines, that simultaneously would be more efficient and reduce harmful emissions (Committee on Environmental Objectives 11).

Furthermore, the use of best possible technology was not only encouraged but mandated by the new Environmental Code. "The best possible technology is to be used for professional operations. The term *best possible technology* applies to the technology used both for the operation itself and for the construction, operation and decommissioning of the plant. An essential condition for using the best possible technology is that it must be feasible in industrial and economic terms in the line of business concerned" (Swedish Ministry, 2000b 12).

The Role of Economic Actors

There is strong evidence from this time period in Sweden supporting Mol's second hypothesis that economic actors were expected to contribute actively to environmental reform resulting in transformations in state-market relations. The government policies promoted greater efficiency in the use of resources, an outcome depending partly on new technology but also the changing habits of resource use by private individuals and corporations. For example, the concept of "Factor 10," as promoted by the United Nations, was a policy guideline. Factor 10 meant "that we must use resources, not only fossil fuels but energy and materials, on average ten times more efficiently within the next generation or two. In the view of many scientists, industrialists, and politicians, this change is essential if we are to be able to face the growing population and reduce pollution without lowering living standards" (Persson and Lindh 6).

The question of efficiency and reduced use of resources related directly to the EMT hypothesis that "ecologizing" economies tend toward dematerialization— the decoupling of production from raw natural resource inputs. There was only mixed evidence in Sweden that such a trend was taking place. One positive indicator was the fact that despite GDP growth of "approximately 43 percent in constant prices over the past 20 years, the total use of energy in industry has declined during this period" (Sustainable Sweden, "Tax Policy"). Over the same period, however, overall energy use in Sweden grew.

Perhaps the most compelling evidence for dematerialization in Sweden consisted of data showing that overall material consumption in industry had been on the decline since the late 1980s (Statistics Sweden 46). The use of nonrenewable materials such as construction minerals, ore and industrial minerals and fossil fuels declined between 1987 and 1997, while the use of renewable resources remained relatively stable.

Data from Statistics Sweden, the Swedish EPA, and other sources indicate an uneven trend over the years, with a mixture of victories and losses on the "dematerialization" front.

For example, although household recycling and composting rates increased and the quantity of waste disposed in landfills decreased between 1985 and 1998, the total amount of annual municipal waste increased during this period (Statistics Sweden 15).

The amount of industrial waste produced in Sweden grew between 1993 and 1998 (Statistics Sweden 17). The per capita use of fuels decreased slightly between 1993 and 1997, but the use of non-fuel hazardous chemicals increased during the same time period, while the annual per capita use of hazardous chemical products remained about even (Swedish Environmental Protection Agency, *De Facto* 24).

There was a trend toward reduced emissions of nitrogen and phosphorus into the sea from coastal point sources between 1985 and 1997. Such reductions were very important for achieving environmental objectives such as mitigated eutrophication and improved wetlands, waterways, and coastal areas (Statistics Sweden 12).

In addition to reduced harmful emissions into Sweden's waterways, there were significant improvements in air pollution. Swedish emissions of sulphur dioxide (SO_2), a prime contributor to acid rain, declined greatly between 1980 and 1998 (Swedish Environmental Protection Agency, *De Facto* 18). Data show that the most dramatic improvements were made in the power/heating and industrial sectors. Reductions in carbon dioxide (CO_2) emissions mirrored those of SO_2 over the same time period; emissions of CO_2 from different sources declined between 1980 and 1999 (Statistics Sweden 52).

Despite these improvements, urban air continued to pose human health risks. Volatile organic compounds and particulate matter emitted by vehicles created smog and ground-level ozone, and caused health disorders such as asthma. Though benzene levels in urban air declined by 50 percent from 1992–1999, the levels were higher than the low-risk limit (Statistics Sweden 11). The fact that VOC emissions remained above acceptable levels was most likely a result of continuing high levels of personal auto use. But while 1998 levels of auto use were more than 10 percent higher than those in 1985, alternative forms of transportation were becoming more popular. In 1998, more than 25 percent of all journeys to and from work or school were made on foot, bicycle or public transportation (Swedish Environmental Protection Agency 13; Swedish Environmental Advisory Council 16, 18, 19).

The Swedish forestry sector appeared already to have begun to balance harvesting with environmental protection. Gustaf Aulén, an ecologist with the Södra association of Swedish forest owners, said his industry had both a "production goal" and an "environmental goal." One "important reason" why forestry companies acted to conserve nature was "increasing pressure from our customers" (Personal correspondence 8 November 2001).

There was evidence that Swedish corporate participation in certification and labeling programs was on the rise thereby suggesting that voluntary action in support of environmental improvement was catching on. The number of firms with EMAS or ISO 14001 certification grew between 1995 and 1999, from zero to over 900 (Statistics Sweden 30). In another example of industry's embrace of green labeling and ecofriendly practices, the amount of land area certified as under sustainable forestry management expanded by more than ten times between 1996 and 2000. This assessment referred to voluntary certification programs administered either by the international Forest Stewardship Council or the Pan-European Forest Certification system. Meanwhile, the amount of forested land area under public protection also grew in the 1990s from about 2.5 percent of all productive forestland in 1990 to more than 3.5 percent in 1999 (Statistics Sweden 31).

Furthermore, Sweden's use of "green taxes" based on the Polluter Pays Principle (PPP) promoted the internalization of ecological costs by corporate and individual decision-makers. The PPP mandated that any entity found to be responsible for environmental degradation, such as oil leaks or toxic emissions, was liable to pay for all costs associated with remediation of the problem. In Sweden, the PPP liability extended in the form of a tax to those who consumed goods and services such as petrol and dry cleaning. Tax reforms in 1991 began the process of imposing consumption taxes on fossil fuels and greenhouse gas emissions. Sustainable Sweden reports, "Energy taxation of motor fuels has led to noticeable improvements in urban air. The sulphur tax has helped our country meet the emissions targets for sulphur. The nitrogen oxide charge has reduced emissions from heating plants, and the energy and carbon dioxide taxes have encouraged the use of biofuels" ("Tax Policy").

. . .

The government argued that implementation of the environmental quality objectives would on balance help the national economy and enhance the performance of Swedish firms (Persson and Lindh 5). One government report asserted, "Conversion to ecologically sustainable development on a wide front can help to create growth and employment and thus promote the conditions for growth even in vulnerable regions" (Sustainable Sweden, "Environmental Quality Objectives" 22).

Thus Swedish officials echoed the EMT claim that economic rationality and ecological values need not be mutually exclusive and that eco-friendly behavior on the part of corporate and individual actors can be profitable. By the early 2000s, Swedes could interpret the energy and materials use trends as an encouraging picture of increasing efficiency and gradual environmental improvements—in other words, ecological modernization. Yet many of the changes in the marketplace were to be achieved not through ecological concern endogenous to the business community, but through economic policy instruments wielded by the state.

The Role of the State

The Swedish government certainly wanted to induce economic actors to consider environmental values. The Swedish Ministry of the Environment declared that "environmental concerns must be integrated into all decision-making. Methods must be developed in order to integrate the costs of environmental impacts into economic and social decision-making models" (Swedish Ministry, 2000c 9). The Committee on Environmental Objectives added, "Powerful policy instruments will be required, of both a normative and an informative nature. The involvement of market forces is important if efforts to reduce risk are to be pursued in a systematic way. Manufacturers' efforts to produce environmental product certificates and positive environmental labeling are examples of this" (28).

Evidence supporting the EMT assertion that the traditional role of the state will be transformed is mixed in Sweden. Mol's third hypothesis is that the government will shift from combative regulator to cooperative facilitator. This change will be accompanied by decentralization and more flexible policy-making and enforcement practices. The Swedish example appeared to validate the prediction of greater flexibility, but instead of becoming decentralized, the state expanded its power over national environmental protection and remained the prime mover in environmental protection. As shown in the previous section about economic actors, much of the impetus for market-based solutions in Sweden was expected to originate in government actions through the use of policy tools to penalize or subsidize certain behaviors.

Prime Minister Persson made perhaps the most straightforward statement of the role of the state in environmental protection: "Central government must lead the way" (Persson and Lindh 7). Sustainable Sweden added, "The task of central government in relation to the market is, inter alia, to establish clear-cut rules. The Environmental Code recently presented by the Government will lay the foundation for this work" ("Environmental Quality Objectives" 29).

The Committee on Environmental Objectives further clarified the government's role in environmental protection: "Central government will assist in this endeavor via measures taken by its sectoral agencies, by providing economic incentives and by creating other beneficial conditions for environmental work. Its task is also to ensure that laws and regulations are enacted and properly implemented. Environmental monitoring and advanced follow-up systems are further responsibilities of central government" (8).

Thus, although county boards and municipalities were responsible for the local implementation of the new policies, the central government remained the strongest force for change. Indeed, while the state sought to work with private enterprise to effect meaningful environmental reform, it also acted as a counterweight to the power of private capital. Svante Axelsson, director of Swedish Society for Nature Conservation, the largest environmental NGO in Sweden, wrote,

> . . . There is considerable potential for change through voluntary action. For example, many multinational corporations are sensitive to environmental issues. They are anxious to protect their brand reputations, of which environmental movement organizations in different regions are well aware. But the politicians are necessary as well. That which is emerging via market forces must be strengthened through political measures.

While the power and responsibility for initiating environmental policy reform remained vested in the central government, there was substantial evidence that the state was becoming more flexible and inclusive in its decision-making and enforcement practices. The green tax policy was a good example of how the state sought to use economic incentives to change behavior. According to the Committee on Environmental Objectives, "The use of economic incentives is an important means of achieving sustainable ecological development. In framing such incentives, it is important to bear in mind that they should encourage technological development and efficiency and hasten the replacement of old technologies by new ones" (19). The state's policy-making toolbox included not only taxes to discourage consumers and producers from harmful activities, but also subsidies to encourage positive changes. This was in contract to the emphasis on "end-of-pipe" pollution mandates in the 1970s and 1980s.

On the whole, the state appeared to be embracing market-based, flexible solutions to meet the environmental quality objectives. There were provisions in the Environmental Code for prosecuting violators, but the new policies emphasized preventative measures.

The Role of Non-Governmental Organizations

Mol's fourth EMT-based hypothesis is that the role and ideology of environmental social movements will change. These "subpolitical" actors will stop acting on a radical, negative "de-modernization" agenda and instead will increasingly seek to work together with the state and corporations to achieve their goals. Not only will these groups' ideologies and activities begin to change, they may become quite influential within the government, leading to the restructuring of power arrangements.

There was evidence of such a trend in Sweden. The government sought the input of environmental groups when formulating the environmental quality objectives, and the Swedish Ministry of the Environment said that "the authorities should pay attention to and support the pro-environment activities of non-governmental organizations" (Committee on Environmental Objectives 9 and Swedish Ministry, 2000c 66).

However, Gunilla Högberg Björck of the Swedish Society for Nature Conservation said that NGOs lacked the influence of business interests (Interview 8 October 2001). This may be explained in part by a trend in Sweden of decreasing public interest in environmental issues by the early 2000s. A social survey found that the Swedish public's concern for environmental issues declined during the 1990s. In 1988, 62 percent of the Swedish public said environmental issues were among their top three policy concerns, compared with only 4 percent in 1999. The study also found that environmental organizations were losing members. In 1989, 15 percent of the population said they were members of an environmental organization, compared with only 5 percent in 1999 (Lövgren and Öhngren 5).

Meanwhile, support for the Swedish Green Party was wavering, casting doubt on whether the environmental movement would continue to have a voice in Parliament. The party was seen by many as obsolete since other parties had incorporated environmental themes into their platforms. Conservative Party advisor Göran Olsson suggested that the Green Party might perish due to its own success: "Actually, it's a paradox because . . . the more they get the Social Democrats or the government to listen to them and actually propose environmental issues, the less need there is for the Green Party" (Interview October 5, 2001).

Thus, the Swedish public's interest in nature and environmental policy was waxing and waning, resulting in uncertain clout for the environmental movement. The variable strength and depth of public environmental concern relates directly to questions about discourse and ideology in Swedish society.

The Role of Discourse, Ideology, and Values

Mol's fifth EMT hypothesis is that the ecological modernization process is characterized by society-wide changes in discourses and ideologies of a proenvironment nature. There was fairly strong evidence that Swedish officials at least "talked the talk" when it came to ecological awareness and their stated intentions to act in a more environmentally friendly way. Examples of the role of discourse and ideology in Swedish environmental decision-making can be found in the political debates over the policy reforms, Swedish notions about their country's international environmental leadership, and consumers' choices to buy eco-labeled products.

While there appeared to be a certain amount of consensus within the Swedish policy arena about adopting the fifteen environmental objectives, debate raged over the question of how best to prioritize and achieve environmental improvements. Most stakeholders, including political parties, environmentalists, and business interests, agreed in principle with the environmental objectives but disagreed about how to implement the reforms.

For example, Conservative Göran Olsson said, "There is no consensus on how to get there or how fast, or other considerations or even the methods—how to do the work to get there. But since the goals are . . . such that it is practically impossible to say no. 'Do we want clean air?' yes. It's laid out in that sense so it's very hard to say that 'No, we are opposed to clean air' " (Interview 5 October 2001).

. . . The question of proper forest management, for example, highlighted ideological differences between those who wanted state control of forests and those who wanted to educate private landowners for eco-friendly management. Genetically modified organisms, Lindbäck said, "presented another question, as did fishery policies, where you could find ideological differences about how to solve the problems. It's seldom difficult to agree on [the existence of] an environmental problem, but it's difficult to agree upon measures how to solve it" (Interview 2 October 2001).

. . . The increasing success of eco-labeled products may serve as a final example supporting Mol's hypothesis that environmental values can be seen permeating society. Gardfjell—in Lövgren and Öhngren—found that the Nordic symbol of eco-friendly products, a green swan label placed on qualifying goods, was well recognized by Swedish consumers. She found that, "The experiences with the environmental label show that voluntary initiatives can be very significant. The environmental label has caused big breakthroughs in certain areas, but there are also commodities—for example textiles—for which the label has completely failed" . . .

Reinhard et al. list the possibilities and limitations of using such eco-labels. The "pros" of using eco-labels include the fact that they complement other policy instruments promoting green consumption and communicate a complex message about green purchasing choices. Updating the labeling criteria leads to continuous improvement in green goods, and the requirements to use the label may be stricter than official legislation. The "cons" of using such labels include the fact that green purchasing does not reduce total consumption, the program is dependent on voluntary action by industry, not all products are suitable for eco-labeling, and products may not qualify for eco-labeling in all stages of their life cycles (Reinhard et al. 65).

Data from Statistics Sweden show that green purchases were on the rise through the 1990s. The purchase of eco-labeled products and services as a percentage of total private consumption rose from less than 1 percent in 1995 to nearly 2.5 percent in 1998. The per capita value of purchases of eco-labeled goods and services rose from less than 500 to more than 2000 Swedish crowns (from about 50 to more than 200 in 2001 US Dollars) (32). While support for eco-labeled products fell in Norway, "the proportion of consumers strongly favouring greener products remained stable in Sweden, Denmark and Finland at around 60 percent" (ENDS Environment Daily "Norway Loses Interest").

A Critical Assessment of EMT and the Swedish Case

Empirical evidence appears to provide compelling validation of many of ecological modernization theory's claims about the characteristics of Swedish environmental policy reform. The information presented here suggests that EMT clarifies the social issues surrounding environmental policy reform in a useful way and is an applicable conceptual framework for better understanding the Swedish environmental reform decision-making process of the late 1990s and early 2000s.

However, the evidence is not an overwhelming confirmation of ecological modernization theory. Obviously, the empirical data regarding materials use and pollution can be interpreted in different ways. While some improvements are obvious and the trend is positive, one may question whether the changes are significant enough to celebrate. The question remains whether dematerialization is a clear and inevitable trend or if perhaps major changes in individual lifestyles and business practices will be necessary to achieve the environmental quality objectives.

Is technology really the magic bullet for Swedish environmental problems? Rather than achieving the vaunted goal of reducing environmental impacts through "dematerialization," critics such as Bunker, Schnaiberg, and Schnaiberg and Gould fear that more efficient production processes will instead simply make possible the use of ever greater quantities of resources (Bunker, Schnaiberg, and Schnaiberg and Gould). Thus, even cleaner and more efficient production may not necessarily lead to environmental improvements.

Even if industry was cleaning up its act and offering more green products during the 1990s, it is unclear whether businesses were acting solely or even mainly out of an ethical concern for the environment. In other words, was "ecological modernization" really taking place, in the sense that there was an emergent new relationship between humans and the natural environment?

Politically, in the Swedish case, the state consolidated its position as the first actor in environmental improvement and did not step aside to let market forces work alone. Indeed, the state choreographed the economic incentives that would make possible the "market solutions" theorized by EMT. It is through such mechanisms as green taxes and government subsidies that many of the environmental objectives are expected to be achieved, but they require substantial centralized control.

Regarding non-governmental organizations, it is possible that instead of being engaged in a collaborative policy-making process among equals, environmental groups that conform to the EMT hypothesis risk being coopted by anti-environmental interests and may even become complicit in the greenwashing of their corporate and government "partners." Rather than taking principled stands for significant reforms and fighting when necessary, environmental groups risk selling out in an effort to gain recognition within the political power structure. It is also interesting to note that EMT does not account for the possibility that anti-environmental NGOs, such as industrial trade organizations, political action committees, and think-tanks, may increase their influence. While there was no evidence for such a situation in Sweden, it is theoretically possible for anti-green NGOs to have as much, or more, power as environmental groups.

Mol's fifth hypothesis addresses changes in social discourse about environmental issues. Lindén shows how young people in Sweden today may be well educated about environmental problems and express ecological values, but their behavior generally has more environmentally negative impacts than that of their parents and grandparents. This suggests that social discourses and personal values may change, but individual and collective behavior does not necessarily become more environmentally friendly as a result. Among economic actors, the discourse still seems to be more about competitiveness and the bottom line than about saving the environment or helping future generations.

To conclude, the process of defining the issues and making decisions in the 1990s and early 2000s in support of new environmental policies did, by and large, reflect an "ecological modernization" of Swedish society. The ecological modernization perspective describes the Swedish environmental reforms, but the Swedish case does not conform one hundred percent with EMT predictions. Future research should track the implementation of the reforms agreed to in the early 2000s and continue to test the empirical evidence against EMT predictions of, and prescriptions for, environmental quality improvement.

Note

Vail, Benjamin. 2008. "Ecological Modernization at Work? Environmental Policy Reform in Sweden at the Turn of the Century." *Scandinavian Studies* 80 (1): 85–108.

References

Aulén, Gustaf. 2001. Personal correspondence. November 8.

Björck, Gunilla Högberg. 2001. Interview. Stockholm. October 8.

Bunker, Stephen G. 1996. "Raw Material and the Global Economy: Oversights and Distortions in Industrial Ecology" *Society and Natural Resources* 9 (4): 419–29.

Buttel, Frederick H. 2000. "Classical Theory and Contemporary Environmental Sociology: Some Reflections on the Antecedents and Prospects for Reflexive Modernisation Theories in the Study of Environment and Society." In *Environment and Global Modernity*, edited by Gert Spaargaren, Arthur P. J. Mol, and Frederick H. Büttel, 17–39. London: Sage.

———. 2000. "Ecological Modernization as Social Theory." *Geoforum* 31: 57–65.

———. 2001. "Environmental Sociology and the Explanation of Environmental Reform." Paper presented at the Kyoto Environmental Sociology Conference.

Committee on Environmental Objectives. 2000. *The Future Environment—Our Common Responsibility: A Summary.* Stockholm.

ENDS Environment Daily. 2000. "Norway Loses Interest in Green Consumerism." *ENDS Environment Daily* 879 (November 21).

———. 2001. "Sweden Legislates for Sustainability." ENDS Environment Daily 984 (May 4).

Karlsson, Kjell-Erik. 2001. Personal correspondence. November 12.

Karlsson, Lars-Ingmar. 2001. Interview. Stockholm. October 8.

Lindbäck, Christina. 2001. Interview. Stockholm. October 2.

Lindén, Anna-Lisa. 1998. "Values of Nature in Everyday Life: Words versus Action in Ecological Behaviour." In *Sustainability the Challenge: People, Power, and the Environment*, edited by L. Anders Sandberg and Sverker Sörlin, 34–41. New York: Black Rose Books.

Lövgren, Kerstin, and Bo Öhngren (eds.). 2000. "Vem bestämmer om var framtida miljö.' Seminarium pa Sigunsborg." Summary from conference at Sigtuna, Sweden November 21–2, 2001.

Mol, Arthur P. J. 2000. "Ecological Modernization and the Global Economy." Paper presented at the Fifth Nordic Environmental Research Conference of the Ecological Modernisation Society.

———. 1997. "Ecological Modernization: Industrial Transformations and Environmental Reform." In *International Handbook of Environmental Sociology*, edited by Michael Redclift and Graham Woodgate, 138–49. London: Elgar.

Mol, Arthur P. J., and Frederick H. Büttel. 2002. "The Environmental State under Pressure." In *The Environmental State under Pressure*, edited by Frederick H. Buttel and Arthur P. J. Mol, 1–11. Oxford: Elsevier.

Mol, Arthur P. J., and David A. Sonnenfeld. 2000. "Ecological Modernisation around the World: An Introduction." *Environmental Politics* 9 (1): 3–14.

Mol, Arthur P. J., and Gert Spaargaren. 2002. "Ecological Modernisation and the Environmental State." In *The Environmental State under Pressure*, edited by Frederick H. Buttel and Arthur P. J. Mol. Oxford: Elsevier.

Nilsson, Thomas. 2001. Interview. Stockholm. October 5.

Olsson, Goran. 2001. Interview. Stockholm. October 5.

Persson, Goran, and Anna Lindh. 1997. Government Communication 1997/98:13—Ecological Sustainability. Sundsvall, Sweden. http://www.sweden.gov.se/content/i/C4/29/75/a7af5i8a.pdf (accessed July 27, 2007).

Reinhard, Ylva, et al. 2001. "Evaluation of the Environmental Effects of the Swan Eco-Label—Final Analysis." Paper presented at the Seventh European Roundtable on Cleaner Production. International Institute of Environmental Economics, Lund University, Sweden.

Schnaiberg, Allan. 1980. *The Environment: From Surplus to Scarcity*. New York: Oxford.

Schnaiberg, Allan, and Kenneth Alan Gould. 1994. *Environment and Society: The Enduring Conflict*. New York: St. Martin's.

Seippel, Ornulf. 2000. "Ecological Modernization as a Theoretical Device: Strengths and Weaknesses." *Journal of Environmental Policy and Planning* 2: 287–302.

Spaargaren, Gert, and Arthur P.J. Mol. 1992. "Sociology, Environment, and Modernity: Ecological Modernisation as a Theory of Social Change." *Society and Natural Resources* 5: 323–44.

Statistics Sweden. 2001. *Sustainable Development: Indicators for Sweden: A First Set 2001*.

Strömdahl, Inger. 2001. Interview. Stockholm. October 8.

Sustainable Sweden. 1999. Swedish Environmental Quality Objectives.

———. 2000. Tax Policy for Ecological Sustainability.

Swedish Commission on Ecologically Sustainable Development. 1998. "Sustainable Sweden: We are on Our Way."

Swedish Environmental Advisory Council. 1999. "Green Headline Indicators: Monitoring Progress towards Ecological Sustainability."

Swedish Environmental Protection Agency. 2000. De Facto 2000: Environmental Objectives: Our Generation's Responsibility.

———. 2001. "Utveklingen av miljömälen: nyheter och bakgnmd."

Swedish Ministry of the Environment. 2000a. The Swedish Environmental Code.

———. 2000b. The Swedish Environmental Code: A Resume of the Text of the Code and Related Ordinances. http://www.sweden.gov.Se/content/1/c6/o2/o5/49/6736cf92.pdf (accessed 27 July 2007).

———. 2000c. The Swedish Environmental Objectives: Interim Targets and Action Strategies. Summary of Gov. Bill 2000/01:130. 2000c.

World Commission on Environment and Development. 1987. *Our Common Future*. Oxford: Oxford Up.

A Tale of Contrasting Trends
Three Measures of the Ecological Footprint in China, India, Japan, and the United States, 1961–2003

6

Richard York, Eugene A. Rosa, and Thomas Dietz

Proponents of ecological modernization theory argue that technological advances, especially ones that decrease the amount of energy and materials used in production, are key to solving many of our environmental problems. However, technological solutions often have unintended consequences that become problematic in themselves. While technologies that allow for higher levels of production with the same energy inputs would appear to be unmitigated successes, Richard York, Eugene Rosa, and Thomas Dietz show that the ecological footprints—the estimated land area required for sustaining use of the environment—of China, India, Japan, and the United States grew since the 1960s despite improved efficiencies. They point to the Jevons Paradox, where technological advances increase, rather than decrease, resource use or consumption.

Introduction

There is a considerable body of research in the world-systems literature examining the structural forces that influence national-level environmental impacts (e.g., Burns, Kick, and Davis 2003; Dietz and Rosa 1997; Dietz, Rosa, and York 2007; Jorgenson 2003; Jorgenson and Burns 2007; Jorgenson and Kick 2003; Jorgenson and Rice 2005; Roberts, Grimes, and Manale 2003; Rosa, York, and Dietz 2004; Rothman 1998; York 2007; York, Rosa, and Dietz 2003, 2004). The importance of tracking the environmental performance of nations is obvious in light of the myriad environmental problems around the globe: global warming, large-scale deforestation, desertification, loss of biodiversity, and disturbances to major geochemical cycles. A substantial variety of national-level environmental indicators have been examined in the environmental social science literature, but there is no consensus regarding which indicators are the most theoretically or substantively important. The reason for this is clear; different indicators reflect different aspects of the complex, hyper-faceted global ecology.

Our primary concern here is not with identifying which indicators are the "best" measures of human pressure on the environment, but with the proper matching of the *form* of the indicator—*total* national ecological consumption and/or pollution emissions, *per capita* ecological consumption and/or pollution emissions, or ecological consumption and/or pollution emissions *per unit of GDP* (the ecological intensity of the economy)—to various theoretical or substantive tasks. Here we use variations of the Ecological Footprint (EF) as our principal indicator for theoretical reasons explained below. Three EF variations (total, per capita, and intensity) tell fundamentally different stories about the environmental performance of nations, and it is, therefore, theoretically important to distinguish among these stories. To illustrate the differences among them, we examine temporal trends in the EF of each variation for four nations which account for a large share of the world's population, economic output, and natural resource consumption: China, India, Japan, and the United States.[1] These four nations combined account for 45% of the total global EF, indicating that what happens in these nations is particularly important for the future well-being of the planet (Loh and Goldfinger 2006: 3). We first explain the EF, the environmental indicator which is our focus here. We then present analyses of the trends in the EF in these four nations, with a discussion of the substantive and theoretical implications of these trends.

The Ecological Footprint

The EF, originated by Wackernagel and Rees (1996) and further developed by them and others (e.g., Chambers, Simmons, and Wackernagel 2000; Kitzes et al. 2007), is designed to assess the demands societies place on the regenerative capacity of the biosphere. The EF has been widely used in the field of ecology and in the environmental social sciences, and is generally regarded as a reliable indicator of anthropogenic pressure on the environment (Dietz et al. 2007; Jorgenson 2003; Jorgenson and Burns 2007; Jorgenson and Rice 2005; Rosa et al. 2004; Rothman 1998; York et al. 2003, 2004). The EF is calculated, much the same way that economic consumption is, by adding up the various forms of consumption in a society—food, housing, transportation, consumer goods, and services—and the waste they generate. That consumption, similar to economic accounting, is converted into a common metric. But unlike economics, which uses prices as its key indicator of value, the EF uses productive land area as its metric. The EF is based on the recognition that land is a fundamental factor on which all societies depend, since it provides living space, products and services, and a sink for wastes. Productive land is, therefore, a justifiable proxy for the demands societies place on the environment. The EF can be interpreted as a measure of the stress a nation places on natural capital and ecosystem services. The EF can be calculated for most nations because flows and consumption of resources and the production of wastes are typically recorded with reasonable accuracy in various national accounts. Human demands on the environment, then, can be converted into the biologically productive land areas necessary to provide these ecological services.

The EF is calculated by "adding up the areas (adjusted for their biological productivity) that are necessary to provide us with all the ecological services we consume" (Wackernagel et al. 1999: 377). The EF is a fairly comprehensive indicator of human pressures on natural resources and ecosystem services, since it does not ignore tradeoffs among different types of environmental exploitation (e.g., wood vs. plastic consumption). National resource consumption is calculated by adding imports and subtracting exports from production.

Because of this matching of impact to locale of consumption, the EF accounting system is particularly suitable for world-systems analyses. World systems theorists and other scholars have focused attention on how resource extraction and consumption, as well as pollution, are geographically distributed in the world-system, where the core nations commonly consume most of the resources, while the environmental degradation associated with resource extraction and polluting industries occurs in the periphery (Arrighi 2004; Brunnermeier and Levinson 2004; Bunker 1996; Frey 1998, 2003; Grimes and Kentor 2003; Hornborg 2003; Podobnik 2002). Due to its consumption-based focus, the EF does not overlook impacts that are externalized by moving production or extraction outside national borders. It, therefore, places environmental responsibility on the nations where resources are consumed—principally in the core—rather than on the ones where they are extracted.

The types of consumption delineated above are converted into the nine types of land area that support that consumption. All nine are aggregated to arrive at the total EF. The land area types are: (1) cropland, (2) grazing land, (3) forest (excluding fuel wood), (4) fishing ground, (5) built-up land, (6) the land area required to absorb carbon dioxide emissions from use of fossil fuel, fuel wood, (8) hydro-power, and (9) nuclear power. The component EF measures are weighted to take into account the fact that different types of land vary in productivity. The weighting for each type of land is scaled to its productivity relative to the worldwide average productivity of all land (including water area). For example, consumption requiring one hectare of arable land would have an EF larger than consumption requiring one hectare of non-arable land, reflecting the high productivity of arable land relative to the average productivity of all land on Earth (Wackernagel et al. 2002: 9268). Built-up land is treated as arable land since cities have historically grown in agriculturally rich areas. The hydro-power component is the area taken up by the reservoirs and infrastructure associated with hydro-electric dams. Each unit of energy from nuclear power is counted at par with one from fossil energy since analyses are inconclusive about the long-term land demands of nuclear power. We emphasize that the calculation of the EF for a nation is not based on its actual land area but on the land area that would be required, at global average productivities, to support its total consumption. Therefore, nations may, as many do, have footprints that are larger than their own land areas. The data in our analyses are from the Global Footprint

Network (http://www.footprintnetwork.org/) 2006 Edition and were gener-ously provided to us by Mathis Wackernagel and are used with his permission.

Understanding Environmental Trends: National Consumption, Eco-Efficiency, and Inequality

The foundational ecological economist William Stanley Jevons (2001 [1865]) identified an important paradox that, although becoming widely known today, still remains underappreciated. He noted, in the early days of England's industrial revolution, that the increasing efficiency of coal use in production was correlated with *increasing*, not decreasing, coal consumption. This is an observation of fun-damental importance since it was commonly assumed then as now that techno-logical improvements in the resource efficiency (eco efficiency) of production (i.e., less energy and/or materials per unit of production) would typically lead to the conservation of natural resources (e.g., Hawken, Lovins, and Lovins 1999; Reijnders 1998). In short, if changes in the production system make it so that a given amount of output can be produced with fewer inputs, it seems "obvious" that the amount of inputs should decline. However, the amount of inputs often does not decline, but instead grows because the scale of production grows faster than efficiency improves. This is Jevons's insight, and it highlights the sharp distinction between efficiency in resource use and total resource consumption. These two different indicators give two different answers to the question of envi-ronmental performance, since the most "eco-efficient" businesses, industries, or economies may be the ones consuming the greatest quantities of resources and generating the most pollution. Jevons's observation raises the possibility that, far from actually contributing to resource conservation, improvements in efficiency in many contexts may actually induce the expansion of resource consumption (Clark and Foster 2001; York 2006; York and Rosa 2003).

The distinction between efficiency and total consumption is of particular relevance to theoretical debates over the effects of economic development and technological change on the environment. Some scholars uncritically assume that advances in the material and energy efficiency of economies (e.g., where more economic capital is generated per unit of energy and/or material consumption), which are often driven by technological refinements, are indicative of environ-mental improvements (Andersen 2002; Fan et al. 2007; Hawken et al. 1999). This assumption is invalid from the point of view of ecosystems and biophysi-cal processes, since the amount of economic capital generated from exploiting the environment is a separate matter from the consequences of environmental exploitation for natural capital and services. Considerable empirical evidence shows that while low-income nations often are the least eco-efficient in the sense that they use a lot of resources or produce a lot of pollution *per unit of GDP*, they are also the nations that consume the least amount of resources in *absolute* and/ or *per capita* terms (Roberts and Grimes 1997; York et al. 2003, 2004).

To illustrate the limitations of focusing on efficiency we turn to an exam-ination of trends in the EF for China, India, Japan and the United States. In Figures 6.1–6.4 we present the trends over time in each of these nations in their

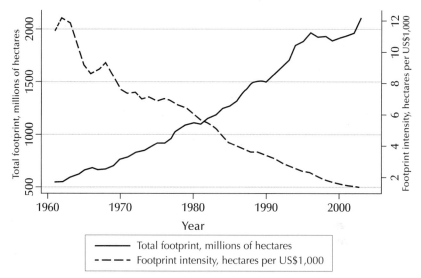

FIGURE 6.1 China's Ecological Footprint.

Global Footprint Network (http://www.footprintnetwork.org/)

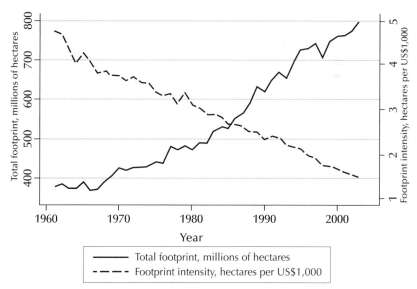

FIGURE 6.2 India's Ecological Footprint.

Global Footprint Network (http://www.footprintnetwork.org/)

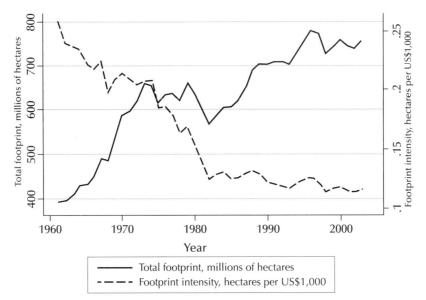

FIGURE 6.3 Japan's Ecological Footprint.

Global Footprint Network (http://www.footprintnetwork.org/)

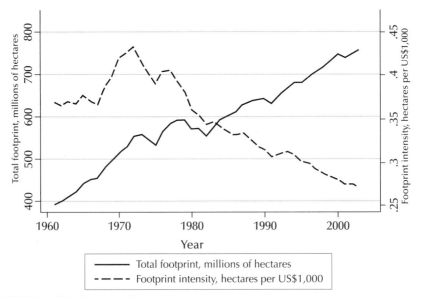

FIGURE 6.4 U.S. Ecological Footprint.

Global Footprint Network (http://www.footprintnetwork.org/)

total EF and their EF per unit of GDP. The latter measure we refer to as "intensity," which is the inverse of efficiency (i.e., high intensity means low efficiency). Strikingly, the trends in total EF and EF intensity illustrate how a focus on efficiency or intensity is misleading. In all nations there is a distinct trend toward declining intensity (i.e., increasing efficiency), which scholars might interpret as a sign of ecological improvements. After all, between 1961 and 2003 the EF per unit of GDP declined by a factor of 8.4, 3.2, 2.2, and 1.4 in China, India, Japan, and the United States respectively.

However, the ecologically relevant observation is that the total EF—which threatens nature's capital and services—increased quite dramatically in all of these nations over this period, by a factor of 3.9, 2.1, 2.9, and 2.9 in China, India, Japan, and the United States respectively. These findings indicate that all of these nations expanded their exploitation of the environment, while simultaneously expanding their economies even more. In other words, they used more resources while simultaneously getting more economic productivity out of each unit of resources. The improvements in efficiency (i.e., declines in intensity) were clearly associated not with reductions but, rather, with steady growth in resource consumption. The two trends are likely connected in a systemic way. Following the logic of Jevons's argument, a declining ratio of EF to economic activity often translates into lower costs per unit of production—since it typically indicates fewer inputs per unit of production—making it more affordable for producers to further expand production and thereby increase their profits. China's trajectory is particularly noteworthy, for it had both the largest improvement in efficiency over the period examined here, while exhibiting the greatest increase in its total EF. In light of these considerations, it might be more appropriate to say that improvements in efficiency are an example of *economic* reform not *ecological* reform and in fact typically indicate rising environmental impacts. Thus, total impact is the most informative measure from an ecological perspective because it signals threats to nature's capital and services. In contrast, efficiency or intensity is perhaps more informative from a profit-oriented perspective. At a deeper level this difference may reflect the differences in two accounting systems: an ecological accounting system where environmental factors are the primary concern and an economic accounting system where, as externalities, environmental consequences are typically ignored.

These observations generally counter the assumption of ecological modernization theory, a prominent theory in environmental sociology, that technological transformations are the key to solving our environmental problems. Ecological modernization theorist Maurie Cohen (1997: 109) argues, for example, that, "a key element in executing this [the ecological modernization] transformation is a switchover to the use of cleaner, more efficient, and less resource intensive technologies. . . ." Similarly, ecological modernization proponents Fisher and Freudenberg (2001: 702) and Milanez and Bührs (2007: 572) agree that the "linchpin" of the ecological modernization argument is the assertion that technological improvements can solve environmental problems. Particularly noteworthy is Andersen's (2002: 1404) statement: "Because ecological modernization by definition is linked with cleaner technology and structural change . . .

we can take changes in the CO_2 emissions relative to GDP as a rough indicator for the degree of ecological modernization that has taken place." In light of the findings we present here, key assumptions in ecological modernization theory appear misguided.

A third way to measure human pressure on the environment is in per capita terms. Examining per capita resource consumption removes the effects of population, assuming consumption and emissions are scaled proportionally by population size.[2] The per capita specification is perhaps most important from the perspective of global inequalities and social justice, since it allows for comparison of demands placed on the environment among the world's people and can highlight the substantial disparity in levels of consumption across nations. Figure 6.5 presents the trends in the EF per capita for the four nations examined here. Two observations stand out. First, while India's per capita EF stayed roughly constant from 1961 to 2003, the per capita EF in each of the other three nations approximately doubled. Second, there is clearly stark inequality across nations in terms of per capita pressure on the environment. For example, in 2003 the EF per capita in the United States was approximately six times that of China and 13 times that of India. Thus, even though China and India each have very large and growing total footprints, it is clearly not because the typical person in each of those nations places a high demand on the environment relative to people in affluent nations.

In 2003, the EF intensity in both China and India was over five times greater than in the United States and about 12 times greater than in Japan. If one focused on intensity, one would conclude that the United States and,

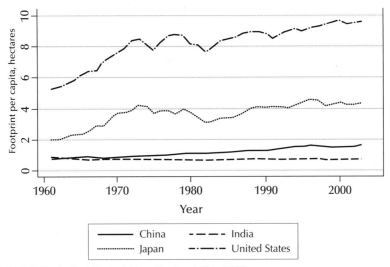

FIGURE 6.5 Ecological Footprint Per Capita, 1961–2003.
Global Footprint Network (http://www.footprintnetwork.org/)

particularly, Japan are good environmental stewards relative to the two larger, poorer nations, but this would mask the fact that the average person in the both the United States and Japan have a much greater effect on the environment than the average person in either China or India.

To further explore the connections between development and environmental degradation, we statistically analyzed the connection between economic production (GDP) and the EF. . . . In all four nations the relationship between the GDP and EF is positive and inelastic (i.e., coefficients > 0 and < 1). This indicates that as the economy of each of these nations has grown, the EF has also grown but not as fast as the economy. This type of relationship is what generates the apparently paradoxical finding discussed above where efficiency improves while the total EF grows. If an inelastic relationship such as this is hypothetically maintained indefinitely, resource consumption will continually grow while intensity declines. Thus, declines in intensity are not necessarily indicative of a move toward reductions in total consumption. It is noteworthy that the GDP coefficient is lowest in China and India, the two low-income nations, and highest in the United States, where it is close to unitary elasticity. This finding counters the common assumption that the most affluent nations will show the greatest improvements in ecological performance due to improvements in technological and other efficiencies as they continue to develop.

Conclusion

We examined trends in measures of human demands placed on nature's capital and services as indicated by three different forms of the EF—total, per capita, and per unit of GDP (intensity)—in China, India, Japan, and the United States. We found that the EF intensity of each nation's economy declined between 1961 and 2003 (i.e., the economies became more efficient in the sense that they each got more economic output per unit of EF). However, interpreting this trend as an environmental improvement is misleading because the total footprint of all four nations rose substantially over this four decade period. Furthermore, the per capita footprint approximately doubled in China, Japan, and the United States over this period, while it remained roughly constant in India. In . . . analyses of the connection between GDP and the EF we found a positive inelastic relationship in all four nations, indicating that improvements in efficiency in each nation over this period were more than counter-balanced by increases in scale, leading the EF of each nation to grow. The key implication of these results is that improvement in efficiency is not a sign of progress toward sustainability but is paradoxically associated with increased threats to the environment.

The ecological fate of the world lies to a growing extent in India and, especially, China. Over the period examined here, China's total ecological footprint expanded four-fold to rival that of the United States. In fact, it appears that China, which builds a new coal power plant approximately every four days, has surpassed the United States to become the single largest emitter of carbon dioxide in the world (although the United States still emits much more carbon dioxide on a per capita basis) due to rapid growth in fossil fuel consumption and cement manufacturing in recent years (Cyranoski 2007; Liu and Diamond 2008). This has occurred while the carbon intensity of China's economy has declined dramatically (Fan et al. 2007), underscoring our key message here: improvements in efficiency typically do not lead to declines in environmental problems. In fact, in footprint terms China is looking more like a core nation since its footprint is larger than its land area. This is due in no small part to its vast and growing appetite for imports, such as fish, tropical wood, and manufactured goods, and to the embodied energy in these goods (Hong et al. 2007; Liu and Diamond 2005). Clearly, China's dramatic improvements in eco-efficiency and its stated commitment to tackling environmental problems have not translated into reductions in its global environmental impacts (Liu and Diamond 2008).

To achieve sustainability and avert environmental crisis the four nations we examined here and the world as a whole need to dramatically scale back their consumption of the Earth's resources. Affluent nations, like the United States and Japan, which have exceedingly high levels of per capita consumption, need to drastically reduce the demands they place on the environment and transform their political, economic, and social systems to meets people's needs without unsustainably devouring natural resources. Less affluent nations, such as China and India, need to shift their development strategies away from relentless economic expansion and focus on strategies that improve people's quality of life without escalating material consumption. It is important to note that such changes—reducing consumption in affluent nations and averting the excessive expansion of consumption in poor nations—do not necessitate inhibiting improvement of social well-being. There is ample evidence that material consumption, once beyond the level required to meet basic needs, does not have a strong association with human well-being. For example, Leiserowitz, Kates, and Parris (2005) note that in nations out of absolute poverty the association between economic development and the subjective well-being of people is very weak. Furthermore, Dietz, Rosa, and York (2007) found that there is no direct connection between environmental degradation and human well-being as measured by education and life expectancy. These findings suggest that we could further human development without worsening environmental quality if we shift our focus away from economic growth as the primary social goal and toward the enhancement of more direct features of well-being.

It is important to recognize that the size and growth of human population plays a major role in the expansion of impacts on the environment. For example, as we show here, while India's per capita EF remained roughly stable, its total EF more than doubled, tracking the growth in its population. Likewise, China has such a large total EF, rivaling that of the United States, despite still having a low per capita EF by global standards, because it has well over one billion people. Addressing the population problem can be an important part of a progressive political agenda aimed at improving quality of life since the most effective ways to reduce fertility rates include improving the status and education of women, reducing infant mortality, eliminating absolute poverty, and providing all people with access to safe, effective and affordable birth control (Cohen 1995). Thus, what is needed to reduce human impact on the global environment—the curtailing of overconsumption and the reduction of fertility rates—can be part of an agenda aimed at directly *improving* human quality of life.

In light of the continuing high levels of consumption in affluent nations, the rapid economic growth in many developing nations in recent years, and the severity of environmental problems the world faces—the most noteworthy of which may be global climate change—averting ecological crisis is arguably humanity's greatest challenge for the twenty-first century. To meet this challenge it is imperative that we move away from the unrealistically optimistic assumption that improvements in the efficiency of economies alone are likely to solve environmental problems. In the face of continued economic and population growth, that assumption is not only misleading, it is also dangerous.

Notes

York, Richard, Eugene Rosa, and Thomas Dietz, "A Tale of Contrasting Trends: Three Measures of the Ecological Footprint in China, India, Japan and the United States, 1961–2003." *Journal of World-Systems Research* 15 (2): 134–46. Reprinted by permission of American Sociological Association.

[1] Based on the most recent year of the data we analyze here, the United States, China, and India have the three largest ecological footprints in the world (in that order) and Japan has the fifth largest, behind Russia. We do not examine Russia here since it has only been an independent nation since 1991. China, India, and the United States have the three largest populations in the world (in that order) and Japan has the ninth. The United States and Japan have the two largest economies in the world (in that order), China the fifth, and India the thirteenth.

[2] A series of analyses have found that the population elasticity of the EF is about (Dietz et al. 2007; Rosa et al. 2004; York et al. 2003) thus justifying the use of per capita measures.

References

Andersen, Mikael Skou. 2002. "Ecological Modernization or Subversion? The Effects of Europeanization on Eastern Europe." *American Behavioral Scientist* 45 (9): 1394–416.

Arrighi, Giovanni. 2004. "Spatial and Other 'Fixes' of Historical Capitalism." *Journal of World-Systems Research* 10 (2): 527–39.

Brunnermeier, Smita B., and Arik Levinson. 2004. "Examining the Evidence on Environmental Regulations and Industry Location." *Journal of Environment & Development* 13 (1): 6–41.

Bunker, Stephen G. 1996. "Raw Material and the Global Economy: Oversights and Distortions in Industrial ecology." *Society and Natural Resources* 9: 419–29.

Burns, Thomas J., Edward L. Kick, and Byron L. Davis. 2003. "Theorizing and Rethinking Linkages between the Natural Environment and the Modern World System: Deforestation in the Late 20th Century." *Journal of World-Systems Research* 9 (2): 357–90.

Chambers, Nicky, Craig Simmons, and Mathis Wackernagel. 2000. *Sharing Nature's Interest: Ecological Footprints as an Indicator of Sustainability.* London: Earthscan.

Clark, Brett, and John Bellamy Foster. 2001. "William Stanley Jevons and The Coal Question: An Introduction to Jevons's 'Of the Economy of Fuel.'" *Organization & Environment* 14 (1): 93–98.

Cohen, Joel E. 1995. *How Many People Can the Earth Support?* New York: W. W. Norton & Co.

Cohen, Maurie J. 1997. "Risk Society and Ecological Modernization: Alternative Visions for Post-Industrial Nations." *Futures* 29 (2): 105–19.

Cyranoski, David. 2007. "China Struggles to Square Growth and Emissions." *Nature* 446 (26): 954–55.

Dietz, Thomas, and Eugene A. Rosa. 1997. "Effects of Population and Affluence on CO_2 Emissions." *Proceedings of the National Academy of Sciences of the USA* 94: 175–79.

Dietz, Thomas, Eugene A. Rosa, and Richard York. 2007. "Driving the Human Ecological Footprint." *Frontiers in Ecology and the Environment* 5 (1): 13–18.

Fan, Ying, Lan-Cui Liu, Gang Wu, Hsien-Tang Tsai, and Yi-Ming Wei. 2007. "Changes in Carbon Intensity in China: Empirical Findings from 1980–2003." *Ecological Economics* 62 (3–4): 683–91.

Fisher, Dana R., and William R. Freudenburg. 2001. "Ecological Modernization and Its Critics: Assessing the Past and Looking Toward the Future." *Society and Natural Resources* 14 (8): 701–9.

Frey, R. Scott. 1998. "The Export of Hazardous Industries to the Peripheral Zones of the World-System." *Journal of Developing Societies* 14: 66–81.

———. 2003. "The Transfer of Core-Based Hazardous Production Processes to the Export Processing Zones of the Periphery: The Maquiladora Centers of Northern Mexico." *Journal of World-Systems Research* 9 (2): 317–54.

Grimes, Peter, and Jeffrey Kentor. 2003. "Exporting the Greenhouse: Foreign Capital Penetration and CO_2 emissions 1980–1996." *Journal of World-Systems Research* 9 (2): 261–75.

Hawken, Paul, Amory Lovins, and L. Hunter Lovins. 1999. *Natural Capitalism: Creating the Next Industrial Revolution*. New York: Little, Brown and Company.

Hong, Li, Zhang Pei Dong, He Chunyu, and Wang Gang. 2007. Evaluating the Effects of Embodied Energy in International Trade on Ecological Footprints in China. *Ecological Economics* 62: 136–48.

Hornborg, Alf. 2003. "Cornucopia or Zero-Sum Game? The Epistemology of Sustainability." *Journal of World-Systems Research* 9 (2): 205–16.

Jevons, William Stanley. 2001 [1865]. "Of the Economy of Fuel." Organization & Environment 14 (1): 99–104.

Jorgenson, Andrew K. 2003. "Consumption and Environmental Degradation: A Cross National Analysis of the Ecological Footprint." *Social Problems* 50: 374–94.

Jorgenson, Andrew K., and Thomas J. Burns. 2007. "The Political-Economic Causes of Change in the Ecological Footprints of Nations, 1991–2001." *Social Science Research* 36: 834–53.

Jorgenson, Andrew K., and Edward L. Kick. 2003. "Globalization and the Environment." *Journal of World-Systems Research* 9 (2): 195–203.

Jorgenson, Andrew K., and James Rice. 2005. "Structural Dynamics of International Trade and Material Consumption: A Cross-National Study of the Ecological Footprints of Less-Developed Countries." *Journal of World-Systems Research* 11: 57–77.

Kitzes, Justin, Audrey Peller, Steve Goldfinger, and Mathis Wackernagel. 2007. "Current Methods for Calculating National Ecological Footprint Accounts." *Science for Environment & Sustainable Society* 4: 1–9.

Leiserowitz, Anthony A., Robert W. Kates, and Thomas M. Parris. 2005. "Do Global Attitudes and Behaviors Support Sustainable Development?" *Environment* 47 (9): 23–38.

Liu, Jianguo, and Jared Diamond. 2005. "China's Environment in a Globalizing World." *Nature* 435: 1179–86.

———. 2008. "Revolutionizing China's Environmental Protection." *Science* 319: 37–38.

Loh, Jonathan, and Steven Goldfinger. 2006. *Living Planet Report 2006*. Gland, Switzer-land: WWF International.

Milanez, Bruno, and Ton Bührs. 2007. "Marrying Strands of Ecological Modernisation: A Proposed Framework." *Environmental Politics* 16 (4): 565–83.

Podobnik, Bruce. 2002. "Global Energy Inequalities: Exploring the Long-Term Implications." *Journal of World-Systems Research* 8 (2): 252–74.

Reijnders, Lucas. 1998. "The Factor 'x' Debate: Setting Targets for Eco-Efficiency." *Journal of Industrial Ecology* 2 (1): 13–22.

Roberts, J. Timmons, and Peter E. Grimes. 1997. "Carbon Intensity and Economic Development 1962–1991: A Brief Exploration of the Environmental Kuznets Curve." *World Development* 25: 191–98.

Roberts, J. Timmons, Peter E. Grimes, and Jodie L. Manale. 2003. "Social Roots of Global Environmental Change: A World-Systems Analysis of Carbon Dioxide Emissions." *Journal of World-Systems Research* 9 (2): 277–315.

Rosa, Eugene A., Richard York, and Thomas Dietz. 2004. "Tracking the Anthropogenic Drivers of Ecological Impacts." *Ambio* 33 (8): 509–12.

Rothman, Dale S. 1998. "Environmental Kuznets Curves—Real Progress or Passing the Buck?" *Ecological Economics* 25: 177–94.

Wackernagel, Mathis, Larry Onisto, Patricia Bello, Alejandro Callejas Linares, Ina Susana Lopez Falfan, Jesus Mendez Garcia, Ana Isabel Suarez Guerrero, and Ma. Guadalupe Suarez Guerrero. 1999. "National Natural Capital Accounting with the Ecological Footprint Concept." *Ecological Economics* 29: 375–90.

Wackernagel, Mathis, and William Rees. 1996. *Our Ecological Footprint: Reducing Human Impact on the Earth*. Gabriola Island, BC: New Society Publishers.

Wackernagel, Mathis, Niels B. Schulz, Diana Deumling, Alejandro Callenjas Linares, Martin Jenkins, Valerie Kapos, Chad Monfreda, Jonathan Loh, Norman Myers, Richard Norgaard, and Jorgen Randers. 2002. "Tracking the Ecological Overshoot of the Human Economy." *Proceedings of the National Academy of Sciences of the USA* 99 (14): 9266–71.

York, Richard. 2007. "Structural Influences on Energy Production in South and East Asia, 1971–2002." *Sociological Forum* 22 (4): 532–54.

———. 2006. "Ecological Paradoxes: William Stanley Jevons and the Paperless Office." *Human Ecology Review* 13 (2): 143–47.

York, Richard, and Eugene A. Rosa. 2003. "Key Challenges to Ecological Modernization Theory: Institutional Efficacy, Case Study Evidence, Units of Analysis, and the Pace of Eco-Efficiency." *Organization & Environment* 16 (3): 273–88.

York, Richard, Eugene A. Rosa, and Thomas Dietz. 2003. "Footprints on the Earth: The Environmental Consequences of Modernity." *American Sociological Review* 68 (2): 279–300.

———. 2004. "The Ecological Footprint Intensity of National Economies." *Journal of Industrial Ecology* 8 (4): 139–54.

PART III RACE, CLASS, GENDER, AND THE ENVIRONMENT

The Du Bois Nexus

7

Intersectionality, Political Economy, and Environmental Injustice in the Peruvian Guano Trade in the 1800s

Brett Clark, Daniel Auerbach, and Karen Xuan Zhang

The four chapters in this section explore how race, class, and gender intersect with environment. Most work in this vein focuses on environmental inequalities and environmental justice activism that has taken place in the past four decades. This chapter by Brett Clark, Daniel Auerbach, and Karen Xuan Zhang examines environmental inequalities as they existed historically. Using an intersectional lens, the authors explore nineteenth-century production and trade practices. To do this, they draw on the frameworks of W.E.B. Du Bois (1868–1963), an African American sociologist, historian, and activist who drew upon Marxist theory to analyze racial inequalities as a product of capitalism. Specifically, Clark and colleagues apply intersectionality to the historical case of the Peruvian guano trade in the nineteenth century. Guano, the excrement of bats and seabirds, was tremendously important as a fertilizer in the early days of large-scale agricultural production. The trade in guano helped fuel capitalist agricultural production in Europe at the expense of indigenous people and labor migrants in Peru.

Introduction

Complex systems of oppression are woven throughout the historic development of capitalism. The intermingled threads of race, class, gender, sexuality, and nationality are ever present in the hierarchy of nations and the international division of labor. They are interlaced in the reorganization of production and labor relations, the historic transfer of value that accompanied international trade, and the ecological contradictions of capitalism (Bunker 1984; Frank 1967; Galeano

1973; Hornborg 2003). Often, these issues are examined separately—no doubt because of the complexity of the social relationships and the different questions scholars ask. Nevertheless, it is important to analyze the distinct historic patterns related to these intersecting realms, deepening our understanding of dynamic systems of power. In this article, we bring together intersectionality, political economy, and environmental analysis through a brief study of the international guano trade in the mid-nineteenth century. Guano was a prized global commodity, used to enrich depleted soils in Europe and the United States. A unique racialized labor regime was imposed to extract the guano from Peruvian islands using "coolie" labor from China (Clark and Foster 2009). This racialized and gendered international division of labor resulted in specific forms of environmental inequalities and injustice, where specific peoples bear a disproportionate burden of environmental exposure and lower "quality of life" due to distinct systems of oppression (Pellow 2007, 15). We examine this historic case, placing intersectionality within an international political—economic context.

We propose that W. E. B. Du Bois's work serves as a fruitful basis for integrating these seemingly disparate traditions. His scholarship provides the basis to integrate race, class, gender, national, and ecological relationships into a coherent, political—economic analysis (Collins 2000b; Clark and Foster 2003). In *The Souls of Black Folk*, Du Bois (2003, xli) wrote, "the problem of the Twentieth Century is the problem of the color line," noting deep historical and racial divisions. Throughout his life, he addressed questions related to "land, labor, and the color line," using "this analysis to explain how the origins of the global capitalist system and imperialism created the hierarchies among nations and the social and racial divisions between peoples," which generated and perpetuated historical patterns of environmental inequities (Clark and Foster 2003, 461; see also Pellow 2007, 38). He contended that addressing the exploitation of women was just as important as confronting racial inequalities, noting that the two must be seen in relation to each other (Du Bois 1999, 105). Du Bois, therefore, contributes significantly to intersectionality analysis by recognizing the importance of distinct and changing historic contexts that shape these systems of oppression. . . .

Intersectionality and Historical Relations

Intersectionality, rooted in black feminist thought and critical race theory, highlights how systems of oppression, along the lines of race, class, gender, sexuality, nationality, etc., do not exist independently of one another. Rather, these systems must be seen as interconnected and mutually constitutive (Collins 2000a, 2000b; Crenshaw 1989, 1991)

Intersectionality is often used to examine individual experiences or as a way to theorize about identity, but it is important to emphasize that this perspective is rooted in illuminating social structures of domination (Acker 2006; Bilge 2013; Davis 2008). It is an approach that attempts to understand a broad variety of experiences and power structures, while placing attention on specific relationships (Carbado, Kimberlé Crenshaw, and Mays 2013). Put differently, there are a range of intersectional powers that are dependent upon specific historical, geographical, and social contexts.

Previous scholarship within this broad, interdisciplinary perspective has run the gamut from examining intersections of race, class, gender, sexuality, nationality, ability, to name but a few (Adib and Guerrier 2003; Bryan 2009; Clarke and McCall 2013; Collins 1998; May 2015; McCall 2005). Precursory writings of intersectionality recognized that during distinct historical periods certain types of oppression became more prominent than others. However, this did not diminish the interlocking nature of these oppressions (Collins 1986, 1993). Ida B. Wells (2002) wrote hard-hitting accounts of lynching in the post-slavery era of the United States, recognizing the intersection of race, class, and gender in maintaining a system of racial oppression. Other pre-intersectionality studies demonstrated the need for a historically grounded analysis that focuses on how systems of racial and gender oppression intersect with the dominant mode of production. Among these historically rooted analyses, Angela Davis argues that the intersections of gender, race, and class must be understood as the result of actual historical relations. Specifically, she contends, "the real oppression of women today is inextricably bound up with the capitalist mode of appropriating and mastering nature" (Davis 1998, 164).

While intersectionality has primarily examined the intersection of race and gender issues, it is able to comprehend other social categories and their relationships to power. Cho (2013, 390) points out that "race and gender intersectionality merely provided a jumping off point to illustrate the large point of how identity categories constitute and require political coalitions." In other words, intersectionality research is useful for comprehending a variety of historical social positions, dimensions of power, and specific forms of oppression. . . .

The Du Bois Nexus: Intersectionality, Political Economy, and Environmental Inequalities

When Du Bois (2003, xli) highlighted the persistence of "the color line," he provided a remarkable historical analysis of race within the United States, taking into account the larger political—economic and ecological context of social relations. Du Bois detailed that the settlement of the country was predicated upon the displacement of indigenous peoples. In particular, in the southern part of the United States, white male landowners helped establish a plantation system on this captured land, whereby Africans were imported as slaves to grow cotton and tobacco. Theories of "racial inferiority" were employed to provide ideological justification for inequalities and a system of domination.

The ranking and ordering of human beings, especially subaltern classes, became an integral part of imperial campaigns (Williams 1994). The triangle trade of humans, rum, and sugar was central to the rise of the global capitalist system, which involved both exploitation of human beings and the expropriation of nature by capital (Du Bois 1947, 56–59). These dynamics, which stratified nations and populations, helped create and maintain the color line.

Du Bois's historical analysis in *The Souls of Black Folk* traced the development of changing relations of production from the time of slavery to the period when he was writing in 1903. After the Civil War (1861–1865), white domination persisted, but it had changed forms. Recently freed African slaves had limited opportunities for employment and housing. While some redistribution of land temporarily took place under the Freedmen's Bureau, landownership remained concentrated in the hands of whites. In the southern part of the United States, blacks farmed their small parcels of land or became tenant farmers, renting from white landowners. The cotton mills and businesses selling farming supplies were owned by whites, who constantly changed prices to take advantage of black farmers. Lacking access to the means of production, black workers remained at the bottom of the class and racial hierarchy (Du Bois 1992).

Black tenant farmers were forced to take loans to obtain food, seed, and other supplies to make a living, leaving them in debt to white capitalists and landowners, maintaining a system of racial domination, despite the end of slavery.

Both Du Bois and Wells detailed how the international cotton trade rested upon a racial division of labor that was, in no small part, enforced through a system of violence via lynching. Such conditions created numerous contradictions and vast inequalities. Du Bois (2003, 138) explained that "the country is rich, yet the people are poor." Blacks massively outnumbered whites, yet whites tended to own the land and the tools necessary to work the land. Blacks sank deeper and deeper into debt, despite the fact they were doing the majority of the work and creating the value, through their labor, that accumulated in distant hands. In this analysis, Du Bois linked social inequalities and environmental injustice. In one example, he detailed these changes, tensions, and broader connections in a rural village:

> It is a beautiful land. . . . The forests are wonderful, the solemn pines have disappeared, and this is the "Oakey Woods," with its wealth of hickories, beeches, oaks, and palmettos. But a pall of debt hangs over the beautiful land; the merchants are in debt to the wholesalers, the planters are in debt to the merchants, the tenants owe the planters, and laborers bow and bend beneath the burden of it all. (Du Bois 2003, 127)

Despite the end of slavery and the potential freedom of black laborers, the class divisions of capitalist production continued to manifest in social inequalities, environmental injustice, and ecological degradation. "The fields . . . begin to redden and the trees disappear. Rows of old cabins appear filled with renters and laborers,—cheerless, bare, and dirty, for the most part" (Du Bois 2003, 127–28). These once beautiful "Oakey Woods" were in the process of being "ruined and ravished into a red waste" (Du Bois 2003, 128). Du Bois (2003, 128–29) explained that the incorporation of these once fertile fields and forests into the capitalist economy steadily degrades the quality of the land as it becomes overworked: "The poor land groans with its birth-pains, and brings forth scarcely a hundred pounds of cotton to the acre, where fifty years ago it yielded eight times as much." Given the racial stratification and the structural organization of capitalist agriculture, the conditions of immiseration were not felt evenly throughout society. The same process that brought degradation to the land also brought continued decimation to the workers. In fact, the exhaustion of the soil deepened the inequalities experienced by black families.

While illuminating specific historical forms of economic and racial domination, Du Bois also examined the positionality of women, such as how even within black families burdens were experienced unevenly. . . . As Du Bois (1999, 98) noted, "the crushing weight of slavery fell on black women. Under it there was no legal marriage, no legal family, no legal control over children."

He recognized that these two issues, race and gender, must not be seen separately. Rather, it is only "when . . . two of these movements—woman and color—combine in one, the combination has deep meaning" (105).

As Du Bois developed his historical-materialist analysis, his critique became even more centered on the structure and operation of the global capitalist system. The germination of his global analysis could be found in *The Souls of Black Folk* as Du Bois insisted that cotton production in the South was foundational to the global market. However, when Du Bois expanded his analysis, specifically examining colonialism, "the problem of the color line" was clearly evident at the global scale and manifested in the hierarchy of nations, the international racialized division of labor, and distinct patterns of environmental injustice and degradation (Clark and Foster 2003). Reflecting upon the racial ideology accompanying these global relations, Du Bois (1999,25) explained, "everything great, good, efficient, fair, and honorable is 'white'; everything mean, bad, blundering, cheating, and dishonorable is 'yellow'; a bad taste is 'brown'; and the devil is 'black.'"

In *Dusk of Dawn*, Du Bois (2007a) described the emergence of the international racialized division of labor. He explained,

> the economic foundation of the modern world was based on the recognition and preservation of so-called racial distinctions. In accordance with this, not only Negro slavery could be justified, but the Asiatic coolie profitably used and the labor classes in white countries kept in their places by low wage. (Du Bois 2007a, 52)

In other words, the wage system helped create and maintain racialized class inequalities as more of the world's people were incorporated into the capitalist system. Slaves and other indentured servants were forcibly moved to colonies where they were "despised and discriminated against" and where the "height of legal discrimination fell upon the Negro and other colored slaves" (Du Bois 2007b, 270). Capitalism depended upon the domination of different racial groups to generate high profits that primarily benefited the white bourgeoisie in Europe and the United States.

Du Bois pointed out that colonial and imperial relationships between the global North and South served as the basis to increase the rate of profit and to transfer important raw materials from the latter to the former. As in the South in the United States, the exploitation of labor and the expropriation of nature by capital worked hand-in-hand. Likewise, colonial nations sought raw materials abroad in part to maintain economic growth and "to retain 'their place in the sun'" (Du Bois 2007b, 270). . . . This usurpation of land and raw materials depended on the exploitation of people of color around the world such as on "coolies in China, on starving peasants in India, on black savages in Africa, on dying South Sea Islanders, on Indians of the Amazon."

The expansionary growth of capitalism took on the form of importing labor from around the world and instilling it in locations where valuable raw materials were located. The extreme conditions of labor exploitation allowed for immense profits that could not be realized in specific countries in the global North due to more stringent labor regulations. Furthermore, the expropriation of nature by capital remained an essential part of capitalism as white colonial powers displaced "indigenous peoples from land while creating labor reserves—both free and forced—on plantations for the production of goods desired by European nations" (Clark and Foster 2003, 461).

These historical developments of capitalism led to a radical reorganization of the relationship between humans and the larger physical world, creating distinct patterns related to the international racialized division of labor and environmental injustice. Du Bois (1992, 15–16) explained, from

> that dark and vast sea of human labor . . . on whose bent and broken backs rest today the founding stones of modern industry . . . despised and rejected by race and color; paid a wage below the level of decent living; driven, beaten, prisoned and enslaved in all but name; spawning the world's raw material and luxury—cotton, wool, coffee, tea, cocoa, palm oils, fibers, spices, rubber, silks, lumber, copper, gold, diamonds, leather—how shall we end the list and where? All these are gathered up at prices lowest of the low, manufactured, transformed and transported at fabulous gain; and the resultant wealth is distributed and displayed and made the basis of world power and universal dominion and armed arrogance.

. . . For Du Bois (1992, 16), the "real modern labor problem" involved the intersection of class, race, gender, nation, and environment through capitalist development, which could only be solved via the emancipation of labor.

The transference of wealth in the global capitalist economy has long involved a racialized hierarchy of labor. The historical development of capitalism generated distinct intersections between race, class, gender, and nation, which influenced the international racialized division of labor, the hierarchy of nations, environmental injustice, and the plundering of the global South. Du Bois recognized the shared burdens of workers, even if there were unique conditions, throughout regions of the world. He focused on how the form and intensity of domination that occurred was refracted through a racial hierarchy, generating distinct patterns of environmental injustice and inequality. In this, Du Bois integrated issues associated with intersectionality, political economy, and environmental justice.

The International Guano Trade: From Soil Crisis to the Racialized Division of Labor

From the 1840s to 1880s, guano was the most prized fertilizer in the world, especially in the global North where it was used to enrich exhausted soils subjected to modern capitalist agriculture. The international trade of guano further integrated the global economy, linking in particular China, Peru, and Great Britain in an elaborate international racialized division of labor. It involved the reorganization of labor and food production, the transfer of resources, and a series of ecological contradictions. The hierarchy of nations influenced the patterns of environmental inequalities and exploitation of labor (Clark and Foster 2009; Cushman 2013; Foster 2000). After providing the historical context that gave birth to the international guano trade, we examine the intersectional relationships associated with the coolie labor system employed in Peru.

Capital, Metabolic Relations, and Soil Science

Marx closely followed scientific debates and discoveries. Of particular interest to him were the discoveries associated with soil science. Justus Von Liebig (1851), a German chemist, incorporated the concept of metabolism into his analysis of the nutrient cycle. He explained that specific nutrients—nitrogen, phosphorus, and potassium—must be present in the soil, in varying quantities, depending on specific plant requirements, in order for crops to grow. The nutrients taken up by the crops must be recycled back to the soil to ensure the productivity of the land. Drawing upon Liebig, Marx put forward a metabolic analysis, arguing that human society, with its own distinct social metabolism, operates within the earthly metabolism, while constantly interacting with it (Marx and Engels 1975, vol. 30, 54–66; see also Foster 2013; Foster and Clark 2016; Saito 2014). Therefore, the historical organization of the social metabolism influenced the soil nutrient cycle.

In many pre-capitalist societies, farm animals were directly incorporated into agricultural production. They were fed grains from the farm and their manure, which contained nutrients, was reincorporated into the soil as fertilizer. Capitalist development, along with the enclosure movement, the division between town and country, and new industrial systems of food production, radically reorganized the metabolic interchange between human beings and the soil. Food and fiber were increasingly shipped to distant markets, transferring nutrients of the soil from the country to the city. These nutrients accumulated as waste, rather than being recycled to farm land (Foster 2000; Mårald 2002). Intensive agricultural practices, including the application of industrial power, increased the scale of operations, expanding the social metabolism while exacerbating the depletion of soil nutrients. Marx (1976, 637–38) explained that capitalist agriculture progressively "disturbs the metabolic interaction between man and the earth," producing a metabolic rift in the soil nutrient cycle.

Soil degradation in Great Britain caused widespread concern. Maintaining agricultural operations required inputs to replenish lost nutrients. Farmers purchased pulverized bones, recovered from battlefields, in an attempt to enrich the soil. But intensive capitalist farming practices continued to create evermore need for fertilizer. In the early 1800s, European explorers brought back samples of Peruvian guano. They shared stories regarding the mountains of bird dung accumulated on Peruvian islands. They explained how indigenous peoples used this fertilizer to enrich the poor soils of the mainland, producing bountiful crops. Scientists in Europe conducted tests, determining that guano had high concentrations of phosphate and nitrogen. It proved to produce high crop yields, setting in motion efforts to secure access to guano deposits (Skaggs 1994).

The largest deposits of high quality guano were on the Chincha Islands off the coast of Peru. Sea birds feasted on anchovies, enriching the dung that accumulated on the rocks. Given that it rarely rained on these islands, the nutrients were retained in the guano. Over thousands of years of accumulation, the guano deposits were hundreds of feet deep (Peck 1854). This fertilizer was used by farmers, who harvested relatively small quantities to support agricultural production on nutrient-poor lands.

Through a series of battles, Peru gained independence from Spain in the 1820s. To wage this campaign, Peru borrowed monies from Britain, resulting in a debt that they continued to carry for decades. In the 1840s, Antony Gibbs & Sons (1843), a British firm, negotiated the first in a series of trade agreements with Lima, gaining exclusive right over the sale of Peruvian guano on the global market (Mathew 1972, 1981; Skaggs 1994). For 40 years, Britain, minus brief interruptions when a French firm held the contract with Lima, controlled this fertilizer trade.

The Peruvian government viewed guano as a potential source of revenue, which could be used to repay debts and gain foreign exchange. Through negotiations, government officials claimed ownership of guano, receiving an agreed upon amount of money per ton of guano shipped. The importance of guano to the national budget dramatically increased. In 1846–1847, guano sales supplied only 5 percent of state revenues, but by 1869 it was 80 percent (Bonilla 1987). From time to time, the trade agreements were renegotiated. Peru hoped to acquire greater financial resources, but the terms of trade continued to decline. Britain forced Peru to accept liberal policies that favored the former (Duffield 1877; Hunt 1973). Peru remained in debt and became increasingly dependent on foreign nations for commodities. Alongside these trade negotiations, national budget considerations, and ecological exchanges, a distinct racialized and gendered international division of labor was organized in relation to the prized commodity.

A Racialized and Gendered International Division of Labor: Guano Work and Chinese Coolies

The guano pits of Peru were notorious during the height of the international guano trade. The "manure theory of social organization," which Du Bois (1999, 69) described, highlights the dynamics of this situation. Guano work was done exclusively by men. It was filthy, as acrid dust coated workers, penetrating their eyes, noses, and mouths. The stench was appalling. In order to keep capital outlay low, extraction of guano depended on physical human labor combined with basic tools, such as picks, shovels, wheelbarrows, and sacks. Workers hacked and shoveled guano into wheelbarrows and sacks, and then they would transport this prized fertilizer to a chute to transfer the commodity to ships waiting to be loaded. This work was known as a process that exhausted human life. As Du Bois (1999, 69) indicated, the worst work is forced upon those not recognized by the dominant society as real human beings, allowing an exclusive class to procure the wealth created. In regard to the global division of labor and wealth, it was white Europe in relation to "the world . . . [which] is black and brown and yellow." And in this hegemonic relationship it was "the duty of white Europe to divide up the darker world and administer it for Europe's good" (23). During the heyday of the international guano trade (1840–1880), the work force changed, but it always consisted of subaltern classes.

As determined in trade negotiations, the Peruvian government retained the right to subcontract the mining and loading process. Domingo Elías, a wealthy cotton plantation owner, who owned African slaves, held the contract for operating the extraction of guano on the islands. In the early 1840s, he organized a workforce that consisted of male convicts, army deserters, and slaves. The size of the workforce varied between 200 and 800 men, who were isolated on the islands. The ruling class in Peru used guano revenues to expand their landholdings. Elías, who was very active in national politics and even served as President of Peru for three months in 1844, followed suit, enlarging his plantations. Together these rich landowners helped transform the agricultural sector into a producer of cash crops, such as sugar, cotton, and cochineal for export to Europe and the United States (Blanchard 1996; Gonzales 1955; Gorman 1979). This system of food production transferred crops (and the nutrients with them) to cities in the global North, while creating a metabolic rift in the soil nutrient cycle in Peru.

The declining availability of slaves and the establishment of the "coolie" labor system ushered in a major transition in Peru. In the 1840s, there was a labor shortage on plantations and in the mines. To remedy this, the government of Peru passed "an immigration law subsidizing the importation of contract labourers" (Gonzales 1955, 390–91). Anyone who imported "at least fifty workers between the ages of 10 and 40" was paid 30 pesos per head. The cheapness of importing labor from China, eventually made it possible for Peru to abolish slavery. Chinese workers replaced slaves in the fields, on the railroad, in the mines, and on the guano islands. When slavery was formally abolished, slaveholders, such as Elías, who was involved in this labor transition, were compensated for the loss of slaves. Like other slave owners, he profited off of the end of slavery and through facilitating the importation of Chinese labor.

The "coolie" labor system was established in the early nineteenth century, in order to supply colonizers with a cheap source of labor for intensive work. While it was officially a system of indentured servitude, it was very much a form of "disguised slavery" (Marx 1963, 112; Marx and Engels 1972, 123). In fact, many slave traders simply shifted to transporting indentured servants. "Coolie" (sometimes spelled "kuli") means hard work. Coolie workers, primarily from China and India, worked on plantations, built railroads, and toiled in mines.

Many factors in China contributed to the expanding coolie trade. Natural disasters, such as droughts and floods, displaced families from their farmland. The British waged the Opium Wars against the Qing Dynasty in China in the name of free trade, in order to expand their hegemonic position in the global economy. The Opium Wars and the Taiping Rebellion created massive social disruption throughout China (Meagher 2008). . . .

These (Natural disasters, war) conditions helped expand the coolie trade, where thousands of Chinese were forced to work in distant lands due to changes in social conditions at home.

Coolie labor was obtained through coercion, deceit, and even kidnapping. This system of contracted labor involved a distinct division between men and women. At the start, Chinese men were conscripted to work abroad building railroads and/or extracting mineral resources from mines. These men were called "piglets," which described the inhumane treatment they received and the conception that they were less than human (Ren 1989). Initially, the coolie trade primarily involved shipping men overseas, who, prior to departure, were stripped down and marked on their chests with letters signifying their destination: C for Cuba, P for Peru, S for Hawaiian Islands (Taku 1991). In some cases, the men who voluntarily signed up for work abroad sold their wives into domestic prostitution to pay the transportation fee. These women became known as "pig flowers" or "pig beauties." As the labor system became more firmly established throughout the world, coolie traders and business operators decided to traffic Chinese women to labor camps as prostitutes, as part of an effort to diminish unrest among the male workers. The Chinese women, who were kidnapped, sold, and forced into prostitution, were generally from the poorest and most destitute families in Chinese society (Luo 1989).

It is clear that the coolie labor system, in the context of the larger political economy, operated as a system of oppression, demeaning and exploiting Chinese men and women, forcing them to lowest stature. Shixiang Mo (2016) explains that the coolie trade was clearly rooted in a political-economic system that pursues endless economic growth, rather than justice.

Hundreds of thousands of Chinese laborers were "contracted" for work through Macao and Hong Kong (Clayton 1980; Hu-DeHart 1989, 2002). European merchants systematically transferred Chinese workers to Cuba and Peru, as well as various British colonies. Reflective of Du Bois's (1999) contention that the racialized and gendered division of labor, under U.S. slavery, created a class of prostitutes and beasts of burden, the coolie labor system generated a historically distinct class of dehumanized workers—"pig flowers" and "piglets." The two groups, although different, remain interlinked within the global story of the Chinese coolie. However, in the specific regional case of guano extraction in Peru, the coolie labor system consisted solely of men, creating a distinct gendered dimension to the division of labor in this context.

The voyage to Peru took approximately five months. Chinese workers were primarily locked away, below deck. From time to time, they were fed small rations of rice on wood pans. Given the conditions, they developed open sores on their bodies. During the first 15 years of the global coolie trade, the mortality rate was between 25 and 30 percent (Wingfield 1873, 4). Occasionally, there were mutinies during passage (Marx and Engels 1972, 123; Marx 1963, 112).

Upon arrival, these male laborers were stripped naked and presented on the open market, where buyers inspected their bodies, determining who would be strong enough for the strenuous work (Chen 1981). Employers freely resold Chinese laborers, without any consent from workers (Chen 1963).

The first Chinese coolies arrived in Peru in 1849. Over 90,000 Chinese coolies were shipped to Peru between 1849 and 1874 (Gonzales 1955, 390–91). Like in other countries and colonies, the majority of the coolies were forced to work on plantations or to build railroads. In Peru, however, an additional, and worse, fate was reserved for select male coolies, as they were sent to the guano pits, where they experienced extreme forms of environmental injustice, often resulting in grievous bodily harm, due to the labor conditions.

On the guano islands, coolie laborers dug through layers of deposits. Each worker was expected to load between 80 and 100 wheelbarrows, close to five tons, each day. They pushed wheelbarrows and carried sacks to chutes used to load boats. British officers reported that often a hundred boats were simultaneously being loaded with guano, while hundreds of other ships waited for their turn ("The Guano Trade" 1855; Matthews 1977; Nash 1857; *New York Observer and Chronicle* 1856). Most years during the international guano trade, over a hundred thousand tons were shipped. From 1866 to 1877, between 300,000 and 575,000 tons were exported to the world to replenish exhausted agricultural fields in the global North (de Secada 1985).

While coolie workers were not legally slaves, they lived in de facto slavery. On the guano islands of Peru, only male coolies lived and worked there. As Alanson Nash (1857) explained, "once on the islands a Chinamen seldom gets off, but remains a slave, to die there." These men were seen as expendable beasts, forced to "live and feed like dogs" (Peck 1854, 207). New male coolies replaced those who died. Failure to meet productivity expectations was met with physical punishment, such as flogging, whipping, and being tied to rocks baking in the sun. Some workers jumped off cliffs to kill themselves to end the misery. From time to time, the workers revolted, only to confront more violence. In an effort to mollify the workers, Peruvian businessmen made arrangements with British traders to import opium (Clayton 1980; Hu-DeHart 1989, 108–9; "Chincha Islands" 1854). Paradoxically, following this strategy, there were newspaper accounts depicting the Chinese workers as degenerates who smoked opium and engaged in homosexual activities. Despite the distance of the guano islands from the major cities in the global North, the suffering imposed on Chinese workers was present in newspapers and magazines in Europe. Writers shared eyewitness accounts, denouncing that Chinese workers are "sold into absolute slavery—*sold by Englishmen into slavery*—the worst and most cruel perhaps in the world" (Chincha Islands 1854). *The Christian Review* detailed how "the subtle dust and pungent odor of the new-found fertilizer were not favorable to inordinate longevity," creating a constant demand for more workers, given that guano labor involved "the infernal art of using up human life to the very last inch" ("The Chinese Coolie Trade" 1862; see also Lubbock 1955, 35). Shipmasters, in 1854, were "horrified at the cruelties they saw inflicted on the Chinese, whose dead bodies they described as floating round the islands" (Wingfield 1873, 5). Duffield (1877, 77–78) wrote,

> No hell has ever been conceived . . . for appeasing the anger and satisfying the vengeance of their awful gods, that can be equaled in the fierceness of [the] . . . heat, the horror of its sink, and the damnation of those compelled to labour there, to a deposit of Peruvian guano when being shovelled into ships.

Commentators also linked the fertile fields of Britain to the exploitation of Chinese workers. In 1856, in the *Nautical Magazine*, it was noted that few probably are aware that the acquisition of this deposit, which enriches our lands and fills the purses of our traders, entails an amount of misery and suffering on a portion of our fellow creatures, the relation of which, if not respectably attested, would be treated as fiction ("The Chincha Islands" 1856).

The complex relationship between intersectionality, political economy, and environmental injustice is evident in this brief case study. The Du Bois nexus serves as a useful means to bridge these distinct literatures, illuminating the general and particular characteristics associated with specific historical systems of oppression.

Conclusion

The Du Bois nexus brings together intersectionality, political economy, and environmental analysis. Intersectionality scholars stress that race, gender, class, nation, region, and environment must be recognized as interconnected and mutually constituted. They also stress that historical conditions, which includes both domestic and international considerations, influence how these relationships are organized. An analysis of coolie labor on the guano islands literally reveals the "manure" theory of social organization. The international racialized and gendered division of labor associated with the guano trade took distinct forms, given the shift from the use of African slaves to Chinese male workers. Both populations experienced extreme forms of exploitation and environmental injustice in the guano pits. The guano islands served as a death sentence for most of these male workers, as the physical labor, meager rations, and environmental conditions shortened their lives. Chinese women, who were left behind in China, were often sold or forced into prostitution. In the latter phases of the general coolie labor system, Chinese women were sent as prostitutes to labor camps in other regions and nations. Wealthy landowners and governmental officials in Peru profited from this international trade and labor system, deepening class divisions. They expanded their agricultural holdings, pursuing cash crops for export, creating a metabolic rift in the soil nutrient cycle. Ironically, they were also profiting off of the most prized fertilizer in the world, which was being shipped to distant parts of the global North, as guano supplies quickly diminished. These relationships were also shaped by the structure of the global economy, given the hierarchy of nations. Peru was in debt to Great Britain. Profits from guano followed the commodity chain, accumulating in England. During this period, Britain was also fighting the Opium War against China, creating much social upheaval, which contributed to the growth in the coolie trade. The international guano trade, itself, was in part a consequence of unsustainable food production in the global North, whereby intensive agricultural practices were creating a metabolic rift, requiring massive fertilizer inputs to sustain crop yields. This case highlights how these historically specific conditions and relationships facilitated the asymmetrical movement of valuable resources, in an unequal ecological exchange, to maintain the growth imperative of capital, involving distinct racial, class, and gendered systems of oppression.

Note

Clark, Brett, Daniel Auerbach, and Karen Xuan Zhang. 2018. "The Du Bois Nexus: Intersectionality, Political Economy, and Environmental Injustice in the Peruvian Guano Trade in the 1800s." *Environmental Sociology* 4 (1): 54–66.

References

Acker, J. 2006. *Class Questions: Feminist Answers.* Lanham, MD: Rowman & Littlefield Publishers.

Adib, A., and Y. Guerrier. 2003. "The Interlocking of Gender with Nationality, Race, Ethnicity and Class." *Gender, Work and Organization* 10 (4): 413–32. doi:10.1111/1468–0432.00204.

Antony Gibbs & Sons. 1843. *Guano.* London: William Clowes & Sons.

Blanchard, P. 1996. "The 'Transitional Man' in Nineteenth-Century Latin America." *Bulletin of Latin American Research* 15 (2): 157–76.

Bonilla, H. 1987. "Peru and Bolivia." *In Spanish America after Independence C. 1820-C. 1870*, edited by L. Bethell, 239–82. Cambridge: Cambridge University Press.

Bryan, A. 2009. "The Intersectionality of Nationalism and Multiculturalism in the Irish Curriculum." *Race Ethnicity and Education* 12 (3): 297–317. doi:10.1080/13613320903178261.

Bunker, S. 1984. "Modes of Extraction, Unequal Exchange, and the Progressive Underdevelopment of an Extreme Periphery." *American Journal of Sociology* 89: 1017–64. doi:10.1086/227983.

Carbado, D. W., V. M. Kimberlé Crenshaw, and B. T. Mays. 2013. "Intersectionality: Mapping the Movements of a Theory." *Du Bois Review* 10 (2): 303–12. doi:10.1017/S1742058X13000349.

Chen, H. 1981. *Collection of Historical Documents Concerning Emigration of Chinese Laborers.* Vol. 4. Beijing: Zhonghau Books.

Chen, Z. 1963. "The Popular Contract Chinese Laborers System in 19th Century." *Historical Research* 10 (1): 161–79.

"Chincha Islands." 1854. *Friends' Intelligencer.* February 11.

"Chincha Islands." 1856. *Nautical Magazine and Naval Chronicle.* April.

"Chinese Coolie Trade." 1862. *The Christian Review.* April.

Cho, S. 2013. "Post-Intersectionality." *Du Bois Review* 10 (2): 385–404. doi:10.1017/S1742058X13000362.

Clark, B., and J. B. Foster. 2003. "Land, the Color Line, and the Quest of the Golden Fleece." *Organization & Environment* 16 (4): 459–69. doi:10.1177/1086026603259095.

———. 2009. "Ecological Imperialism and the Global Metabolic Rift." *International Journal of Comparative Sociology* 50 (3–4): 311–34. doi:10.1177/0020715209105144.

Clarke, A. Y., and M. Leslie. 2013. "Intersectionality and Social Explanation in Social Science Research." *Du Bois Review* 10 (2): 349–63. doi:10.1017/S1742058X13000325.

Clayton, L. A. 1980. "Chinese Indentured Labor in Peru." *History Today* 30 (6): 19–23.

Collins, P. H. 1986. "Learning from the Outsider Within." *Social Problems* 33 (6): S14–S32. doi:10.2307/800672.

———. 1993. "Toward a New Vision: Race, Class, and Gender as Categories of Analysis and Connection." *Race, Sex & Class* 1 (1): 25–45.

———. 1998. "It's All in the Family: Intersections of Gender, Race, and Nation." *Hypatia* 13 (3): 62–82. doi:10.1111/j.1527–2001.1998.tb01370.x.

———. 2000a. *Black Feminist Thought*. New York: Routledge.

———. 2000b. "Gender, Black Feminism, and Black Political Economy." *Annals of the American Academy of Political and Social Sciences* 568: 41–53. doi:10.1177/000271620056800105.

Crenshaw, K. 1989. "Demarginalizing the Intersection of Race and Sex." *University of Chicago Legal Forum* 1: 139–67.

———. 1991. "Mapping the Margins: Intersectionality, Identity Politics, and Violence against Women of Color." *Stanford Law Review* 43 (6): 1241–99. doi:10.2307/1229039.

Cushman, G. T. 2013. *Guano and the Opening of the Pacific World*. Cambridge: Cambridge University Press.

Davis, A. 1998. *The Angela Y. Davis Reader*. Malden, MA: Blackwell Publishers Limited.

Davis, K. 2008. "Intersectionality as Buzzword: A Sociology of Science Perspective on What Makes a Feminist Theory Successful." *Feminist Theory* 9 (1): 67–85. doi:10.1177/1464700108086364.

De Secada, C., and G. Alexander. 1985. "Arms, Guano, and Shipping." *Business History Review* 59 (4): 597–621. doi:10.2307/3114596.

Du Bois, W.E.B. 1947. *The World and Africa: An Inquiry into the Part Which Africa Has Played in World History*. New York: Viking.

———. 1992. *Black Reconstruction in America, 1860–1880*. New York: Atheneum.

———. 1999. *Darkwater*. Mineola. New York: Dover Publications.

———. 2003. *The Souls of Black Folk*. New York: Modern Library.

———. 2007a. *Dusk of Dawn*. New York: Oxford University Press.

———. 2007b. *The World and Africa and Color and Democracy*. New York: Oxford University Press.

Duffield, A. J. 1877. *Peru in the Guano Age*. London: Richard Bentley and Son.

Foster, J. B. 1994. *The Vulnerable Planet*. New York: Monthly Review Press.

———. 1999. "Marx's Theory of Metabolic Rift." *American Journal of Sociology* 105 (2): 366–405. doi:10.1086/210315.

———. 2000. *Marx's Ecology*. New York: Monthly Review Press.

———. 2013. "Marx and the Rift in the Universal Metabolism of Nature." *Monthly Review* 65 (7): 1–19.

Foster, J. B., and B. Clark. 2012. "The Planetary Emergency." *Monthly Review* 64 (7): 1–25.

———. 2016. "Marx's Ecology and the Left." *Monthly Review* 68 (2): 1–25.

Foster, J. B., B. Clark, and R. York. 2010. *The Ecological Rift*. New York: Monthly Review Press.

Foster, J. B., and H. Holleman. 2012. "Weber and the Environment." *American Journal of Sociology* 117 (6): 1625–73. doi:10.1086/664617.

Frank, A. G. 1967. *Capitalism and Underdevelopment in Latin America*. New York: Monthly Review Press.

Galeano, E. 1973. *Open Veins of Latin America*. New York: Monthly Review Press.

Gonzales, M. J. 1955. "Chinese Plantation Workers and Social Conflict in Peru in the Late Nineteenth Century." *Journal of Latin American Studies* 21: 385–424. doi:10.1017/S0022216X00018496.

Gorman, S. M. 1979. "The State, Elite, and Export in Nineteenth Century Peru." *Journal of Interamerican Studies and World Affairs* 21 (3): 395–418. doi:10.2307/165730.

"Guano Trade." 1855. *Friends' Intelligencer*. August 4.

"Guano Trade." 1856. *New York Observer and Chronicle*. July 24.

Hancock, A.-M. 2005. "W.E.B. Du Bois: Intellectual Forefather of Intersectionality?" *Souls: A Critical Journal of Black Politics, Culture, and Society* 7 (3–4): 74–84. doi:10.1080/10999940500265508.

Hornborg, A. 2003. "Cornucopia or Zero-Sum Game?" *Journal of World-Systems Research* 9 (2): 205–16. doi:10.5195/JWSR.2003.245.

Hu-DeHart, E. 1989. "Coolies, Shopkeepers, Pioneers." *Amerasia Journal* 15 (2): 91–116. doi:10.17953/amer.15.2. b2r425125446h835.

LaDuke, W. 1999. All Our Relations. Boston: South End Press.

Liebig, J. 1851. *Familiar Letters*. 3rd ed. London: Taylor, Walton, and Maberley.

Lubbock, B. 1955. *Coolie Ships and Oil Sailers*. Glasgow: Brown, Son & Ferguson.

Luo, H. 1989. "The Discussion on 'Pig Flower.'" *Journal of Jinan University* (Philosophy and Social Science Edition) 8 (4): 78–84.

Marx, K. 1963. *The Poverty of Philosophy*. New York: International Publishers.

———. 1976. *Capital*. Vol. 1. New York: Vintage.

———. 1991. *Capital*. Vol. 3. New York: Penguin Books.

Marx, K., and F. Engels. 1972. *On Colonialism*. New York: International Publishers.

———. 1975. *Collected Works*. New York: International Publishers.

Mathew, W. M. 1972. "Foreign Contractors and the Peruvian Government at the Outset of the Guano Trade." *The Hispanic American Historical Review* 52 (4): 598–620. doi:10.2307/2512783.

May, V. M. 2015. *Pursuing Intersectionality, Unsettling Dominant Imaginaries*. London: Routledge.

McCall, L. 2005. "The Complexity of Intersectionality." *Signs* 30 (3): 1771–800. doi:10.1086/426800.

Meagher, A. J. 2008. *The Coolie Trade*. San Bernardino, CA: Xlibris Corporation.

Mo, S. 2016. "The Kuli Trade in Hong Kong and Macau and the Issue between Britain and Portugal." *Guangdong Social Science* 33 (2): 80–89.

Nash, A. 1857. "Peruvian Guano." *Plough, the Loom and the Anvil* 10 (2): 69–76.

Peck, G. W. 1854. *Melbourne and the Chincha Islands*. New York: Charles.

Pellow, D. N. 2007. *Resisting Global Toxics: Transnational Movements for Environmental Justice*. Cambridge: The MIT Press.

———. 2016. "Toward a Critical Environmental Justice Studies." *Du Bois Review*. doi:10.1017/S1742058X1600014X.

Peluso, N. L. 1992. *Rich Forests, Poor People*. Berkeley: University of California Press.

Ren, D. 1989. "The Research on the Origin of the Appellation of 'Piglet.'" *Overseas Chinese History Studies* 3: 61.

Saito, K. 2014. "The Emergence of Marx's Critique of Modern Agriculture." *Monthly Review* 66 (5): 25–46. doi:10.14452/MR-066-05-2014-09_2.

Skaggs, J. M. 1994. *The Great Guano Rush.* New York: St. Martin's Griffin.

Taku, S. 1991. "Slavery System of Colony and Coolie Trade in China." *Southeast Asian Studies—A Quarterly Journal* 61 (1): 60–70.

Wells, I. B. 2002. *On Lynching.* Amherst, New York: Humanity Books.

Williams, E. 1994. *Capitalism and Slavery.* Chapel Hill: University of North Carolina Press.

Wingfield, C. 1873. *The China Coolie Traffic from Macao to Peru and Cuba.* London: British and Foreign Anti-Slavery Society.

Ruin's Progeny
Race, Environment, and Appalachia's Coal Camp Blacks

Karida L. Brown, Michael W. Murphy, and Appollonya M. Porcelli

The authors of this article point out that the literature on race and environment has largely focused on the unequal distribution of environmental burdens, the mobilization of communities of color against those burdens, and, to a lesser extent, exploring how those inequalities are produced. At the same time, the literature on environmental concern and meaning making has focused almost entirely on white communities while ignoring communities of color. Instead of asking questions about inequalities, these authors explore how African Americans formed individual and collective identities from their experiences living in the coal mining mountains of Appalachia. In focusing on meaning making and its complexities, these authors are complicating the field, revealing how previous research has presented an incomplete story of race and environment.

> *[It would] remain impossible for the majority to conceive of a Negro [being] stirred by the pageants of Spring and Fall; the extravaganza of summer, and the majesty of winter.*
> —Zora Neale Hurston, "What White Publishers Won't Print," *Negro Digest* (1950, p. 87)

Introduction

The vibrant and meaningful relationship between communities of color and their surrounding environments has been historically disregarded. . . .

Moreover, the majority of literature on environmental concern and meaning-making centers on White communities (Agyeman and Spooner, 1997; Deluca and Demo, 2001; Finney 2014) thereby buttressing the dominant narrative that ties environmental awareness to Whiteness. Our research therefore seeks to rectify this significant omission in existing literature by asking: (1) How is the environment implicated in the conditioning of racialized subjectivities? And (2) How do landscapes and environment impact the formation of collective identity and sense of belonging for African Americans?

In this article, we focus on the lived experience of a generation of African American coal miners and their families, who migrated into and out of central Appalachia, during the twentieth century Great Migration. Our analysis shows that their subjectivities were largely conditioned in and through their relationship to the landscape and environment—from their deep understanding of agriculture, the mountainous landscape, and the bituminous coal that symbolized life and death to them. Although the vast majority of this population migrated over half a century ago, this generation of African Americans continues to call these coal towns in Appalachia "home"—a place of origin and belonging that deeply informs their collective identity. This changing "landscape of meaning"— referring to historically specific and particular landscapes upon which the social emerges (Reed 2011, p. 92)—is expressed, refashioned, and sustained through a variety of ongoing cultural formations and invented traditions instantiated by this group. This study offers an empirical investigation of the way in which collective identities can emerge out of the shared experience of racialized displacement from land and environment. This paper complicates the process of *ruination*—the resultant wreckage of the inevitable decline of imperial and industrial formations, and the ensuing displacement of the people and places that are subsequently rendered debris (Mah 2012; Stoler 2013)—by focusing on the productive tensions that exist between the duality of oppression and resilience. In documenting . . . what humanist geographer Yi-Fu Tuan (1974, p. 4) refers to as "the affective bond between people and place or environment," among this group of African Americans, this study thereby counters the otherwise dominant narrative that portrays the people of Appalachia as *hopeless, helpless, homeless,* and *White.*

Previous Literature on Race and the Environment

Environmental sociology has addressed questions of race and environment in three primary ways thus far. The first explores the unequal distribution of environmental hazards, which disproportionately trouble low-income communities and communities of color. Secondly, scholars have analyzed the mobilizations of the racially marginalized around these environmental inequalities, as seen in the Environmental Justice Movement(s). Finally, though much less prevalent, some of the literature has focused on how these environmental inequalities are produced. . . . By focusing solely on the (unequal) distribution of environmental burdens and the mobilizations of people of color against the pollution in their communities, this literature is blinded to ways in which even the most degraded landscapes figure into the identity formation of marginalized groups. Below, we briefly review this literature.

Opposition to the placement of a toxic, polychlorinated biphenyl (PCB) landfill in Warren County, North Carolina, in 1980 initiated the unfolding discourse of what would come to be known as Environmental Justice (EJ). As a social movement, EJ is centered on addressing disparities in environmental health, hazards, and risks. Within sociology, Environmental Justice is a very productive field, with deep roots and linkages to activism and advocacy. In the past, the study of Environmental Justice has been conflated with the study of environmental inequality. Here, we would like to distinguish the two within the academic literature of sociology. Environmental inequality is a field of inquiry concerned with the disparity of environmental health, between communities, especially based on categories of race, class, nationality, and gender, whereas Environmental Justice refers simultaneously to an ethical principle and the socio-logical study of the social movement formed around it. Simply put, Environmen-tal Justice literature is the study of the social movements that emerge to combat environmental inequalities.

Environmental justice is a normative concept that emerged in response to environmental racism, and can be thought of as the remedy for environmental racism. Robert Bullard, a founding figure in Environmental Justice scholarship, defines environmental justice as the principle that "all people and communities are entitled to equal protection of environmental and public health laws and reg-ulations" (Mohai et al., 2009, p. 407). Environmental justice takes an ethical stance that calls for both positive and negative rights: All people have the right *to* live in clean environments in which they can thrive, and they should have free-dom *from* exposure to toxins. . . .

Early studies of environmental inequality focused on demonstrating the phenomena through descriptive study. Beginning in the 1970s, a small group of researchers used data from the Environmental Protection Agency (EPA) to study exposure to air pollution. They found that poorer neighborhoods were exposed to more polluted air than wealthier communities (Berry 1977; Kru-vant 1975; Zupan 1973). From here, a new research enterprise blossomed as scholars and activists began to share accounts of environmental injustice (Bullard 1990, 1993; Hofrichter 1993; Mohai and Bryant, 1992). These scholars were primarily focused on the disproportionate impact of environmental hazards on communities of color—or environmental racism—which is "racial discrimination in environmental policy making, the enforcement of regulations and laws, the deliberate targeting of communities of color for toxic waste facilities" (Mohai et al., 2009, p. 407).

While important, these studies often focused solely on the *existence* of unequal environmental outcomes and the struggles against them without paying much attention to the mechanisms that created them (Pellow 2000; Szasz and Meuser, 1997; Weinberg 1998). David Pellow (2000) suggests that social scien-tists instead focus on the mechanisms that produce environmental inequalities, and some scholars have taken up this suggestion (Crowder and Downey, 2010; Pastor et al., 2001; Saha and Mohai, 2005). Kyle Crowder and Liam Downey (2010), for example, focus on inter-neighborhood migration, race, and environ-mental hazards, in an attempt to specify the mechanisms that create environmen-tal inequalities. . . . Their findings . . . provide evidence that racial discrimination plays a stronger role than class in that when socioeconomic status is controlled for, even the highest-income Blacks and Latinos end up moving into neighborhoods with higher exposures to pollution than Whites. . . .

Overall, the dominant discourse about the environment is heavily racialized, linking Whiteness to concern and attention to environment (Agyeman and Spooner, 1997; Deluca and Demo, 2001; Finney 2014), and non-Whiteness to environmental decay. Unfortunately, the literature within environmental sociology perpetuates a narrative of urban environmental decay in communities of color by overlooking the ways in which people of color think about the environments in which they are embedded. This paper makes an important intervention in this literature by disrupting this meta-narrative of inequality and degradation by offering a thick interpretation of the ways in which collective identities are shaped by the inextricable relationship with racialization, landscape, and environment. . . .

The Setting: Eastern Kentucky

This study is set high in the rugged, verdant, mountainous, industrial region of Central Appalachia. The last region in the United States to be swept by the boom and bust coal mining industry, Central Appalachia became a niche layover for Black migrants who were early movers between 1880 and 1930, a period that came to be known as the first wave of the African American "Great Migration" (Lewis 1987, 1989; Trotter 1990). These migrant streams primarily initiated from the mineral districts of Birmingham and Bessemer, Alabama, and to a lesser extent, the agricultural regions of Virginia and North Carolina (Lewis 1989). A part of the cadre of the estimated six million Blacks who fled the oppressive and violent context of the deep South in the early twentieth century, these tens of thousands of early migrants became "Black Appalachians" (Turner and Cabbell, 1985), if for only one or two generations. . . .

Methodology

The data for this paper come from the Eastern Kentucky African American Migration Project (EKAAMP), an archival collection housed at the *Southern Historical Collection* (SHC) at the University of North Carolina at Chapel Hill. The collection is comprised of 215 oral history interviews that were primarily conducted with African Americans who share roots in the coal towns of Eastern Kentucky. In addition to the oral history interviews, EKAAMP also holds thousands of vernacular photographs and hundreds of documents and objects relating to the history and culture of this population of Black Appalachians and their descendants.

In this study, we examine the ways in which the intersections between race and environment bear on the formation of the social self and collective identity. In the following sections we explore the myriad meanings that emerge from interaction with the landscape as we follow the life history of southeastern Kentucky's coal camp Blacks.

Ruin's Progeny: Race, Environment, and Appalachia's Coal Camp Blacks

To understand life in a coal camp you must know this: there is no separation between home, environment, and industry. It is a place where the mountainside provides berries, peaches, and apples; prime spots for hide-and-go-seek; and the perfect glade for the perfect tree house. At the same time, it is a place where the soles of mountains are etched in the dark, dank, and dangerous crawlspaces that most of the men in the town call their workplaces. Like Atlas, the Titan god condemned to hold up the sky for eternity, their back-scraping work supported the entire economy of the town.

What follows is a narrative that traces the lives of the African American families who migrated to these coal camps in southeastern Kentucky, and who were eventually displaced upon the closure of the mines. We explore these families' landscapes of meaning in a double sense, both conceptually—examining a life-world constructed and contested through the tensions between life and death, freedom and oppression, ruin and progeny, that emerges from the subjective experience of living in these coal camps—and in the literal sense—drawing attention to the very trees, creeks, and creatures that make up the physical landscape and that have meaning for Appalachia's coal camp Blacks.

Brenda Thornton grew up in a bowl, or at least that's the way she describes it, a town "buried in the mountains," where you couldn't see anything "as far beyond those mountains."[1] For Black children in southeastern Kentucky, the edges of the world fell along the contours of the Appalachian Mountains. Their parents had moved to this region in search of new economic opportunities, fleeing the oppressive structure of opportunity in the Deep South—sharecropping, convict labor, and the ever-present threat of unmediated racial violence. What they found in southeastern Kentucky were towns built solely for the purpose of supporting the extraction of coal—towns entirely owned and operated by the coal company. Because of this totalizing, built environment everything seemed to have a purpose and everyone seemed to know his or her place. The Blacks lived on one side of the town, Whites on the other, coal mining managers along the main street, and the elites tucked away in villa-like estates on what was called "Silk Stocking Row." All necessary goods could be bought at the company store—from rice to dish soap to a custom tailored suit for that year's Easter Sunday church service. The company even provided certain amenities to keep the miners and their families entertained within the confines of the mining town—a movie theater, a town pool (for Whites only), two school houses (one for the Whites and the other for the "colored" children), and a number of churches to serve the various racial and ethnic immigrant groups in the community. In a town so contrived, however, Cynthia Brown-Harrington remembers the "freedom of it all":

> I remember the mountains, friends and families and mothers and fathers. Everybody's father was a coal miner, everybody's mother stayed home unless she was

a nurse or a teacher. I remember snowy days, having snow rides on 2nd Street, stealing tires from the White people's yard to make us a fire. I remember going to the poolroom to warm up when it snowed. I remember summers, playing in the rain, walking in the ditch, playing baseball in the street. We made our own ball from newspaper and wrapped a tape around it. We made hot rods, we raided apple trees at night, especially Ms. Almon's tree. I remember going to 3rd Street, 4th Street, 5th Street just walking up there, to the graveyard behind our house . . .[2]

Childhood memories full of such bounty and self-sufficiency were not unique to just Cynthia Brown-Harrington, but rather pervasive among all those who grew up in Harlan County at this time. It was a cradle of natural abundance—some called it a utopia, others called it a "little garden of Eden." Jack French reminisces, "the good Lord provided everything; he really did. Because we could go three or four miles into the mountains; we would find all types of apple trees, blackberries, blueberries."[3] Weekday afternoons and on Saturday mornings children would hike up the mountains to fetch fruit for their mothers' pies and jams, while helping in their fathers' gardens was a daily responsibility—planting tomatoes, cabbage, and peppers. "We never went hungry," Lee Arthur Jackson recalls:

We raised food in the garden and we raised hogs and some chicken, and daddy would always buy a side of beef. And mama canned all the vegetables and toma-toes and you know cabbage and greens and make apple jelly. Everybody had, I should say about a lot of people in Lynch, had gardens, and they had their little spot and nobody else messed with anybody's spot or anything like that. And we raised the hogs and when it was hog-killing time everybody in the com-munity came and slaughter the hogs all at the same time, and that's just the way it was.[4]

Part of what made Black life in Harlan County at this time so abundant was the convergence of rural and industrial living. The generation of African Americans who migrated to Kentucky largely came from the mineral district of Alabama, where agriculture was not only a way of life, it was embodied. Intuitive knowl-edge about the science of farming, tilling soil, raising livestock, and living off of the bounty of the land is woven into the tapestry of the African American tradition through generations of enslavement and forced labor in the United States. The men and women who journeyed to southeastern Kentucky brought this knowledge with them and continued these agricultural practices in the coal camps, and transmitted it to their posterity.

Given the time spent interacting with the land, whether in their garden plots, rearing livestock, or trekking through the mountains for fruits and game, the Black families of Harlan County applied their rich agricultural knowledge base towards the local flora and fauna. Brenda Wills-Nolan recalls going for walks with her grandmother: "I mean every now and then when you go pick blackber-ries you would see a green snake, but my grandmother was very familiar with the snakes, different snakes and animals, so she would tell you which one was poison and which one wasn't."[5] Because people knew which snakes were poisonous and which ones were not, which paths through the mountainside would lead to fruit orchards, where the abandoned coal mines were, and which springs had toxic coal mine effluence, the forest and the mountains around the towns were not to be feared. Through this embodied knowledge came freedom—the freedom to move about and engage with the land.

Knowledge of the natural landscape also lent itself to strategies of re-use, further accentuating the bounty of the land and the ability to be self-sufficient:

> Now we had cows and we had chickens and we would take the cow manure and the chicken stuff there and we would put them in the middle of the [garden] rows and that's like fertilizer . . . that's why we had, our garden grew bigger than anybody else because we used that cow mess and that chicken mess, and all we could say to anybody was "my daddy had a green thumb." Yes I know what made that green thumb, because we put all this cow manure and chicken mess on there.[6]

The gardens, livestock, and the mountains provided not only biological, but also cultural vitality. The seasonal slaughtering of hogs brought the community together as each social group had their own role. Each fall, the Black men in the community would gather on a Saturday to slaughter their fattest hogs, which would supply meat for the winter and early spring. In recalling hog-killing season, Clara Smith reminisced:

> [A]nd so in the fall, my father would bring two hogs down. They had a big iron pot. They would make a fire and put this pot over the fire and boil the water and I believe those hogs knew that it was "slaughtering" time because they went "oink, oink, oink" all around the yard there. And so they'd wait until we went to school and my dad and maybe four or five other men in the community, they'd stand around smoking their cigars and whatever, after we'd left and gone to school, they would shoot the hogs right in the middle of the forehead . . . right between the eyes. And they slit them open and they'd hang them up so that the blood would drain down, and put them in a big tub of—a big iron pot— and they would you know . . . pull the hair off. But I would be in school and I wouldn't know anything about it so when I would come home from school my mother and maybe four or five ladies in the neighborhood, they would be there at the kitchen table cutting up all that fresh meat, and it was red and fresh and smelling like blood.[7]

These cyclical traditions—daily, weekly, and annual—were the backbone of a vibrant African American culture in Appalachia at this time.

That said, life in the coal camps was not only characterized by freedom and vitality, but also death, as men toiled in the bowels of the earth to extract coal for the corporation. The occupational hazards that coal miners faced were manifold, from respiratory illness to injury from falling objects and collapsing infrastructure. It was a daily sacrifice that only reified the precarity of the Black male body. One coal miner's daughter said her father worked from "caint to caint [*sic*]"— can't see when they start work and can't see when they end it—as they worked from before sunrise to after sunset five to six days a week. Most men would make sure to shower in the bathhouse before returning home covered in coal dust. This was an effort to keep their own homes clean, but also to mark a clear mental separation between their work and home. Most fathers wanted to prevent their children from knowing how gruesome, how tiring, and how terrifying it was to work in a coal mine every day.

However, it was impossible to maintain this separation between, what was supposed to be, the interior sphere of home and the exterior world of industry. On one occasion, as a little girl, Clara Smith accompanied her mother to meet her father at the coal mine:

> And we were standing there and she was talking to him and I was a little kid. I was looking up at this man and I was wondering who is this man my mother's talking to? . . . he was pitch black from head to toe. All you could see was his eyes you know, and he was talking and she was talking, and I guess he saw me staring at him. And he looked down at me and my mother looked at me and she said, "Clara, do you know who that is?" and I said, "no," she said "that's your daddy." I looked at him and I said to him, "Daddy that's you?!" And then my hand was black, and my mother had to get some Kleenex and scrub my hand because I couldn't believe that was him. That's just how dirty they were . . . when they came out of the mines. You couldn't see anything on him that would make me know that that was father.[6]

Although the mines were segregated, with Black workers working on one side, and White workers on another, at the end of the day, everyone came out of the mines covered in the same black dust. For Jeff Turner, this produced a certain sense of evenness, in that you couldn't differentiate between White and Black men when both were smothered in dust and enduring similar conditions.[8] Reminiscing about conversations he would have with his father, Turner notes: "When you go in that hole every day you go in even and you come out, you are dark. . . . You couldn't tell from the face of a White man and you can't tell the face of a Black man that's when you are even." The common fear and fate that defined a lifetime of coal mining had an equalizing effect on Black and White relations, to some extent.

Coal mining was a death sentence. Coal dust is associated with a host of health threats, including coal workers' pneumoconiosis (CWP) and chronic obstructive pulmonary disease (COPD), as well as silicosis from crystalline silica dust. Commonly referred to as "black lung," CWP was a common ailment for miners in eastern Kentucky. CWP is caused by the continual build-up of dust within the lungs, from constant inhalation, and results in inflammation, fibrosis, and in many cases death. Children of Black coal miners knew all too well the inevitable fate of their fathers. Black coal miners, especially, were forced into precarious conditions—working longer hours and taking dangerous jobs such as setting detonators—which made their bodies more susceptible to injury and ailment than their White counterparts. Upon learning their diagnosis, coal miners were shipped from the local hospital to Knoxville General where death was inevitable.

> [T]hey took us there [Knoxville General], kids really couldn't visit, but they took us there and I remember it being an old hospital, it wasn't you know like Oakridge was clean and bright and cheery, it was dingy looking. And they took us in the room where my dad was, I didn't know he was dying, and they had this pump pumping this black stuff out of his body and they had an IV in him, and he was laying there you know looking really really bad.[6]

Once the father died, families were given 90 days to move out of company housing so that the company could find a replacement.

The eventual degradation of the Black male body from a lifetime of working in the coal mine mirrored that of the industry itself. After decades of exploiting the land, the coal industry began to sell off property and lay off its workforce in the early 1950s, beginning with the Black workers. Because of this, young Black men and women began to leave Harlan County in droves, seeking work in the growing metropolises of Chicago, Detroit, Cleveland, and New York. Upon emigrating from the rural coal towns of southeastern Kentucky to new urban landscapes, Black teenagers were met with new forms of racism and oppression. Yvonne McCaskill recalled being called the slur, "country," during her first year of college and equating that with the more recent derogatory nature of the term "ghetto":

> I think they [today's youth] replaced ghetto with country and stuff like that. I think that's the new term now and we said "country." But you know when we first came here because the landscape is drastically different from where we came from, it was culture shock in the sense that we lived around Black people, with Black people, but it was a different kind of living. You know our doors were opened all the time; we didn't have to lock our door—as I would say and everybody knew everybody. You know you didn't fear the White people, you didn't fear Black people.[9]

The condescending and offensive connotation of being called "country" contradicted her upbringing, which characterized the country, and rural areas like those in Appalachia, as in fact being rich in community, equality, and vitality. The pejorative "country," rather, echoed the vestiges of slavery within contemporary vernacular at that time, invoking images of ignorance, poverty, and backwardness which served as a way to promote both Whites over Blacks and certain Blacks over others. Vyreda Davis-Williams remembers, "We didn't know we were poor," because they always had "food to eat and clothes to wear."[10] However, the opinions of her peers were shaped by popular media that cast Appalachia as a place where old men have moss for teeth, no one is literate, and everyone is White. Thus, even in migration out of the semi-rural environment in the coal camps in southeastern Kentucky, the Appalachian landscape continued to define them both in their response to their new urban contexts as well as the way in which others, Black and White, characterized them.

As the young, working-age Blacks moved out, the aging fathers succumbed to black lung as their weary wives did their best to keep them comfortable through their inevitable transition to death. As the industry continued to dry up, towns like Lynch, Kentucky, began to fall apart. When the progeny of Harlan County's coal camp Blacks would come back to their family homes, they found abandoned lots, overgrown gardens, and most recently, bears roaming through the streets. Coming home meant walking among the ruins of what once was a vibrant community, or what Clara Smith calls a "ghost town":

> And see like when I was down there we lived on 2nd Street. . . . Well they [the Black neighborhood] had six streets. Now all those streets are gone and the forest . . . has come down . . . to the back of 2nd Street, yes. It's hard to look at that and imagine there were six streets going up there. They were all full of people; there were houses and people in those places. . . . And it breaks your heart really to see how Lynch has gone down and it's almost like a ghost town as far as I'm concerned . . . But nobody's buying anything down there [in Lynch], nobody's moving in down there and coming back home to live. I know I wouldn't ever even consider going back down there [to live] because there's nothing there. There's no entertainment, there's no work, there's just nothing there.[7]

For those that still call Appalachia home, this ruination is more than just physical or economic—it is also deeply symbolic. Jeff Turner explains, "We're seeing the mountains die, we're seeing the streets being swallowed up with the trees, we're seeing death which is imminent to all of us."[7] The very visible decay is a reminder that the oasis of their youth—a place of self-sufficiency and cultural vitality—is gone. It mirrors the disintegration of their fathers' bodies and the shriveling of the coal mining industry. The trouble for communities like Lynch and Benham, that emerge for the sole purpose of one industry, is the residue left after the boom and bust—the people and places that get caught in the gravity of the explosion. Once the boom is over, there is a sense of emptiness where there was once abundance—an implication that those people and places were simply kindling for the corporate fire. However, in the case of Appalachia's coal camp Blacks, we see resistance to that hopelessness.

Despite the physical degradation of the town, families continue to come "home" every year for Memorial Day weekend. The companies gave many families the opportunity to buy their homes, often for as little as five hundred or one thousand dollars, unbeknownst to the new homeowners that the coal companies planned to exit the region. Instead of selling their homesteads once they realized they were deceived, many families kept these properties and they still remain in the family today. Several families still pay for electricity and running water, if only for that one weekend a year when their family and so many other families return. The Browns, Davis's, Garners, Clarks, and so many others come home to spruce up their houses—sweeping the front porch, putting on a fresh coat of paint, and fixing a broken shutter. The air smells of barbecue and firecrackers, and sounds of chatter, honking cars, and church hymnals circulate throughout.

When asked why they continue to return, many attribute it to "love," "family," and a sense of "home." There is something special about growing up in the mountains, they say. Cynthia Brown-Harrington argues that early summer in Appalachia is one-of-a-kind. "Kentucky at that time of the year is beautiful. And when you're young and you're growing up there you don't really see all that but once you left and came back you saw the beauty of it. I was like, oh my god, look at these mountains, look at these trees."[2] The mountain is not just a physical place to return to, but also a spiritual place that defines who they are:

> When I go home it gives me an opportunity to fill my tank up. Sometimes in the city your tank gets so empty and you have to refuel. So when I go back home to rejuvenate the mountain spirit, it's all through the concept of home . . . home is where mama and daddy raised us in that little house. So I sort of call it our condo in the mountains.[7]

It is a mutual relationship—between the people and the land. As the Black families of Harlan County continue to make their annual pilgrimage to the mountains to "fill up their tanks," they too revive the mountains, giving new meaning to Appalachia. For them, this region has become one of recreation and nostalgia, instead of a place of self-sufficiency and economy as it was in their youth.

Because of this deep affection for the mountains, many desire to return to Harlan County not only as an annual social destination, but also as a burial place. Odell Moss, for example, says that he wants to be buried in Harlan County because that is where the rest of his family has been laid to rest:

> Why, I was born and raised there, all my family was born and raised there. And before my people passed away there was home there. And when people begun to passed away that's where they are buried right at Benham, now I have an older brother who was not buried there but the rest of them they were buried right there in Benham. And my baby sister, she was buried there because she was cremated but we buried, at least I did buried her right down between mom and daddy, you see. And you know, because my oldest brother was the only one who wasn't buried there I've already prepared things for myself to be buried there, everything is this over set and lined up and everything.[11]

The fact that most people want to return to Harlan County in death shows us just how important it was during their lives. Throughout their life histories, children of Appalachia's coal camp Blacks continue to come back to the place of their youth, creating new memories through re-engagement. In this way, we see that memory is embodied in the physical scars on their backs from crawling through the mines, in the cut on their foot from chasing their brother through the apple orchard, and in the burn on their finger from the firecrackers set off every Memorial Day weekend. At the same time, the shared experiences of racism, both in and beyond Harlan County, solidify the subjective experience of being a Black Appalachian.

Race and Landscapes of Ruin: Towards an Interpretive Approach

> Here we take the charge to be a vital one: to refocus on the connective tissue that continues to bind human potentials to degraded environments, and degraded personhoods to the material refuse of imperial projects.
> —Ann Stoler, *Imperial Debris* (2013, p. 7)

What does the case of Appalachia's coal camp Blacks tell us about race and the environment? Most importantly, our analysis points to the ways in which a slight change in perspective can illuminate previously unexplored dimensions of the intersection of race and environment. This is partially a matter of method and methodology. Most studies of race and environment proceed by asking questions about the unequal distribution of environmental burdens along racial lines, and therefore employ spatial and statistical methods to explore this phenomenon. Yet, when scholars focus solely upon these negative dimensions of how race and environment intersect, they are often blind to questions of meaning and identity formation within these same contexts. This paper shows how despite living in a socionatural context fraught with problems, from social segregation to death from occupational hazards, people created meaning and value through engagement with myriad forms of non-human nature within the coal mining landscape. They found freedom in the same mountains that would later cause many community members an early death. In spite of the collapse of the coal economy and the subsequent ruination of the built landscape, Harlan and Letcher County's coal camp Blacks continue to return in celebration of their persistent connection to the land. . . .

Notes

Brown, Karida L., Michael W. Murphy, and Appollonya M. Porcelli. 2016. "Ruin's Progeny: Race, Environment, and Appalachia's Coal Camp Blacks." *Du Bois Review* 13 (2): 327–44.
Authors are listed in alphabetical order and made equal contributions.

[1] EKAAMP Interview #153, Brenda Thornton. August 2014, SHC.
[2] EKAAMP Interview #7, Cynthia Brown-Harrington. June 2013, SHC.
[3] EKAAMP Interview #74, Jack French. June 2014, SHC.
[4] EKAAMP Interview #34, Lee Arthur Jackson. August 2013, SHC.
[5] EKAAMP Interview #105, Brenda Wills-Nolan. July 2014, SHC.
[6] EKAAMP Interview #117, Ernest Pettygrue. July 2014, SHC.
[7] EKAAMP Interview #54, Clara Smith. August 2013, SHC.
[8] EKAAMP Interview #126, Jeff Turner. August 2014, SHC.
[9] EKAAMP Interview #3, Yvonne McCaskill. June 2013, SHC.
[10] EKAAMP Interview #70, Vyreda Davis-Williams. June 2014, SHC.
[11] EKAAMP Interview #10, Odell Moss. June 2013, SHC.

References

Agyeman, Julian, and Rachel Spooner. 1997. "Ethnicity and the Rural Environment." In, *Contested Countryside Cultures: Rurality and Socio-cultural Marginalization*, edited by Paul Cloke and Jo Little, 197–217. New York: Routledge.

Berry, Brian. 1977. *The Social Burdens of Environmental Pollution: A Comparative Metropolitan Data Source*. Cambridge, MA: Ballinger Publishing.

Bullard, Robert. 1990. *Dumping in Dixie: Race, Class, and Environmental Quality*. Boulder, CO: Westview Press.

Bullard, Robert. 1993. *Confronting Environmental Racism: Voices from the Grassroots*. Boston, MA: South End Press.

Crowder, Kyle, and Liam Downey. 2010. "Inter-Neighborhood Migration, Race, and Environmental Hazards: Modeling Micro-Level Processes of Environmental Inequality." *American Journal of Sociology* 115 (4): 1110–49.

Deluca, Kevin, and Anne Demo. 2001. "Imagining Nature and Erasing Class and Race: Carleton Watkins, John Muir, and the Construction of Wilderness." *Environmental History* 6 (4): 541–60.

Eastern Kentucky African American Migration Project. Southern Historical Collection (SHC). University of North Carolina at Chapel Hill.

Finney, Carolyn. 2014. *Black Faces, White Spaces: Reimagining the Relationship of African Americans to the Great Outdoors*. Chapel Hill: The University of North Carolina Press.

Hofrichter, Richard. 1993. *Toxic Struggles: The Theory and Practice of Environmental Justice*. Philadelphia, PA: New Society Publishers.

Hurston, Zora Neale. 1950. "What White Publishers Won't Print." *Negro Digest*, April, 85–89.

Kruvant, William. 1975. "People, Energy, and Pollution." In *The American Energy Consumer*, edited by Dorothy Newman and Dawn Day, 125–67. Cambridge, MA: Ballinger Publishing.

Lewis, Ronald. 1987. *Black Coal Miners in America: Race, Class, and Community Conflict, 1780–1980*. Lexington: University Press of Kentucky.

Lewis, Ronald. 1989. "From Peasant to Proletarian: The Migration of Southern Blacks to the Central Appalachian Coalfields." *The Journal of Southern History* 55 (1): 77–102.

Mah, Alice. 2012. *Industrial Ruination, Community, and Place: Landscapes and Legacies of Urban Decline*. Toronto, ON: University of Toronto Press.

Mohai, Paul, David Pellow, and J. Timmons Roberts. 2009. "Environmental Justice." *Annual Review of Environment and Resources* 34 (1): 405–30.

Mohai, Paul, and Bunyan Bryant. 1992. "Environmental Racism: Reviewing the Evidence." In *Race and the Incidence of Environmental Hazards: A Time for Discourse*, edited by Paul Mohai and Bunyan Bryant, 163–76. Boulder, CO: Westview Press.

Pastor, Manuel, Jim Sadd, and John Hipp. 2001. "Which Came First? Toxic Facilities, Minority Move-in, and Environmental Justice." *Journal of Urban Affairs* 23 (1): 1–21.

Pellow, David. 2000. "Environmental Inequality Formation." *American Behavioral Scientist* 43 (4): 581–601.

Reed, Isaac. 2011. *Interpretation and Social Knowledge: On the Use of Theory in the Human Sciences*. Chicago: The University of Chicago Press.

Saha, Robin, and Paul Mohai. 2005. "Historical Context and Hazardous Waste Facility Siting: Understanding Temporal Patterns in Michigan." *Social Problems* 52 (4): 618–48.

Stoler, Ann. 2013. *Imperial Debris: On Ruins and Ruination*. Raleigh, NC: Duke University Press.

Szasz, Andrew, and Michael Meuser. 1997. "Environmental Inequalities: Literature Review and Proposals for New Directions in Research and Theory." *Current Sociology* 45 (3): 99–120.

Trotter, Joe. 1990. *Coal, Class, and Color: Blacks in Southern West Virginia, 1915–1932.* Champaign, IL: University of Illinois Press.

Tuan, Yi-Fu. 1974. *Topophilia: A Study of Environmental Perception, Attitudes, and Values.* New York: Columbia University Press.

Turner, William H., and Edward J. Cabbell (eds.). 1985. *Blacks in Appalachia.* Lexington: University Press of Kentucky.

Weinberg, Adam 1998. "The Environmental Justice Debate: A Commentary on Methodological Issues and Practical Concerns." *Sociological Forum* 13 (1): 25–32.

Zupan, Jeffery. 1973. *The Distribution of Air Quality in the New York Region.* Baltimore, MD: Johns Hopkins University Press.

Environmental Apartheid
Eco-Health and Rural
Marginalization in South Africa

Valerie Stull, Michael M. Bell,
and Mpumelelo Ncwadi

Since the 1980s, grassroots activists have responded to environmental inequalities by launching an effort to change community, governmental, and corporate practices that endanger the health of people of color and low-income communities. These activists have experienced success in calling attention to environmental inequalities and framing this problem in terms of environmental justice. The literature on environmental inequality argues that the ecological "goods" and "bads" are unfairly distributed. This reading goes a step further to suggest that degraded environments can be used to marginalize racially defined groups and, at the same time, can be a result of that marginalization; the authors call this process "environmental apartheid." This reading deconstructs "environmental apartheid" processes using a case study of KuManzimdaka, a communal land in South Africa's Eastern Cape.

Introduction

Resilient. That's the first word that comes to mind when meeting Mildred Ncapayi, known in her community as MamBhele. She will shake your hand with a firm squeeze and a calloused palm. While her colorful red skirt and bright eyes might suggest an easier life, tilling the earth and working with her hands is the glue that holds together her meager possessions and extended but tight-knit family. Near MamBhele's *kraal*—the corral for her small herd of sheep and cattle chickens scurry to and fro as someone sprinkles maize. Behind the house, potatoes and cabbages emerge like jewels from the crusty red soil. Here, Mam-Bhele plants vegetables, along with other women from the small cooperative she organized.

MamBhele lives in a section of KuManzimdaka, a village of about 300 people in South Africa's Eastern Cape Province. Positioned atop high bluffs at the base of the Drakensberg Mountains, KuManzimdaka rests among vast rolling pastures. Warm summer breezes brush miles of lush grassland, dotted with

143

traditional amaQwathi huts and cattle grazing under the seemingly endless South African sky. Women collect water and firewood. Young boys herd cattle along ridges, and you can just make out the sound of a hoe striking the ground in someone's home garden. The pandemonium of modern South Africa seems far removed from this intoxicating calm.

Such a quick snapshot, however, obscures stark realities. Hundreds of *dongas*—the local word for erosion gullies—wound KuManzimdaka's green pasturelands, chiseling red gashes deep into the soil. The sward on the pastures is short, patchy, and increasingly overtaken by water-sucking invasive species, especially black wattle. A hard pan crusts the land's surface, baked beneath the overgrazed and often dry grass. Rainfall runs off the thirsty ground, and streams are brown with soil that used to be on the hills. Local springs and wells yield little or not at all. And the people are poor, terribly poor. Unemployment is rampant, as is malnutrition. Some 12 percent of the province suffers from HIV/AIDS, including nearly 20 percent of adults (Shisana et al., 2014). Household vegetable gardens sit neglected and abandoned. Crop ground lies fallow or sprouts only patchy rows of stunted maize.

It's not a pretty picture. And it didn't come about accidentally. South Africa's legacy of apartheid is well known, if still not well understood. In this paper, we contribute to a better understanding by examining the intertwined ecological, social, and health implications of what we term environmental apartheid and by showing the instrumental role of rural space in the implementation of South Africa's inequalities. These inequalities are not only social, political, and economic; they are also environmental. Social injustice and environmental injustice feed on each other in a continuing cycle of immiseration of people and land.

By environmental apartheid we *mean the deliberate use of the environment to marginalize racially defined groups,* as well as the subsequent consequences of that marginalization. We present South Africa as the paradigmatic example of environmental apartheid, investigating the use of the environment in the apartheid government's efforts to marginalize the majority of the population. Our focus will be on the use of rural space as an environmental means for marginalizing groups, what we term rural marginalization. As we will show, the apartheid government of South Africa wielded rural space as a means to deny most South Africans their political rights, relegate them to the least healthy and least productive ecological contexts, and leave them economically dependent upon distant White-owned capital. This forced many South Africans into slave-like employment in faraway mines and factories, as well as in services for whites.

Environmental apartheid is a manifestation of the more general phenomenon of environmental racism, which we define as Bullard (2001) did: "any policy, practice, or directive that differentially affects or disadvantages (whether intended or unintended) individuals, groups, or communities based on race or color." Environmental racism, then, is a critical term that highlights environmental framings which disproportionally negatively affect people of color (Dickinson, 2012) and advantage whites (Bullard, 2001).

The conventional use of the term environmental racism points out the environmental abuse of a racially-defined marginalized group. Environmental apartheid is the reverse logic of power. It commits environmental abuse in order to marginalize a racially defined group. As stated above, environmental apartheid is not accidental. Moreover, in practice the two logics—environmental abuse of the racially marginalized and environmental abuse in order to racially marginalize—often work in consort to varying degrees. In this sense, environmental apartheid is both cause and consequence.

Many aspects of the environment might potentially be made use of in environmental apartheid. In the case of South Africa, marginalizing forces mobilized a variety of facets of the environment to implement apartheid, keeping Black South Africans apart from the resources of livelihood, well-being, and political power. Our focus here, though, is on the use of the rural to marginalize racially defined peoples. We trace the nesting of three levels of rural marginalization, what we term first order, second order, and third order rural marginalization. By first order rural marginalization, we mean the forcible location of Black South Africans in rural spaces distant from the economic and cultural advantages controlled by Whites. By second order rural marginalization, we mean how Black South Africans were generally relegated to the worst lands within these distant rural spaces. Lastly, by third order rural marginalization we mean the continued isolation and neglect of Black South Africans within first and second order rural marginalization. These three orders of rural marginalization have had major eco-health implications, continuing consequences that cannot be separated from an understanding of the social, political, and economic repercussions of apartheid policies.

This paper explores environmental apartheid through a case study of KuManzimdaka where we have been working since 2011 on a participatory approach to agroecological development.

The Roots of Apartheid and the Three Orders of Rural Marginalization

Although sometimes seen as a mid-twentieth century offense, apartheid has old roots. Ever since colonization, South Africa has faced severe racial tension. In 1652, the Dutch East India Company established the Cape Colony at the site of what is now Cape Town as a kind of refueling station for its ships in need of food and water to make it all the way to the East Indies and back. . . . By the mid-eighteenth century, the Dutch had taken control of nearly all the agriculturally productive lands of the Khoikhoi and San. Further north, the Bantu-speaking peoples resided, including the two largest groups: the more warlike amaZulu and the amaXhosa. In the late eighteenth century, the Dutch advance into those lands began as well (Thompson, 2014).

Then in 1795, the surging British Empire swept into the Cape Colony and forced the Dutch to capitulate. In the decades to follow, the British also swept north, especially along the east coast, forcing back the Bantu speaking people onto ever higher and ever less desirable ground, through a series of bloody wars. Dutch "voortrekkers," however, disgruntled with British rule, preceded the British in vanquishing the local people in the dryer lands in the interior. The British were initially content to let the Dutch—who were coming to be known as the Afrikaners—have the interior. But between 1866 and 1886, a series of discoveries made plain the region's wealth of diamonds and gold, attracting British interest. The ensuing British-Afrikaner conflict culminated in Britain's victory in the South African War of 1899–1902, and the founding of the modern South African state—albeit as a unit of the British Empire, subject to legal override by the British parliament (Thompson, 2014).

After the war, white farmers returned to the business of developing and expanding their properties. They received a mighty boost with the passage of the Natives Land Act of 1913 which banned "any person, male or female, who is a member of an aboriginal race or tribe of Africa" from owning or renting land in 93 percent of South Africa (Thompson, 2014; Wotshela, 2004). The Natives Land Act made explicit the latent first order rural marginalization of South Africa's earlier colonial period, for the remaining seven percent was hardly the best land. Rather, it primarily included the last refuges of South Africa's various indigenous peoples and the final frontlines of the many wars with the British and the Dutch over the course of the 18th and 19th centuries. It was generally land too high, dry, rocky, and rural to be worth the time and blood of Whites to pursue further. Here, indigenous peoples continued to live in largely traditional manners, communally managing what land they had left through customary land rights awarded based on tribal association.

The Natives Land Act established this seven percent of land as "reserves," under control of tribal authorities—a fact that helped the South African government win the allegiance of tribal authorities to the scheme. But as well, the allocation of these poor lands as reserves assured that Black South Africans would struggle to maintain secure livelihoods there, virtually guaranteeing that many would be willing to accept low wage work in White-owned farms, mines, factories, and homes. The first order rural marginalization of the Natives Land Act did not merely represent the outcome of competition for land but the creation of a population with little social, economic, and political power in the wider society.

The Natives Trust and Land Act of 1936 expanded the area of land available to Black South Africans to 13.5 percent of country, purchasing unprofitable lands back from White South African farmers (Beck, 2000). Even allocating 13.5 percent of the land to 60 percent of the population, though, could hardly be considered generous. Whites hand-picked the most productive land for themselves, furthering the second order rural marginalization that was already evident in the Natives Land Act of 1913. While Blacks were forced to use marginal land and to live in economically less productive areas (Durning, 1990), Whites formerly living on these disadvantaged lands were bought out.

Nor was there much concerted effort to alleviate the disadvantages faced by Blacks segregated by first and second order rural marginalization. Villages and towns in these low-productivity areas lacked basic services like running water, sanitation, electricity, and decent roads. Hospitals were few and poor. Schools as well. This third order rural marginalization of continuing isolation and neglect was not inevitable. South Africa's mineral riches meant it long held the status of being the wealthiest country and largest economy in Africa, although it has now been surpassed by Nigeria in both categories. Of course there are limits to what even a relatively wealthier country can afford. But it is a political choice whether to allocate what wealth a nation has towards the top or the bottom. South Africa pursued the former course of action. Consequently, its relative wealth was also accompanied by being the world's most economically unequal country, a status it retains to this day (World Bank, 2016).

Apartheid's Deepening Shadow

. . . Some of the pillars of legalized racism included the Mines and Work Act of 1911 (part of broader job reservation practices), the Prohibition of Mixed Marriages Act of 1949, the Population Registration Act of 1950, the Group Areas Act of 1950, and the Bantu Education Act of 1953. In parallel, a series of moves stripped away voting rights until only White South Africans had suffrage.

It was the Bantu Authorities Act of 1951, however, that put the capstone on rural marginalization. This is the act that established political entities based on the rural areas where Black South Africans had been confined by first and second order rural marginalization, and gave further legal license to their continued third order rural marginalization. Based on the terminology used by the British in 1947 for partitioning Pakistan from India, White leaders called these areas "Bantustans." Later, trying to make the claim that these were the indigenous locations for the different Black tribes, the White authorities took to calling them tribal "homelands" in complete neglect of the two centuries of war that had disenfranchised Blacks from the regions now owned by Whites.

Ultimately, apartheid policies were legal sanctions by which the government could control labor, movement, freedom, and economic status according to race. The Bantu Authorities Act established ethnic governments in the homelands. Outside of the homelands, Black South Africans were required to carry passbooks designating their race and where they could travel and live. Millions were arrested for pass law violations. Alongside homeland creation came forced removal, whereby at least 3.5 million people were pushed from their land and resettled in the homelands between 1960 and 1983—land that was largely barren and uninhabitable (Platzky and Walker, 1984).

Beginning in the late 1970s, South Africa began to spin off the homelands as independent states. Four homelands—Ciskei, Transkei, Bophuthatswana, and Venda, the so-called "Bush Republics"—eventually gained this status before the end of official apartheid. As a result, Black South Africans were no longer considered citizens of South Africa and had no citizenship rights inside it. In reality, however, these were quite obviously puppet states controlled by South Africa, and no other nation ever recognized them. Meanwhile, the homelands received little economic development and remained mired in poverty and lack of opportunity.

Direct evidence that this was all a deliberate plan of marginalization—such as memos and tape recordings of leaders discussing plans—is hard to come by. The politics of marginalization are rarely in full public view, although the proceedings of South Africa's unique Truth and Reconciliation Commission following apartheid subsequently gave much disturbing evidence of deliberate intent, especially the policy known as "forced removals" that relocated millions of Blacks to the homelands, as well as the overall shocking brutality of the apartheid era. Few Black South Africans who lived through this period doubt its deliberateness.

The result was quite effective for the South African government, which aimed to: Create a large population of poor, desperate people who would therefore be essentially forced to migrate to the White areas of South Africa to accept low-wage, menial work with virtually no rights. First, second, and third order rural marginalization kept labor cheap and without rights, while also ensuring that blacks were subordinate and disorganized (MacDonald, 2006). In other words, the point of apartheid was not to keep Blacks and Whites separate. Whites and Blacks interacted closely on a daily basis during apartheid. Rather, the point was to keep Blacks powerless and willing to work for Whites for very little. Separateness was the means; marginalization was the ends.

Recognizing that it was impractical to expect people to show up at work in the morning if they had to sleep hundreds of kilometers away in their designated homelands, Black South Africans were given leave to live on the rural outskirts of cities in vast slums called "townships," or to live in huts on White farms and barracks at mines and factories, or in some cases to live in small shacks in the back gardens of White houses if they held domestic jobs there. But permission to stay outside the homelands had to be granted by, and registered with, the White authorities. It could be terminated at any time, should a Black worker show signs of being insufficiently compliant to White control. The maintenance of first and second order spaces of rural marginalization gave White authorities a continued legal and economic whip to keep Blacks living outside of the homelands in line. Moreover, employment opportunities in the homelands deteriorated throughout apartheid thanks to on-going third order marginalization, as the regime aimed to guarantee a secure stream of cheap labor into White-controlled cities, keeping the homelands economically dependent on the Republic of South Africa (Percival and Homer-Dixon, 1998).

With the eventual success of the anti-apartheid movement and the enfranchisement of all South Africans, the homelands ceased to exist as of April, 1994, when they were reincorporated into a new democratic Republic of South Africa that had nine provinces. Sadly, the end of apartheid has not brought relief to much of South Africa. National income inequalities have actually worsened since 1994 (Seekings and Nattrass, 2005).

Apartheid and the Environment

There is little dispute regarding the deleterious impact of apartheid on the South African environment (Beinart, 2003; Beinart and Hughes, 2007; Department of Environment Affairs and Tourism 1999; Durning, 1990; McDonald, 2002; Percival and Homer-Dixon, 1998; Steyn, 2005). The Worldwatch Institute's influential 1990 text, "Apartheid's Environmental Toll," provided stark evidence of the overlap between apartheid policies and environmental degradation. South

Africa has a well-established environmental movement, dating back to the 18th century. But the focus of concern was long upon wildlife conservation, and upon correcting other aspects of what Whites regarded as the "environmental profligacy of African farmers" (Beinart, 2003: 355), which resulted in soil erosion, tree removal, and over-grazing. Before the Worldwatch report, South African conservationists rarely contextualized these problems within the history of Black rural marginalization onto the worst land with high population densities—lands that are easily degraded, especially when farmers are poor and desperate. Our village of interest, KuManzimdaka, is no exception; apartheid policies twenty years gone still impact the quality of land available to local farmers, their agricultural productivity, their environmental impact, their access to markets, as well as their access to healthcare, as we will describe.

The very word apartheid—Afrikaans for "separateness" or "apartness"—itself points to the environmental character of the policy through its political use of space. Of course, apartheid was also founded on an ideology of race and racial hierarchy, as well as motivated by efforts to build and enforce economic and political inequality. Numerous other social settings have also seen ideologies of racial hierarchy with economic and political motivations. This paper emphasizes a distinctive feature of apartheid in South Africa, which is not characteristic of all such ideologies: how it was enacted through the environment, emphasizing the powers of the rural to marginalize racially defined populations.

Environmental apartheid always has a spatial dimension. Powerful populations have often relocated or confined their enemies to the places and spaces they consider to be least valuable or most precarious. Humans draw arbitrary boundaries around swaths of land to form district, state, or national boundaries that often reinforce racial divisions. They cultivate identities and freedoms by way of citizenship acquired from a geographic separation. Although it is outside the scope of this paper to provide a global review, the widespread prevalence of discrimination reinforced using spatial divisions is undeniable. In the case of South Africa, the apartheid government commonly isolated Black South Africans through the strategic use of rural space, forcing them into confined, desolate, and underdeveloped regions of the country as a means to preserve a cheap labor force and to retain racial control.

Blaming the Environmental Apartheid Victim

Following Bell et al. (2010), we note two general forms of the use of the rural in environmental apartheid: the material practices of "rural power" and the symbolic practices of the "power of the rural." Physical removal of Black South Africans from spaces of economic opportunity ensured their migration to work for low wages. Forcing Blacks into economically undeveloped regions was dependent on the material power of rural space to isolate people. Additionally, the spatial association of Black South Africans with rural homelands, separate from White spaces, symbolically propelled a sense of social difference, essential for creating the categories of hierarchy. It also served to put poor Black people in a "blind spot" for Whites with more power and privilege. The notion of separate citizenship in rural spaces served to reinforce this difference and the political disenfranchisement it created.

The Village Road

. . . Early during our work in KuManzimdaka, the extent of true isolation felt here became evident. Despite the many improvements in the South African economy, infrastructure has not reached this place.

Elliot itself is a small town, but the ways in which it resembles larger South African cities are striking. Occupied by many Afrikaner and English families even today, the town is surrounded on its formerly rural outskirts by informal settlements and townships inhabited by poorer Black South Africans. Moving just one block in almost any direction outside of the town's main grid and you will find overcrowded dwellings, many of them mere shacks, with narrow streets and uncollected garbage. Like the countrywide first order marginalization, the Black population, even in the rural Eastern Cape, was pushed out of urban centers into surrounding rural spaces where infrastructure was poor. It is as though there is an invisible barrier between the town and its rural surroundings where those seen as inferior were placed. Once there, and once given little opportunity to improve their condition, the mental imagery of Blacks as different from Whites and happy to live in squalor became a widespread symbolic, power-of-the-rural cultural trope. "I don't know how they can stand to live like that," one of us heard a White South African woman remark at an elite bed and breakfast. The answer to this question—because they are forced to—seems hard even now for some White South Africans to comprehend.

The road to KuManzimdaka from Elliot shows a tortured history of rural marginalization that still seems stuck to the land. After passing the peri-urban townships, lush green fields appear again where the ground is relatively flat and the soil rich. Commercial farmers—most of whom are White, with a small number of emerging Black farmers who obtained land through South Africa's post-apartheid redistribution policies—and large agribusinesses grow impressive stands of maize. With irrigation, fertilizer, pesticides, high-yield seed, and plenty of machinery, these lush plots on the gentle plateau just below the Drakensberg mountains resemble Nebraska in July. But where the road veers right to the even more rural KuManzimdaka, the topography steepens as the heads of small valleys begin to cut into that plateau. The crest is still high—up to 1600 m, or about 5200 feet—but ever more deeply incised. Dark groves of invasive black wattle (*Acacia mearnsii*) begin to obscure the view. Behind the evergreen foliage rests a rock-strewn landscape stippled with trees, sparse crops, and anemic grass. Black wattle is an aggressive Australian tree that invades bare spots in over-grazed grasslands or over-harvested woodlands, and it successfully competes with indigenous species to the point of take-over. It forms dense thickets that reduce grazing areas for live-stock and wild animals, all while hoarding available water. Black wattle is one of the most pervasive and damaging invasive alien trees in South Africa, posing a major threat to local water catchment yields (Dye and Jarmain, 2004). Nearly every household in KuManzimdaka is plagued by the tree.

The road and the view from the road represent environmental apartheid's first order rural marginalization compounded by second and third order rural marginalization. The amaXhosa and amaQwathi who live here were historically pushed to the rural fringes through first order marginalization. Yet even within these margins they were relegated to the worst possible land. Apartheid-era surveyors carefully drew the line of the former Transkei along the heads of the valleys cutting into the plateau that forms an apron of rich crop ground at the base of the Drakensberg—a second order line of rural marginalization that separated ground Whites wanted to own from less valuable land. To north and west of the line, the land could be bought and sold on the open market if you were White, as Black South Africans were then not allowed to own land. To the south and east, the rougher ground was designated as communal land, governed by Black tribal authorities. It remains so today. . . . Having access to land without having to buy or rent it is fundamental to local people's livelihoods and is one advantage that they do have amid their poverty. But managing such marginal land is difficult. The margin for error is tight, and over time it gets harder and harder as bad luck compounds. Excessive grazing in a dry year, for example, could perhaps be followed by a veld fire. The ground heats in the sun, the grass does not reestablish fully, and opportunistic black wattle along with other weeds and invasives leap into the holes in the grass sward. And sometimes before they do, a donga starts to form, carving the land away and sending it into the muddy streams and rivers.

Today, the wattle stands tall and the dongas cut deep in KuManzimdaka, mocking signposts of stifled productivity where verdant grasslands could have flourished. Black wattle yields serious negative impacts on biodiversity, water resources, and the cohesion and stability of riparian ecosystems throughout rural South Africa, while also increasing erosion (Wit et al., 2001). The extensive veld fires that raged outside Cape Town in 2000 have also been blamed at least in part on the presence of invasive species (Neely, 2010), which pull moisture from the ground and thereby encourage wildfires. In August of 2014, a veld fire decimated almost all the pastureland in KuManzimdaka, exactly in the depths of winter when both the rains and grass growth subside. Charred grass is inedible, and for a time the community livestock herds were on the verge of collapse. Without the generosity of neighboring villages who gave KuManzimdaka access to unburned grazing lands, the people might have lost every single animal. No help came from central government. Third order rural marginalization still pervades.

The journey to KuManzimdaka also exemplifies the symbolic practices of the power of the rural. The village's physical distance and difference from any "modern" town makes it feel outdated and foreign to several of the White South Africans we have brought to the village as a part of our participatory development work, resonating with past notions that Black South Africans are somehow inferior and uncivilized. "I didn't know people still lived this way in South Africa," one of them remarked, struggling with her cultural expectations. Traditional round huts abound, many with thatched roofs. Women cook on outdoor fires, except in the winter, and can often be seen carrying bundles of fuel wood on their heads. In the winter, people cook inside on open fire pits so as to also warm their houses, leading to dangerous levels of indoor air pollution. About half the homes do not have electric wires leading to them. One might expect that such realizations would motivate more urgent efforts to improve local livelihoods, but as yet they have not. We suspect that some of the reason for this is likely the cultural feeling among South African elites that local people are comfortable living their "traditional" ways.

But environmental apartheid's third order rural marginalization is not only due to symbolic constructions of rural life in KuManzimdaka. The material practices of rural power make such inattention to local needs easier for government to contemplate. It is indeed a long and difficult road from KuManzimdaka. But that road could be tarred. And because it is not, employment opportunities in KuManzimdaka remain a major problem, pushing locals to migrate to cities and mines in search of work. Employment outside of the area has yet to bring much wealth or prosperity to KuManzimdaka itself, but the practice lingers, as the people have few options. Meanwhile, those left behind—mainly women—are tasked with managing the degraded landscape in addition to all other responsibilities.

Eco-Health in KuManzimdaka

One warm January afternoon, an elderly woman of 84 is walking up a steep hillside to fetch water from a spring. She moves slowly but meticulously up the gradient, taking calculated steps to avoid holes and stones. Her traditional headpiece is large and blue, indicative of her age and marital status. The spring she seeks is meager, where a slow trickle of water emerges from the hillside to pool in a muddy puddle on a flat ledge. Upon reaching the spot, she pulls aside the sheet of metal someone has thrown over the source of the water, perhaps to serve as a bridge for passersby or to protect the pool from rain. . . . But the sheet of metal provides very little real sanitation. A cluster of hoof prints encloses the area, and there are a few in the mud in front of the pool, evidence that cattle and other livestock share this spring and defecate close by. With no running water, no system for keeping out cattle, and no alternative reliable source nearby, this is the only option for this elder. She comes here a couple of times every day to carry this dirty water back home. In recent years, she tells us, the spring has become unreliable. "There were other springs on this hill before," she says, "but now they are dry." It seems the water is disappearing.

Grass on the hill is patchy; the soil is hard and chalky. The veld has been overgrazed for a long time, and even the cattle appear despondent when feeding here. The degraded hillside itself contributes to the lack of water, as overgrazing has led to soil compaction, reduced plant growth, and high levels of erosion. Sufficient rainwater no longer seeps in to recharge the groundwater levels. There are scarce plant roots to hold the soil in place and collect moisture. Instead, water runs off down the valley, carrying with it rocks and sand that amass in the river. From here, one can see the river's brown color far below. The land is rocky, the hillside sharp, due to second order rural marginalization's continued legacy of environmental apartheid. It would be nearly impossible to farm decent field crops here, especially without advanced equipment; livestock grazing is the only realistic option. But there is no fencing or training to help people institute a technique like rotational grazing that could restore the grasslands and ground water levels. Nor is there help with establishing even the most basic of water supply improvements, such as a spring box or a drilled well. Signs of NGOs and international aid agencies like USAID are as scarce as the South African government here, in part because of the feeling that South Africa, as one of the richest countries on the continent, ought to be able to handle simple investments like these. But it doesn't. Environmental apartheid persists as third order rural marginalization compiles with first and second order rural marginalization to leave people with little option but to continue the cycle of livestock over-grazing.

MamBhele and other women in the village voice three primary concerns regarding their community. First, there is limited access to clean water and healthcare. The wells are drying up and the water is contaminated. Plus, water sources are far away and water must be carried by hand. Boys, girls, and women walk kilometers to fetch heavy buckets every day. Once home, there are few options for water purification. Even boiling it is a challenge because of the added firewood or other expensive fuel required—resources that take precious time or money to procure. Not surprisingly, children get sick, frequently. A mobile clinic comes through the area every once in a while, but its focus is vaccinations and young child wellness appointments. It does not provide acute care or antibiotics, and it does not visit the village on a consistent schedule. Isolation has left KuManzimdaka in a "healthcare desert" of sorts, cut off from both government and private healthcare services. While primary healthcare provisions are available for free in South Africa, they aren't truly free for the people here. Just getting to a clinic can be a prohibitive expense.

Second, the women note that there are few jobs or opportunities for them in the village. "We need income generating activities," one woman says. "Without money, we can do nothing," they say. Where are the jobs? Farming barely meets subsistence levels for most; it is hardly lucrative. And transporting crops grown here to Elliot is as difficult as transporting people. Without jobs, people leave. When they return, they don't stay long.

Third, "the soil is dead," they say. "There is not enough to eat." Farmers struggle to grow and sell crops, especially during the dry season. With little water and limited access to fertilizer or training in organic production techniques, agricultural productivity is low in KuManzimdaka. Community members face periods of food shortage throughout the year. The Eastern Cape province is South Africa's second poorest, with almost 60% of its 6.5 million people living in poverty (ECSECC, 2012; Statistics South Africa, 2014), a problem compounded by the fact that nearly 20% of adults in the Eastern Cape are living with HIV and AIDS—and likely far more in KuManzimdaka, were a local survey to be done. (In our experience, nearly every weekend there is at least one funeral in local villages for someone rumored to have died of AIDS.) Malnutrition is still a major problem here too, with rates of stunting in children under-five around 32%, the highest in the country (Zere and McIntyre, 2003). Close to 50% of the population was food insecure in 2008 (Labadarios et al., 2011), making the Eastern

Cape the most food insecure province in the country. And yet there is little done to help. The effects of environmental apartheid are everywhere, plain to be seen, as health declines are visible in tandem with environmental degradation.

Majola

If it wasn't for Majola, this research likely would not have started in KuManzimdaka. But one of us was fortunate enough to meet him at a local agricultural event. Majola occupies an iconic space among village leadership. He is wise, relatively affluent, and unendingly generous. His home overlooks the only school in the village and is nestled up against the hillside where the elder described above climbs twice a day to fetch her water. It is as though he lives on a perch, whereby he sees the entire community. Not only has Majola adopted six village children orphaned by AIDS, but he has also adopted the school, at least metaphorically, investing resources into improving education and access to healthy food for students. Majola knows the community inside and out, and he recognizes both its challenges and opportunities. It was Majola who first helped Mambhele begin a chicken rearing project. Without using our terminology, Majola has recognized and highlighted the need to address environmental apartheid in KuManzimdaka.

Majola, like others in the village, is a pastoralist; cattle are his most esteemed possessions. Historically, cattle have been invaluable to amaQwathi and ama-Xhosa people, serving as a sign of wealth, means of exchange, and symbol of status (Afolayan, 2004). But the modern worth of local cattle in a post-apartheid economy is astonishingly low. Because of environmental apartheid, the current population of KuManzimdaka is almost entirely isolated and thus cut-off from regional markets. They lack the necessary connections with the predominately White-controlled supply chains and do not possess marketing teams or sales experts to support their businesses. Rural marginalization—the base of environmental apartheid—has incentivized many young people to seek work outside the community. Some travel to and from mines for work. Others move to cities in seek of employment. But still others remain in KuManzimdaka, attempting to succeed as farmers—almost all by rearing cattle alone or in addition to other crops. The only hope of successful farming is better management of natural resources.

The situation for Ncedisizwe farmers in KuManzimdaka is grim. Again, this grimness is not the result of accidental or localized activities. It is born from a history of environmental apartheid and its policies of rural abuse that trampled on the people of the former Transkei. As a result, pastures were overgrazed, agrarian land overcultivated, and soils depleted to the extent that traditional methods of production were no longer viable. Meanwhile, agricultural industrialization in South Africa resulted in high yields on White-owned farms. While this productivity protected food security, it came at considerable environmental and social costs (Fakir and Cooper, 1995); many farmers began to rely heavily on synthetic fertilizers, other chemicals, and mechanization. White-farmers have been historically protected by government drought insurance potentially increasing their willingness to take risks with land (Mather, 1996). Environmental apartheid ultimately hindered the productivity and ecological vitality of all rural lands because it forced Blacks to farm on already marginal soils; encouraged White farmers to carelessly industrialize with over-application of pesticides, commercial fertilizers, and heavy machinery; and perpetuated inequality through unequal access to resources. Today, pastures in the former homelands—including KuManzimdaka— are largely overgrazed, facing extreme erosion, poor water sanitation, and lowered river flows (Cooper, 1991; Weiner and Levin, 1991). Consequently, in the late 1990s, South Africa was identified as having the highest extinction rate of plants and animals globally, abnormally high air pollution levels in some areas, and an unsafe waste and hazardous waste disposal program, along with widespread soil erosion and water contamination (Department of Environment Affairs and Tourism 1999). And yet almost no attention is given to the desperate situation of most rural people in South Africa today, especially in the former homelands. Official apartheid may be over, but environmental apartheid is not.

Rural and Urban Consequences

The negative impact of apartheid on the environment and human health has not been limited only to rural areas. Colonial control of South Africa prompted a shift away from self-sufficiency for African smallholders by creating a wage laborer system through the use of taxation and land expropriation, among other strategies (Mitsuo, 1996). Thus, apartheid policies led to a collapse of rural livelihoods, and concomitantly massive migration to the outskirts of cities as policies restricted Black residence inside the cities and controlled employment. People flocked to cities because of the lack of jobs in rural areas, and once they got there things did not improve. Consequently, the use of the rural environment to impose racial segregation actually led to an increase in the population of urban areas.

But wage laborers migrating to cities were still constrained in their work options and residency, as well as their movement to and from the homelands where many still had legal access to farmland and had families. Overcrowding, poor sanitation, and pollution led to environmental degradation and poor health for non-Whites in urban areas. Even those that originated from cities were pushed into townships by racial segregation and enforced curfews. Historically, townships were designed to house non-White laborers who worked in the cities, preserving separate living spaces. Today, the cost of living at the heart of South African cities remains too high for most migrant laborers. Racial spatial divides are still apparent in all major cities. Thus, environmental apartheid's use of the rural to marginalize had and continues to have urban effects.

By forcing Black South Africans back to live in homelands where their citizenship supposedly resided, but where there was little opportunity, urban populations in the homelands grew rapidly as well, especially beginning in the 1970s (Jooma, 1991). These urban areas in the rural and fragmented homelands lacked an economic base, however, so they could not function as typical cities (Mitsuo, 1996). Whether it be on the fringes of White cities or the population centers of the homelands, the environmental impact of apartheid on the urban was also devastating. Thus, environmental apartheid's use of rural marginalization has spatial consequences that continue across South Africa today.

Challenges and a Brighter Future

Although this paper outlined the ways in which KuManzimdaka has long been marginalized, scholars should not overlook some recent efforts to improve the lives of rural peoples made by the South African government, however inadequate they may be for the scale of the problems. Specifically, the ANC has worked to redistribute land to Black South Africans since the end of apartheid, buying land from willing White sellers and making it available to emerging Black commercial farmers. The fruit of that policy can be seen in much of the privately owned lands on the plateau between KuManzimdaka and the mountains. Other initiatives have supported Black farmers through grants and training, although most of that work has focused on the emerging Black commercial farmers, not the smallholders on communal land. There has also been a recent surge in legally recognized cooperatives in South Africa, as cooperative agricultural business models were identified as a potentially powerful mechanism for simultaneously empowering previously marginalized groups and jump-starting the economic engine. Government incentives were introduced to achieve this purpose including grants to promote co-op development. Beginning with the end of official apartheid in 1994, the South African government implemented the Reconstruction and Development Programme (RDP), the National Growth and Development Strategy (NGDS), and the Growth Employment and Redistribution strategy (GEAR) to name a few (Alemu, 2012).

But progress in most, if not all, of these areas has largely fallen short—as is plainly evident in KuManzimdaka and elsewhere. Throughout the Eastern Cape, much of the land that was redistributed is unused or under-produces because of lack of skills and capital held by new owners, enduring deficiencies in support services, and continued exclusion from White-controlled supply chains. Cooperative development efforts have been stymied by a lack of fundamental financial literacy, foundational agricultural knowledge, and working capital. Government agricultural policy, while encouraging cooperative growth and smallholder farmers through grant making, often prioritizes cheap food and has been more effective at supporting large, predominantly White, farm operations. Black smallholders and emerging commercial farmers struggle to compete with large, established, White-owned farms.

Rural powers can push two ways, however, as we note above and as the people of KuManzimdaka are starting to discover. Indeed, the leadership of the African National Congress (ANC) developed out of the rural isolation of the Transkei. Nelson Mandela was born in a Transkei village about a hundred kilometers away. Local people are not passive actors. MamBhele and Majola are both examples of the current and necessary movement away from rural oppression, building on the strengths of their rural history and location. KuManzimdaka, like other rural places in South Africa, has substantial potential. The slow progress in these areas underscores that the impact of environmental apartheid runs deep. But with collaborative and participatory efforts that strive for holistic and agroecological approaches to justice and development, South Africa's rural margins can become spaces of hope and well-being.

Note

Stull, Valerie, Michael M. Bell, and Mpumelelo Ncwadi. 2016. "Environmental Apartheid: Eco-health and Rural Marginalization in South Africa." *Journal of Rural Studies* 47: 369–80.

References

Afọlayan, F. S. 2004. *Culture and Customs of South Africa*. Westport, CT: Greenwood.

Alemu, Z. G. 2012. "Livelihood Strategies in Rural South Africa: Implications for Poverty Reduction." International Association of Agricultural Economists Conference, August 18–24, 2012, Foz do Iguacu, Brazil, No. 125411. https://ideas.repec.org/p/ags/iaae12/125411.html.

Beck, R. B. 2000. *The History of South Africa*. Westport, CT: Greenwood.

Beinart, W. 2003. *The Rise of Conservation in South Africa: Settlers, Livestock, and the Environment 1770–1950*. Oxford: Oxford University Press.

Beinart, W., Hughes, L. 2007. *Environment and Empire*. Oxford: Oxford University Press.

Bell, M., Ashwood, L. 2016. *An Invitation to Environmental Sociology*, 5th ed. Thousand Oaks, CA: Pine Forge Press (Sage).

Bell, M. M., Lloyd, S. E., Vatovec, C. 2010. "Activating the Countryside: Rural Power, the Power of the Rural and the Making of Rural Politics." Rural Sociology 50 (3): 205–24.

Bullard, R. D. 2001. "Environmental Justice in the 21st Century: Race Still Matters." *Phylon* 49 (3/4): 151–71. http://dx.doi.org/10.2307/3132626 (1960–).

Cooper, C. 1991. "From Soil Erosion to Sustainability: Land Use in South Africa." In *Going Green: People, Politics and the Environment in South Africa*, edited by J. Cock and E. Koch. Cape Town: Oxford University Press.

Dickinson, E. 2012. "Addressing Environmental Racism through Storytelling: Toward an Environmental Justice Narrative Framework." *Communication, Culture & Critique* 5 (1): 57–74. http://dx.doi.org/10.1111/j.1753-9137.2012.01119.x.

Durning, A. R. 1990. *Apartheid's Environmental Toll*. Washington, DC: Worldwatch Institute.

Dye, P., and C. Jarmain. 2004. "Water Use by Black Wattle (*Acacia mearnsii*): Implications for the Link between Removal of Invading Trees and Catchment Streamflow Response." *South African Journal of Science* 100 (1/2): 40–44.

ECSECC. 2012. Eastern Cape Development Indicators, 2012. Eastern Cape Socio Economic Consultative Council, p. 39. http://www.ecsecc.org/files/library/documents/EasternCape_withDMs.pdf.

Fakir, S., and D. Cooper. 1995. *Sustainable Development, Agriculture and Energy: NGO Renewable Energy Summit*. New York: Land and Agriculture Policy Centre.

Jooma, A. 1991. *Migrancy after Influx Control*. Johannesburg: South Africa Institute of Race Relations (SAIRR).

Labadarios, D., Z.J.-R. Mchiza, N. P. Steyn, G. Gericke, E.M.W. Maunder, Y. D. Davids, and W. Parker. 2011. "Food Security in South Africa: A Review of National Surveys." *Bulletin of the World Health Organization* 89 (12): 891–99. http://dx.doi.org/10.2471/BLT.11.089243.

MacDonald, M. 2006. *Why Race Matters in South Africa*. Cambridge, MA: Harvard University Press.

Massey, D. S. 1990. "American Apartheid: Segregation and the Making of the Underclass." *American Journal of Sociology* 96 (2): 329–57.

Mather, C. 1996. "Towards Sustainable Agriculture in Post-Apartheid South Africa." *GeoJournal* 39 (1): 41–49. http://dx.doi.org/10.1007/BF00174927.

McDonald, D. (ed.). 2002. *Environmental Justice in South Africa*. Athens: Ohio University Press and University of Cape Town Press.

Mitsuo, O. 1996. "Urbanization and Apartheid in South Africa: Influex Controls and Their Abolition." *The Developing Economies* 34: 402–23.

Neely, A. H. 2010. " 'Blame it on the Weeds': Politics, Poverty, and Ecology in the New South Africa." *Journal of South African Studies* 36 (4): 869–87.

O'Farrell, J. 2005. "Apartheid." *New Statesman* 134 (4768): 14–17.

Percival, V., and T. Homer-Dixon. 1998. "Environmental Scarcity and Violent Conflict: The Case of South Africa." *Journal of Peace Research* 35 (3): 279–98.

Platzky, L., and C. Walker. 1984. *The Surplus People: Forced Removals in South Africa*. Johannesburg: Ravan Press.

Seekings, J., and N. Nattrass. 2005. *Class, Race, and Inequality in South Africa*. New Haven, CT: Yale University Press.

Shisana, O., T. Rehle, L. Simbayi, K. Zuma, S. Jooste, N. Zungu, and D. Onoya. 2014. *South African National HIV Prevalence, Incidence and Behaviour Survey, 2012*. Cape Town: Human Sciences Research Council Press.

Statistics South Africa. 2014. *Poverty Trends in South Africa an Examination of Absolute Poverty between 2006 and 2011* (No. 03–10–06), Pretoria, South Africa.

Steyn, P. 2005. "The Lingering Environmental Impact of Repressive Governance: The Environmental Legacy of the Apartheid Era for the New South Africa." *Globalizations* 2 (3): 391–402. http://dx.doi.org/10.1080/14747730500367983.

Thompson, L. 2014. In *A History of South Africa*, 4th ed., edited by L. Berat. New Haven, CT: Yale University Press.

Weiner, D., and R. Levin. 1991. "Land and Agrarian Transition in South Africa." *Antipode* 23 (1): 92–120. http://dx.doi.org/10.1111/j.1467-8330.1991.tb00404.x.

World Bank. 2016. GINI Index (World Bank Estimate) Data Table. http://data.world bank.org/indicator/SI.POV.GINI (accessed April 3, 2016).

Wotshela, L. 2004. "Territorial Manipulation in Apartheid South Africa: Resettlement, Tribal Politics and the Making of the Northern Ciskei, 1975–1990." *Journal of South African Studies* 30 (2): 317–37.

Zere, E., and D. McIntyre. 2003. "Inequities in Under-Five Child Malnutrition in South Africa." *International Journal for Equity in Health* 2 (7). http://dx.doi.org/10.1186/1475-9276-2-7.

Zreik, R. 2004. "Palestine, Apartheid, and the Rights Discourse." *Journal of Palestine studies* 34 (1): 68–80.

Turning Public Issues into Private Troubles
Lead Contamination, Domestic Labor, and the Exploitation of Women

10

Lois Bryson, Kathleen McPhillips, and Kathryn Robinson

This reading examines government efforts to deal with contamination from lead smelters in some Australian provinces. Lead is a dangerous environmental toxin, which can cause an array of physical problems, from kidney damage to hyperactivity. Lead is especially dangerous to young children and can cause, among other things, mental retardation, brain damage, and behavior problems. However, rather than address the source of the problem (the smelters themselves), government policies have focused on individual behaviors, specifically on housecleaning, to minimize the amount of lead dust in homes. The authors, pointing to the fact that most of the homes near smelters are occupied by working-class families and most of the housecleaning is done by women, argue that the government policies are not only ineffective but are also embedded in, and reproduce, unequal social class and gender relations.

Using a case study of state intervention in the industrial contamination of a residential community, this article offers a contribution to feminist analysis of the role of the state. Residents of three Australian lead smelter towns, with high levels of toxic pollution, have been given a strong message by state health authorities that their children's health could be irreparably damaged unless they adopt a strict regimen of housecleaning and child management to reduce the ingestion of lead particles by their children.

This case study of state action on a site that is classically women's domain provides insight into a "state gender regime" (Connell 1990) through examining the state "doing gender" (West and Zimmerman 1987). . . .

Unraveling the complexity of the state's role in developing and implementing its health strategy for dealing with the effects of lead pollution within smelter towns allows us to tease out some of the complexities of state intervention. How do they identify and address this health issue? Whose interests do the interventions serve? What are the impacts for different groups of women? Because the intervention is focused on mothering, we start by examining some relevant features of motherhood in contemporary Australia and its place within the wider scheme of gender relations.

Motherhood and the State Gender Regime

. . . While schools of feminist thought account in different ways for women's position and gender relations, they do not contest that motherhood involves a responsibility for family work, which falls unequally to women. Graham (1983) pointed to the bifurcated nature of caring as involving "caring about" and "caring for." In terms of parenthood, she suggested that fathers are expected to "care about" their children, and this may involve taking some responsibility for overseeing that care is provided. For mothers, the two aspects are firmly fused: They are expected to both "care about" and "care for."

Empirical studies in Australia and elsewhere of perceptions of motherhood clearly expose a dominating ideology that reflects such views of caring (Dempsey 1997; McMahon 1999; Russell 1983). . . .

Decades of research into time use confirm women's disproportionate share of the work involved in both domestic cleaning and child care relative to their male partners (Bittman and Matheson 1996; Bryson 1997; Fenstermaker Berk 1985). Women in Australia, as elsewhere, are far more likely to undertake cleaning chores and physical care of children. The Australian Bureau of Statistics (1994) found that women's share of laundry was 89 percent (90 percent if in full-time employment), their share of cleaning 82 percent (84 percent if in full-time employment), and of physical care of children 84 percent (76 percent if in full-time employment) (Bittman and Pixley 1997, 113). These tasks are central to the housecleaning regime devised by state authorities that we examine here. . . .

The Smelter Communities: Method and Background

We first became interested in the public health intervention processes in lead-affected towns when one of us (McPhillips) became caught up with the effects of lead contamination in her residential community, which bordered on Boolaroo, a smelter town in NSW [New South Wales]. McPhillips's personal experience of dealing with lead contamination became the subject of spirited debate and theorizing in our (then) shared work context—the sociology department of the local university. This led to the development of the research reported in this article.

Method

. . . In investigating the case study of Boolaroo, we collected qualitative data through direct engagement in the community; through participant observation in community activities; and through interviews with eight residents of Boolaroo, including female members of community groups (both for and against the smelter). We also interviewed personnel in the local health authority who had been involved in testing children's blood levels and in designing and implementing subsequent intervention measures.

Historical materials relating to the genesis of the political conflict over contamination in Boolaroo were available to us. These included reports in local newspapers (which had been systematically filed by the municipal library and collected by activists), reports and scholarly articles produced by the public health authorities, and a television documentary produced for the state-owned public broadcasting authority that critically recorded an intervention in 1991 intended to remove historic contamination from Boolaroo residences (Australian Broadcasting Corporation 1992). The smelter company produced regular community newsletters, which were available to us, as were the public health information brochures.

In interpreting the data relating to Boolaroo, we used a comparative perspective, drawing on the reports of similar interventions in the other Australian lead smelter towns. There are three major sites of lead production in Australia: Port Pirie in South Australia, and Broken Hill and Boolaroo in NSW [New South Wales]. . . .

. . . The critical features of the populations in all three areas affected by the pollution is that they are of low socioeconomic status with manufacturing providing predominantly male employment.

The Household Cleaning Regime

Public health authorities have systematically turned to an approach dealing with lead effects that is focused on the ways in which it is ingested, particularly by children, rather than with stopping pollution. In the vicinity of the Boolaroo smelter, for example, an attractive poster was distributed to residents, with the title "Lead: Lower the levels & protect your child" (PHU n.d.). It mentions soil, food, household dust, old paint, and the lead worker as potential sources, identifying interventions that can be made, with most of them involving an intensification of domestic labor. To avoid exposure of children to lead in household dust, parents are advised to use a wet mop instead of a broom; keep dust from children's play areas, including under beds and in closets; and remember to dust corners, along sides of windows, and behind furniture and doors. Advice to the lead worker includes the following: Keep kit bags out of reach of children; keep children away from work clothes; and clean dust from inside and outside of car, especially if driven to work. To avoid lead in food, the parents are advised to intensify domestic labor by preventing children from sucking dirty hands, fingernails, or objects; wiping surfaces before preparing food; and covering food and utensils to prevent lead dust from settling on them. The problem of contaminated soil is addressed by advice to wash children's hands before eating, especially if playing outside; to provide clean soil or sand for children to play in; and to use a nail brush under nails. The poster does not represent the source of the lead contamination in any way.

The focus on domestic-based interventions deflects attention away from the source of the pollution, which is clearly in the interests of capital. We argue that the burden of activities to ameliorate the effects of pollution falls disproportionately to women and must be counted as an aspect of the state's gender as well as class regime. The communities, which already bear the burden of toxic contamination and its attendant health, social, and economic effects, are further burdened with the responsibility of "putting things right."

The spotlight in the three smelter towns in recent years has been very much on children, even though evidence suggests that lead is a health hazard for all age groups (Alperstein, Taylor, and Vimpani 1994; Centers for Disease Control and Prevention 1991). In earlier years, the focus of official programs was on workers and also in a manner that deflected attention away from the smelting companies' practices. . . .

The 1980s and 1990s: Developing the Domestic Cleaning Regime

In the early 1980s, the South Australian health department responded to U.S. research linking impaired intellectual development with lead by undertaking a survey of the factors implicated in elevated blood lead levels of young children living within the vicinity of the smelter at Port Pirie (Landrigan 1983). Using a control sample of children with low blood lead levels, researchers found that 5 of the 16 evaluated behavioral factors were significantly associated with high lead levels. These factors were a history of placing objects in the mouth, nail biting, dirty hands, dirty clothes, and eating lunch at home rather than at school (Landrigan 1983, 8). The survey also found that higher lead levels in children were associated with the number of persons in the household working in the lead industry (cf. Donovan 1996).

However, the research also showed that none of the factors were as important as living in a contaminated environment and that living near the smelter was three times more important than anything else (Landrigan 1983, 9). Because past emissions still contaminate the atmosphere and the soil, reducing emissions was acknowledged as insufficient for dealing with the problem. More recently, public health officials recognized that the only effective way of dealing with lead contamination at smelter sites is relocating residents (Galvin et al. 1993, 377), although to our knowledge no public body has ever implemented a relocation program.

In 1984, the South Australian health department and the smelter management started an education program about the importance of personal hygiene and household cleaning. The focus was on "pathways" through which lead is ingested by children and the ways this can be minimized within the home. The general manager of smelter operations expressed the view that "given reasonable care and hygiene, then you can live with the levels of contamination from past emissions." Fowler and Grabosky (1989, 153–57) suggested that the company was "defensive and cautious" lest remedial action imply admission of legal responsibility. The company maintained public health to be a government responsibility and invested far less than the government. The South Australian government's response was restrained as well, illustrating the power of major capital by showing an unwillingness to "antagonise one of the state's largest employers" (Fowler and Grabosky 1989, 150). The government was prepared to burden working-class women, rather than business.

In 1986, another case control study examined children's blood lead levels in Port Pirie (Wilson et al. 1986). The study focused mainly on environmental issues and the implications for the smelter. It assessed behavioral differences between "cases" and "controls" and again pointed to the "pathways" for ingestion such as "biting fingernails" or "dirty clothing/hands at school." In 1988, the public health program at Port Pirie also undertook external decontamination in the yards of 1,400 homes at highest risk, and adjacent vacant blocks and footpaths were sealed (Heyworth 1990, 178). Subsequently, a "partial decontamination" of the same area was instituted, in recognition that contamination is continuous. Such a program requires a permanent cycle of treatment of the homes and vacant blocks and a continuing awareness within the community of the need for vigilance in terms of personal and home hygiene (Heyworth 1990, 183).

In 1993, after several months of remediation, including replanting tailings dumps and the implementation of housecleaning regimes, the blood lead levels had elevated in a part of the town deemed to be a nonrisk area and hence not subject to the campaign. Further investigation found the source to be stacks of lead ore left on the wharves in open piles, with dust being carried by the winds into residential areas not previously considered at risk from the smelter.

The South Australian Health Commission published a decade review of the Port Pirie program in 1993 that concluded, "Given the amount of lead contamination to which these households are exposed, changes in dust hygiene would not seem to be a realistic way to ensure lower exposures to lead by the child" (Maynard, Calder, and Phipps 1993, 25). This point, that domestic strategies are unsustainable in the long term (Maynard, Calder, and Phipps 1993, 5), had previously been made by Landrigan in 1983 and by Luke in 1991 after an extensive search of the world literature on the topic (Luke 1991, 161–67). Many people are prepared to modify their behavior in the short term, but sooner or later they revert to more comfortable habits. The report notes that because there is a constant process of recontamination, it is unrealistic to expect continuing voluntary participation of residents in programs for lead remediation. Another barrier to ongoing participation is the stigmatization that is involved if a child's blood levels are high. Parents are reluctant to have their child labeled as potentially intellectually impaired, and they themselves feel stigmatized by the implication that they have dirty houses.

Over time, parents tend to withdraw from participation in blood-monitoring programs. This can be seen as a form of resistance in situations where parents are expected to comply with surveillance of a problem that has a source external to the home and family, but they are expected to deal with the consequences. Their concerted efforts at housecleaning fail to bring the promised results, and they continue to suffer the stigmatization of the threats to their children's healthy development and the implication that they are poor housekeepers.

During the evaluations of cleaning regimes by health authorities, little attention was paid to the women involved as the cleaners, which indirectly provides us with some insight into the state gender regime. Apart from the effort and the responsibility, there is evidence that the cleaning process itself can be contaminating (Australian Broadcasting Corporation 1992; Luke 1991, 99).

Fowler and Grabosky (1989, 148) suggested that the "regulatory orientation to pollution control has been characterized by negotiation and compromise, rather than strict enforcement." Governments have been reluctant to antagonize business, and business has been motivated by a "desire to avoid the loss of a marketable product rather than environmental concern" (Fowler and Grabosky 1989, 147). State recognition of business interests as more important than smelter community residents has ultimately resulted in greater recourse to the cleaning regime. Furthermore, it is unlikely that the public health care system has the long-term capacity to maintain the level of involvement and monitoring that is required. This suggests the program is "window dressing," a project of state legitimation, which rests on the state's class and gender regime and which masks the interests that are served by de facto tolerance of levels of industrial pollution damaging to a small and easily ignored segment of the population.

Domestic Cleaning Regimes—The 1990s

A similar household cleaning regime to that developed and applied in the 1980s by public health workers in the vicinity of the Port Pirie lead smelter was subsequently put in place in the NSW smelter towns of Boolaroo and Broken Hill. The regime was formalized in 1994 by a federal agency, the Commonwealth of Australia Environmental Protection Authority (EPA), and outlined in a document published by the EPA titled *Lead Alert: A Guide for Health Professionals* (Alperstein, Taylor, and Vimpani 1994).

Lead Alert set out the "steps to minimise exposure and absorption" of lead by children. Health professionals are told to give the following advice "to parents if a child's blood lead is more than 15 µg/dL" (Alperstein, Taylor, and Vimpani 1994, 17). Parents should ensure that children's hands and face are washed before they eat or have a nap, discourage children from putting dirty fingers in their mouths, encourage children to play in grassy areas, and wash fruit and vegetables. In terms of housecleaning, they are advised to wet dust floors, ledges, window sills, and other flat surfaces at least weekly or more often if the house is near a source point for lead; to clean carpets and rugs regularly using a vacuum cleaner; to wash children's toys, especially those used outside; and to wash family pets frequently and discourage pets from sleeping near children. The parent should ensure that the child does not have access to peeling paint or chewable surfaces painted with lead-based paint and that the child's diet is adequate in calcium and iron, which helps to minimize lead absorption. Children should be provided with regular frequent meals and snacks, up to six per day, because more lead is absorbed on an empty stomach (Alperstein, Taylor, and Vimpani 1994, 17).

The document was widely disseminated, not just in smelter communities but in situations where people had elevated blood lead levels from any source. The housecleaning regime has been promoted as a primary intervention strategy, despite the evidence from evaluations of the Port Pirie experience (as well as studies in other countries) that show it to be ineffective.

This extremely detailed cleaning regime involves an implicit threat for non-compliance, the threat of adversely affecting one's child's health and intellectual development. Because of the nature of the tasks, the responsibility for most of this work falls to mothers rather than all parents.

For communities in the vicinity of lead smelters in Australia, public health authorities recommend an even more stringent regime than that implied in *Lead Alert*. The regime suggests not vacuuming with a child in the room since the cleaning raises dust. Dusting should ideally cover what one Boolaroo mother described as "bizarre places" such as the fly wire in screen doors. Other suggested strategies include moving children's beds from under windows and putting away soft toys because they cannot be easily washed (Gilligan 1992, 4). To add to this intensification, some anxious parents intensify the regime further. The "wet dusting" or mopping over of all horizontal surfaces is done more than once a day by some mothers. In a television documentary on the subject of lead poisoning and children, women who had implemented the regime in both Boolaroo and Port Pirie expressed their frustration. "They tell you to run round with a washer after them. They are not allowed to put their fingers in the mouths"; "You do things that you just would not do"; "We must be the only housewives in New South Wales that do these tasks" (Australian Broadcasting Corporation 1992). Another mother verbalized the worry and guilt associated with such responsibility for her children's health, "you feel guilty if you just don't want to do the work . . . you think your child does not have a normal life . . . should you have more children?" (Gilligan 1992, 45).

The lead abatement program also involves the monitoring of the child's blood lead levels as the most significant means of measuring the levels of lead absorption. This means subjecting the child to frequent blood sampling and a constant measuring as to whether the family's efforts have been successful. There is constant contact for both mother and child with health and other professionals and often researchers (Australian Broadcasting Corporation 1992; Gilligan 1992, 4). The constancy of the monitoring creates the impression that the responsibility for the public health problem falls on the residents themselves (and especially on mothers), rather than the government or the corporation (see also McGee 1996, 14). The monitoring of blood lead levels also gives the impression that something is being done but actually does nothing to address the source of the contamination. In some cases, children's blood lead levels have risen after the implementation of household cleaning regimes.

A study of the effects of the lead issue on Boolaroo residents found that families of children with high blood lead levels experienced "feelings of guilt, stigma, anxiety, stress and powerlessness . . . that the difference may be due to or to be seen to be due to some action or inaction of them as parents" (Hallebone and Townsend 1993, 17). We found that in Boolaroo, health professionals regarded parental failure to present children for monitoring as evidence of irresponsibility. As with smelter workers in the 1920s, stigma is attached to those who do not conform to the cleaning standards. . . .

Not only are women enslaved by the domestic cleaning, but their children's psychological and/or emotional development is put in question. In the education booklet written by the education department for use in Boolaroo schools, the family unit is presented as the most important element in dealing with lead exposure: "Children who are supported and confident in their family unit will be better able to deal with any problems associated with the lead issue" (quoted in Mason 1992, 41).

Despite evidence of limited success in the long term, domestic cleaning regimes are becoming popular for dealing with lead contamination from gasoline. As has occurred with the smelter communities, the effect again is to shift responsibility from the polluting source to the private sphere of the family and women.

Resistance to State Interventions

The focus on housecleaning regimes as a response to children's elevated blood lead levels has the effect of stigmatizing parents and calling into question their capacity to care for their children. As noted above, mothers who have implemented the regime, however, find it very oppressive and anxiety provoking, hence many cease doing it (McGee 1996, 14). There is also resistance to the surveillance of children's blood lead levels in situations where the health authorities are not offering an effective response to the problem. The proportion of smelter-town families who continue to present their children for blood testing has declined considerably over time (Isles 1993; Maynard, Calder, and Phipps 1993, 9; Stephenson, Corbett, and Jacobs 1992; Western NSW PHU 1994).

Nevertheless, women in smelter towns have responded with active as well as passive resistance. Following the PHU's 1991 revelation of elevated blood lead levels in Boolaroo children, local residents, mainly women, formed the North Lakes Environmental Action Defence Group (No Lead). It has campaigned for action by state and local government and by the smelter to counter the high levels of lead pollution. . . .

No Lead has addressed the issue of lead contamination by refocusing public attention on the responsibilities of the government to regulate the industry and the industry's responsibility to reduce toxic emissions. They have found themselves in direct conflict with the company's public relations strategists who have attempted to downplay their political significance as legitimate representatives of community interest. No Lead has been able to successfully work with environmental groups like Greenpeace and has achieved some success in refocusing government attention on Boolaroo. For example, a Parliamentary Select Committee set up in 1994 resulted in a management plan that placed more responsibility on the smelter and the local government in limiting the effects of plant emissions, although it reinforced the emphasis on housecleaning regimes (NSW Parliament 1994).

Other women in Boolaroo have actively resisted government interventions intended to ameliorate the exposure of children to lead, which they see as stigmatizing their children as intellectually limited or themselves as poor housekeepers. The most significant of these was the 1992 resistance by the Parents and Citizens Association to a temporary closure of the local school for remediation. The parents questioned whether the school was indeed contaminated or if the contamination was any more significant than that which they experienced in their homes. They rejected the implication that the school posed a threat to their children's intellectual development, citing cases of local children who had succeeded academically. The women involved in this public protest expressed a fear that the school would be closed forever, once the children had been relocated. McPhillips (1995, 48) commented that the issue brought out the "deep-felt suspicion that this section of the local community held for government bureaucracies." That is, they do not see the public authorities as acting in their interests and resist the authorities' efforts to intervene in the situation. In this case, the women's resistance was successful, and the remediation was carried out while the school was routinely closed for summer holidays.

Conclusion: Lead Levels, Public Health Strategies, and State Gender Regimes

The strategy promoted by public health authorities for smelter towns, rather than dealing with the source of the pollution, turns this public issue into a private family matter. In appearing to do something, the state selected a remediation strategy that has been repeatedly proven to be ineffective. Research from many countries (Luke 1991) persistently shows that the major issue is one of proximity to the smelter and the level of emissions, with past pollution also a critical factor. There is a continuing history of failure of the smelting companies to own the pollution problem they cause and to deal with it. Profit levels, rather than health concerns, have historically taken precedence, with the state mediating the corporations' interests. The state historically has facilitated the shifting of the focus of responsibility for the problem to the relatively powerless. Now the blame is laid on working-class women, whereas early in the twentieth century, it was directed toward recently arrived male immigrant workers, although women were indirectly implicated.

The official remediation strategy does not fall in a gender-neutral way on both parents. It relies for its implementation on additional daily caring labor being undertaken by mothers of children deemed "at risk" and thus on the basis of a traditional understanding of the responsibilities associated with motherhood.

In addition, it relies for compliance on the mothers' emotional commitment to their children's health, a situation with great potential to engender feelings of guilt and to stigmatize those who "fail." The exploitation of these working-class women's unpaid caring labor is not only an example of the state "doing" gender but a recent form of "doing" class as well. Brown and Ferguson (1995, 161) noted that:

> when activists discover that local industry values its bottom line or international reputation more highly than it does the health of children in the community, this realization violates the trust that the women toxic waste activists have placed in the ideal of corporate citizenship and governmental protection.

Working-class mothers are the target of a burdensome housecleaning regime that absorbs their labor but at the same time effectively shifts responsibility for ensuring the pollution does not damage their children's health away from the corporation and the state. This strategy reflects a state gender regime that involves material exploitation in the form of reliance on women's labor in a manner that serves powerful interests and ideological exploitation through the manipulation of the women's sense of maternal responsibility.

The structural features of class and gender do not, of course, account for the whole story. Residents are not duped nor have they have been silent on the matter. The women as individuals resist or reject the recommended regime and have been at the forefront of organized community resistance strategies. Women organized into community-based pressure groups have had limited success in directing attention back onto the smelter as the source of the toxic pollution and away from their responsibility as domestic carers for the health of their children.

This case study raises specific questions about where the responsibility lies for the health of residents of smelter communities. But it also illuminates class and gender relations in contemporary Australia and the manner in which the state is engaged in the reproduction of class and gender difference.

Note

Bryson, Lois, Kathleen McPhillips, and Kathryn Robinson. 2001. "Turning Public Issues into Private Troubles: Lead Contamination, Domestic Labor, and the Exploitation of Women's Unpaid Labor in Australia." *Gender & Society* 15 (5): 755–72.

References

Alperstein, G., R. Taylor, and G. Vimpani. 1994. *Lead Alert: A Guide for Health Professionals.* Canberra, Australia: Commonwealth Environmental Protection Agency.

Australian Broadcasting Corporation. 1992. *Living with Lead.* Television documentary screened on *Four Corners*, September 14.

Australian Bureau of Statistics. 1994. *How Australians Use Their Time.* Canberra: Australian Bureau of Statistics.

Bittman, Michael, and George Matheson. 1996. *"All Else in Confusion": What Time Use Surveys Show about Changes in Gender Equity.* SPRC Discussion Paper Series No. 72. Sydney: Social Policy Research Centre.

Bittman, Michael, and Jocelyn Pixley. 1997. *The Double Life of the Family: Myth, Hope, and Experience.* Sydney: Allen and Unwin.

Brown, P., and F. Ferguson. 1995. " 'Making a Big Stink': Women's Work, Women's Relationships, and Toxic Waste Activism." *Gender & Society* 9: 145–72.

Bryson, Lois. 1997. "Citizenship, Caring and Commodification." In *Crossing Borders: Gender and Citizenship in Transition,* edited by Barbara Hobson and Anne Marie Berggren. Stockholm: Swedish Council for Planning and Co-ordination of Research.

Centers for Disease Control and Prevention. 1991. *Preventing Lead Poisoning in Young Children.* Atlanta, GA: U.S. Department of Health and Human Services.

Connell, R. W. 1990. "The State, Gender, and Sexual Politics." *Theory and Society* 19: 507–44.

Dempsey, Ken. 1997. *Inequalities in Marriage: Australia and Beyond.* Melbourne: Oxford University Press.

Donovan, John. 1996. *Lead in Australian Children: Report on the National Survey of Lead in Children.* Canberra: Australian Institute of Health and Welfare.

Fenstermaker Berk, Sarah. 1985. *The Gender Factory: The Apportionment of Working American Households.* New York: Plenum.

Fowler, R., and P. Grabosky. 1989. "Lead Pollution and the Children of Port Pirie." In *Lead Pollution: Fourteen Studies in Corporate Crime or Corporate Harm,* edited by P. Grabosky and A. Sutton. Milson's Point, NSW: Hutchinson.

Galvin, J., J. Stephenson, J. Wlordarczyk, R. Loughran, and G. Wallerm. 1993. "Living near a Lead Smelter: An Environmental Health Risk Assessment in Boolaroo and Argenton, New South Wales." *Australian Journal of Public Health* 17: 373–78.

Gilligan, B., ed. 1992. *Living with Lead: A Draft Plan for Addressing Lead Contamination in the Boolaroo and Argenton Areas, NSW.* Lake Macquarie, Australia: Lake Macquarie City Council.

Graham, Hilary. 1983. "Caring: A Labor of Love." In *A labor of Love: Women, Work and Caring,* edited by Janet Finch and Dulcie Groves. London: Routledge and Kegan Paul.

Hallebone, E., and M. Townsend. 1993. "Who Bears the Weight of Lead in Society? A Social Impact Assessment of Proposed Changes to the National Guidelines for Blood Lead Levels. Technical Appendix 3." In *Reducing Lead Exposure in Australia: An Assessment of Impacts.* Vol. 2, edited by Mike Berry, Jan Garrard, and Deni Greene. Canberra, Australia: Department of Human Services and Health.

Heyworth, J. S. 1990. "Evaluation of Port Pirie's Environmental Health Program." Master of Public Health diss., Adelaide University, Adelaide, Australia.

Isles, Tim. 1993. "Boolaroo Blood-Lead Tests Show No Change." *Newcastle Herald,* July 13.

Landrigan, P. J. 1983. *Lead Exposure, Lead Absorption and Lead Toxicity in the Children of Port Pirie: A Second Opinion.* Adelaide: South Australian Health Commission.

Luke, Colin. 1991. "A Study of Factors Associated with Trends in Blood Lead Levels in Port Pirie Children Exposed to Home Based Intervention." Master of Public Health diss., Adelaide University, Adelaide, Australia.

Maguire, Paul. 1994. "Green Buffer 'Best Way' to Cut Lead Poisoning in Young Children. *Newcastle Herald,* June 3.

Mason, Chloe. 1992. "Controlling Women, Controlling Lead." *Refractory Girl* 43: 41–42.

Maynard, E., I. Calder, and C. Phipps. 1993. *The Port Pirie Lead Implementation Program: Review of Progress and Consideration of Future Directions (1984–1993).* Adelaide: South Australian Health Commission.

McGee, T. 1996. "Shades of Grey: Community Responses to Chronic Environmental Lead Contamination in Broken Hill, NSW." PhD diss., The Australian National University, Canberra.

McMahon, Anthony. 1999. *Taking Care of Men: Sexual Politics in the Public Mind*. Cambridge: Cambridge University Press.

McPhillips, K. 1995. "Dehumanising Discourses: Cultural Colonisation and Lead Contamination in Boolaroo." *Australian Journal of Social Issues* 30: 41–55.

NSW Parliament. 1994. *Report of the Select Committee upon Lead Pollution*. Vol. 1. December.

Public Health Unit. n.d. Hunter Area Health Service. Poster.

Russell, Graeme. 1983. *The Changing Role of Fathers?* St. Lucia, Australia: University of Queensland Press.

Stephenson, J., S. Corbett, and M. Jacobs. 1992. "Evaluation of Environmental Lead Abatement Programs in New South Wales." In *Choice and Change: Ethics, Politics and Economies of Public Health. Selected Papers from the 24th Public Health Association of Australia Conference, 1992, Canberra*, edited by Valerie A. Brown and George Preston. Canberra: Public Health Association of Australia.

West, C., and D. H. Zimmerman. 1987. "Doing Gender." *Gender & Society* 1: 125–51.

Western NSW Public Health Unit. 1994. *Risk Factors for Blood Lead Levels in Preschool Children in Broken Hill 1991–1993*. Dubbo, Australia: Western New South Wales Public Health Unit.

Wilson, D., A. Esterman, M. Lewis, D. Roder, and I. Calder. 1986. "Children's Blood Lead Levels in the Lead Smelting Town of Port Pirie, South Australia." *Archives of Environmental Health* 41: 245–50.

PART **IV** MEDIA

Media Framing of Body Burdens
Precautionary Consumption and the Individualization of Risk

Norah MacKendrick

Because many of us get our information about health- and environment-related issues from the mainstream media, it is important to think critically about how newspapers cover these issues. Sociologists may look for trends and patterns in reporting—What types of information are routinely discussed? What is left out? What types of people are featured in the news? In this chapter, Norah MacKendrick examines how the Canadian press reports on "body burdens"—chemical contaminants stored up in our bodies—from 1986 to 2006. She finds that news articles focus increasingly on green consuming and self-protection (as opposed to protection by the government or business); these have become dominant media frames in reporting on body burdens.

Introduction

Bioaccumulation—the gradual accumulation of environmental contaminants in biological organisms—has long been the concern of natural scientists, but has only recently surfaced in the sociological literature. Attention has now turned to "body burdens," an aspect of bioaccumulation that refers to the internal contaminant load carried by most organisms in the industrialized world. Of special concern are the health effects of body burdens in human populations, particularly children. As individuals, we are not aware of our own internal contaminant load, as detection of body burdens requires specialized technology. But exposure to contaminants is universal and largely involuntary, as chemicals are present in air, water, food, soil, and indoor environments. . . .

As expert systems have considerable control over the definition and subsequent management of chemical hazards (e.g., Cable, Shriver, and Mix 2008), the communication of risk around body burdens to the lay public largely depends on the popular news media. By encouraging certain problem frames, the media shapes public understandings of key causal factors and attributions of responsibility in the management of risk (Brown et al. 2001; Stallings 1990). Consequently, the media plays a vital role in framing body burdens as both a risk and a social problem, and we can assume that public knowledge of body burdens is significantly influenced by the news media. The importance of the news media in interpreting local or regional contamination events is acknowledged in the literature (e.g., Griffin, Dunwoody, and Gehrmann 1995; Robinson 2002; Zavestoski et al. 2004), although to date no study has systematically examined news media coverage of body burdens arising from chronic and universal exposure to environmental contaminants. An understanding of how the news media frames body burdens as a social problem is critical for understanding how this issue is taken up in the public sphere, by policy makers, communities in close proximity to polluting industries, health professionals dealing with environmental illnesses, and individual readers who become aware of their bodies and their environments as "polluted."

Drawing on the sociology of risk literature, this article uses frame analysis to examine the social problem frame around body burdens in the Canadian news media from 1986 to 2006, and links these findings to broader theories of risk and society, particularly the individualization of risk. Most significantly, recent news articles underscore the individual's responsibility to avoid contaminant exposure through what I call "precautionary consumption" behavior. The frame of precautionary consumption communicates a sense of individual empowerment and control through acts of "green" consumption and chemical avoidance. While it transforms body burdens into a manageable risk, it directs attention away from the failure of states and industries to properly manage chemical production and dispersal. . . .

Background: Body Burdens

The internal contaminant load carried by living organisms or "body burden"[1] is associated with intensive industrialization that has released multitudes of persistent chemical compounds into the environment. These chemicals accumulate up the food chain, and predatory organisms, such as polar bears, whales, seals, and fish, contain chemical concentrations far higher than those of their external environments. Chemical bioaccumulation is now a prominent environmental issue for both national and international environmental groups. Biomonitoring technology—methods for analyzing animal tissue for biological markers of contaminant exposure—has improved significantly in recent years, allowing the detection of compounds not previously known to bioaccumulate (Sexton et al. 2004).

Biomonitoring data suggest that the body of nearly every person living in an industrial society contains minute concentrations of contaminants from exposure to a polluted environment, food treated with pesticides, and consumer goods treated with brominated flame retardants, stain repellents, and other compounds (e.g., CDC 2005). Health effects from chronic chemical exposure are uncertain and controversial, although some studies link exposure to certain forms of cancer, fertility problems, behavioral disorders in children, and thyroid disorders (Grandjean and Landrigan 2006). In other words, body burdens represent the contamination of the environment and are a serious health issue.

The industrial production and use of chemicals is regulated in Canada, as in the United States, by state institutions that employ risk assessments to determine the toxicity, persistence and bioaccumulative potential of chemicals, and to remove problematic compounds from production. Human body burdens are frequently cited as evidence of the failure of these risk assessments to prevent universal exposure to bioaccumulative chemicals (CELA 2006; Steinemann 2004).

Such regulatory failures, along with catastrophic localized pollution events, have given rise to a broad-based toxics movement comprised of advocacy groups and community groups fighting for pollution prevention (Szasz 1994) and, more recently, to movements concerned about environmental illness at both the individual and community levels (Brown et al. 2004; Kroll-Smith, Brown, and Gunter 2000; McCormick, Brown, and Zavestoski 2003). Greater accessibility to biomonitoring technology has allowed these groups to track community exposure to environmental contaminants and to raise public awareness of chemical bioaccumulation (Altman et al. 2008; Morello-Frosch et al. 2009).

Technology, Risk and the Environment

Like other forms of contamination and pollution, the bioaccumulation of contaminants in the environment and living organisms represents an environmental and technological risk. This section discusses two separate risk literatures: the institutional and organizational perspective and the risk society and individualization perspective.

Institutional and Organizational Factors

The body burden phenomenon has a clear institutional component in that government agencies regulate chemical production and are responsible for minimizing adverse impacts of chemical exposure on environmental quality and human health. As Perrow (1984) demonstrated in his study of nuclear accidents, risk is not necessarily the product of human error but can be an inherent property of industrial operating systems where sections are so tightly coupled that minor problems quickly develop into systemic failures. Organizational characteristics and configurations are keys to the production of risk, making the institutional perspective on risk a useful analytical tool, as it examines how features of organizations, bureaucracies, and policy-making processes lead to environmental hazards and technological failures (e.g., Clarke 1993) . . .

The "Risk Society" and Individualization

Pointing to widespread ecological degradation (e.g., DDT in breast milk, transboundary pollution, and radioactive fallout), Beck (1992, 1995, 1999) argues that we are moving toward a "risk society" characterized by universal and high-consequence risks that threaten human and ecological existence and transcend geographic and social divisions. He observes that risks have ceased to be managed by traditional institutions associated with the welfare state and are managed instead by institutions with little visibility and public accountability, such as the market, and science and technology. Thus, the production of risks cannot easily be associated with an identifiable actor or institution, a condition he calls "organized irresponsibility" (Beck 1999: 6). Elsewhere, Beck has written extensively on individualization and modernity (Beck and Beck-Gernsheim 2002) and the anxiety associated with negotiating universal risks at the individual level (Beck 2006). He suggests that individualization is exacerbated by the inability of state institutions to manage modern risks, such that the market becomes a key institution for individuals to manage their exposure to risk (Beck 2006).

A related, but distinct, body of risk literature has studied the individualization of risk, with a strong focus on risk as a technique of governance related to the dismantling of the welfare state. These authors argue that the individualization of risk is associated with principles of neo-liberalism, emphasizing individual choice, personal responsibility, and the market as an efficient institution for bringing about social and political change (Ericson, Doyle, and Barry 2003; Rose 1999). They note that with the expansion of neo-liberalism, universal risks—such as crime and economic instability—are increasingly managed through the purchase of private insurance and other commodities of self-protection (Doyle 2007; Ericson, Doyle, and Barry 2003; O'Malley 2004). Here, risk is used as a technique of governance, with risk management transferred to "citizen-consumers" (Scott 2007: 37) who negotiate their exposure to risk by modifying their consumer choices. While not explicitly embedded in risk theory, Szasz (2007) also observes the growing commodification of risk protection, citing gated communities and the increase in consumption of bottled water and specialty non-toxic products.

Media and Risk

Risks are communicated through the mass media, an important source of information on science, health and environmental issues (Carvalho and Burgess 2005; Nelkin 1987; Seale 2003), and a critical actor in the social construction of risk (Allan 2002; Driedger 2007). The media help to construct phenomena as social problems, identifying which issues are cause for concern, outlining responsibility for these problems, and recommending courses of action (Gamson and Modigliani 1989; Hilgartner and Bosk 1988; Kitsuse and Spector 1973). Public risk perception is shaped by news stories that frame and construct events in ways that amplify or down-play risks. The media can amplify risks through an increase in the volume of news stories about a risk event (Kasperson et al. 2005) or through dramatic language and images (Eldridge et al. 2005). News stories can exacerbate fear and anxiety (Altheide 1997) or offer reassurance to the public by highlighting the actions of "responsible" public institutions purporting to have risks under control (Freudenburg et al. 1996).

Frame Analysis

. . . Frames in news stories do several things: define a problem in a particular way, interpret causal relationships, provide moral evaluation, and recommend courses of action or treatment (Entman 1993). Importantly, frames in news texts represent a choice between what messages to communicate (Tuchman 1978) by drawing attention to certain aspects of reality and ignoring others (Entman 1993; Oliver and Johnston 1999). This article examines the content of frames rather than the conditions behind the construction of frames or the interaction between frames and public opinion formation (see D'Angelo 2002). This approach is useful for identifying general cultural frames and ideological standpoints (Binder 1993; Oliver and Johnston 1999), as news frames deliberately draw on these to increase their resonance or "potency" (Gamson and Modigliani 1989; Scheufele 1999).

It is important to pay attention to the primary sources selected by journalists to define issues and their terms of reference (Coleman, Hartley, and Kennamer 2006). A shift in these "primary definers" (Antilla 2005: 344) can coincide with a shift in how risks are framed in the news media. In the case of climate change, for example, what was once almost entirely a scientific issue (with scientists as primary definers) is now also framed as a financial risk because of the involvement of insurance industry representatives as primary definers (Carvalho and Burgess 2005). Journalists seek out "official" sources (often publications or spokespeople from government or industry) from organizations that hold some responsibility for a given event (Stallings 1990), or select sources that lend credibility to the journalist's own viewpoint (Ericson, Baranek, and Chan 1989). Organizations with opposing viewpoints must, therefore, be well organized and persuasive to appear in news articles, and many rarely gain access to reporters (Coleman, Hartley, and Kennamer 2006; Freudenburg et al. 1996). Importantly, frames are not determined by primary definers but are the outcome of an interaction between journalists and their sources (Miller and Reichert 2000).

The Framing of Causal Mechanisms

A key component of the social problem frame in media accounts of risk is the representation of causal mechanisms and attributions of responsibility (Altheide 1997). Media discourse may offer a selective view of causation by failing to identify systemic causes of hazards. In his seminal study of media framing, Stallings (1990) observed that news reports of a bridge failure focused more on the immediate problems associated with environmental conditions and engineering quality than on systemic factors, such as the design of transportation networks or the reliance on cars for transportation. Importantly, Stallings made this observation by noting the absence of systemic causal frames in media representations. In a similar study, Brown et al. (2001) found that references to systemic causes of breast cancer—particularly environmental pollution—decreased over time in the popular print media, while individual-level factors, such as personal lifestyle and genetics were presented as primary causes of the disease.

While the media's influence over public discourse is "neither trivial nor decisive" (Gamson and Modigliani 1989: 80), in the case of highly technical risks such as body burdens, we can assume that it exerts a tremendous influence over our knowledge and understanding of risk. Contact with chemicals is part of everyday life, and each of us has a chemical body burden; nevertheless, the ability to detect and observe body burdens is limited to expert systems. Consequently, the news media play a vital role in communicating the risks of chemical bioaccumulation to the lay public—a role that has not yet been examined in the sociological literature . . .

Methods

Over the past 20 years, chemical bioaccumulation has received considerable attention in the Canadian news media, as both an environmental and health topic. Canadian research institutions—including federal government agencies—actively study the ecological and human health effects of chemical bioaccumulation, and the Canadian government has pushed for the ratification of international treaties to control the production and disposal of persistent organic pollutants as a result of concerns about Arctic pollution.[2] In the last 5 years, several prominent Canadian environmental groups (e.g., Environmental Defense and Pollution Probe) have launched campaigns to raise public awareness of chemical pollution and human body burdens.

The first step in the research process was collecting background information to inform content and frame analysis. This included searching through Canadian newspaper articles from 1986 to 2006, and a careful reading of Centers for Disease Control (CDC) reports on human exposure to chemicals, as well as online publications from several environmental groups. . . . This background research identified the main environmental and health concerns of chemical bioaccumulation, particularly the concern that body burdens are a *universal* and *unavoidable* risk because of frequent exposure to contaminants through normal everyday experiences.

The next step was to find news media articles that reflected these concerns, particularly exposure to multiple chemical compounds through normal everyday experiences, such as breathing, eating, drinking, and using common household goods. Articles therefore had to mention *more than one* environmental contaminant, so that coverage would not be biased toward issues specific to an individual compound. Because topics around bioaccumulation vary significantly over time (see Figure 11.1), it was not possible to track one single topic in sufficient depth over a 20-year time period.

Newspaper selection was largely determined by circulation size and geographic representation, but was also influenced by the political economic realities of newspaper ownership in Canada. *The Globe and Mail*, a national paper, and the *Toronto Star*, a paper serving Canada's largest metropolitan area, were selected as they have the largest circulation in Canada . . .

An electronic database of articles from the four papers was generated through systematic searches in the ProQuest Canadian Newsstand database. All searches were limited from January 1, 1986, to December 31, 2006. . . .

. . . Only news articles and special features were selected for analysis, since these articles are guided by the journalistic norm of objectivity with the purpose of providing information to readers, in comparison with editorials and opinion columns that explicitly state an opinion or take a position in a broader debate

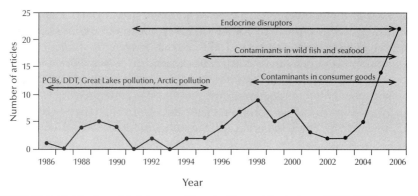

FIGURE 11.1 Distribution of Articles over Time by General Topic Area: 1986–2006.
Note: Arrows indicate the temporal distribution of the major topic areas related to bioaccumulation.

(Gamson and Modigliani 1989), and are therefore assumed to be more representative of the media's role in a risk society—as an intermediary between expert and lay systems of knowledge.

Several exclusionary criteria were used to further refine the articles selected for analysis. Articles that focused entirely on bioaccumulation in nonhuman organisms, for example, were excluded unless there was some mention of human consequences of contamination. Articles were rejected if they covered only accidental or specific point-source pollution (e.g., a chemical spill, contamination from a fire) or exposure to contaminants from certain occupations. Also excluded were those covering "single-instance" food safety scandals (e.g., carcinogens in soft drinks or the contamination of eggs with dioxin) and articles discussing farmed fish.[3] These articles were rejected because single-instance food safety scandals and specific point-source pollution represent a different form of risk from the everyday, normal exposure to contaminants associated with the body burden phenomenon. . . .

Findings

In total, 102 articles were selected for analysis. The greatest proportion came from *The Globe and Mail* (45%), with 25 percent from the *Vancouver Sun*, 22 percent from the *Toronto Star*, and 9 percent from the *Calgary Herald*.

Figure 11.1 illustrates the distribution of articles and specific topics over the 20-year period. Topics vary according to the period of coverage. PCBs and DDT were two important contaminant issues in the 1980s because of extensive concern over the presence of these chemicals in human breast milk and in large mammals. News reports of chemical bioaccumulation in the Arctic and Great Lakes region were steady throughout the 1980s and early 1990s, but declined after the mid-1990s. Contaminants in wild fish and seafood are consistently covered in the news since the mid-1990s. Coverage of endocrine disruptors began in the early 1990s and became more prevalent in the mid-1990s after the publication of the popular book *Our Stolen Future* (Colborn, Dumanoski, and Myers 1996). News reports of contaminants in consumer goods began in the late 1990s after the publication of CDC reports and scientific studies identifying these compounds in humans and large mammals such as polar bears. The spike in news coverage around 1998 seems to be explained by a convergence of topics, particularly concerning endocrine disruptors and consumer goods. The considerable increase in articles after 2004 reflects discussions of human body burdens specifically, with an emphasis on health effects in children and consumer goods as contaminant sources.

Content and Frame Analysis

Content analysis is used to identify: (1) the specific pathway of contaminant exposure identified in news articles, and (2) the primary definers used in the article. Frame analysis is used to identify aspects of the dominant social problem frame.

The pathway of chemical exposure reveals the characterization of contaminant origins. Pathways are identified by searching for references to the presence of contaminants in: (1) the external environment, defined as water, soil, and air (including indoor air), as well as reference to bioaccumulation in non-human organisms; (2) food, including wild game and wild-caught fish; and (3) consumer goods, such as electronics and furniture. Primary definers are identified in each article by counting the number of people or organizational representatives quoted, including quotations taken from written statements or organizational publications. . . .

In the first stage of frame analysis, the dominant problem frame is identified, that is, *how* contamination is described as a social problem. Frames in textual materials "are manifested by the presence or absence of certain keywords, stock phrases, stereotyped images, sources of information, and sentences that provide thematically reinforcing clusters of facts or judgments" (Entman 1993). Therefore, frames are identified by examining how the articles used concepts, themes, or metaphors to describe body burdens as a "concern" and assign meaning to this concern. Three different aspects of the problem frame are identified:

1. The overall representation of body burdens as a problem. This includes identifying the specific *characterization* of the problem—whether body burdens are cited as an example of environmental degradation (e.g., references to contaminants as a form of pollution or a threat to nature and wildlife) or a threat to human health (in terms of illness, disease or disorders).
2. The way in which responsibility or blame for contamination is articulated and whether aspects of blame are missing from the article. Here, the particular actors and organizations held responsible for creating contaminant risks are documented (e.g., industry or government).
3. The kinds of solutions to the body burden problem proposed by the article (e.g., changes to regulation, cleaning up the environment, or suggestions for avoiding exposure to contaminants).

In the second stage of frame analysis, the presence of these aspects of the problem frame is documented for each article, making it possible to track variation over time using basic descriptive statistics. Examining annual variation, however, reveals only subtle changes in framing largely because of the small number of articles published each year before 1996. For this reason, coverage is divided into two separate decades to more clearly capture shifts in framing. When divided by decade, 28 percent of the articles (N = 29) appear between 1986 and 1996 and 72 percent (N = 73) appear between 1997 and 2006. *T*-tests are used to test for significant differences in averages across decades. . . .

Pathway of Exposure and Primary Definers. Most articles note more than one pathway of exposure. . . . [B]etween 1986 and 1996, 86 percent of articles mention an environmental pathway of exposure, and 52 percent mention food as a pathway. In the second decade, there is a drop in articles noting an environmental pathway, and *t*-tests of statistical difference (p < .05) show that this difference is significant. Moreover, 43 percent of the articles in the first decade connect environmental pollution to the presence of contaminants in food, but this decreases sharply in the second decade (17%). The proportion of articles identifying consumer goods as an exposure pathway, in contrast, increases significantly in the second decade, jumping from 10 percent to 60 percent (p < .001).

Several categories of primary definers are used in the news articles . . . Regardless of time period, academic scientists (those affiliated with universities rather than government) appear most frequently as primary definers. Their presence in the articles contributes to a consistent focus on the "scientific" aspects of body burdens. Although representing a small proportion of primary definers across decades, when individual lay people appear in articles, they typically express fear of contamination or provide information on how others can protect themselves from exposure. With one exception, all of these individuals are women who are presented as mothers (or women wishing to have children) and their advice applies to the social reproductive domain. Some articles, for example, feature women who fear that contaminants in their breast milk are "poisoning" their babies (McInnes 1988), women dealing with infertility problems (Bueckert 1991) or warning against using plastics in the microwave (Ubelacker 2005). These individuals support the frame of precautionary consumption discussed below.

Frame Analysis

Two Major Problem Frames: Health and Environmental. All articles frame body burdens as a "problem" and communicate a sense of fear and anxiety around this issue. They note, for example, the "surprising level of foreign substances" in breast milk (Bergman 1989), express "fears over cancer" (McAndrew 1998), and describe endocrine disruption in aquatic ecosystems as "ominous and frightening" (Simpson 2001).

Across both time periods, all articles identify potential health problems from contaminant exposure, including cancer, birth defects, decreased sperm counts, reproductive disorders, and learning and behavioral disorders. Over time, these health dimensions overshadow the environmental dimensions of body burdens. . . . [F]rom 1986 to 1996, about 83 percent of the articles note environmental problems associated with contaminants (such as compromised ecological integrity or the reduced reproductive success of mammals, fish, and amphibians), but this drops dramatically to 34 percent from 1997 to 2006, and this drop is statistically significant (p < .001). In this later decade, even articles identifying an environmental pathway of exposure often do not explicitly reference environmental concerns associated with chemical bioaccumulation. This shift reflects a broader movement toward seeing environmental degradation through a health lens (Brown and Zavestoski 2004), and attests to the increasing dominance of the health frame in the narrative of environmental problems. The health frame marks a shift in how environmental problems are conceptualized: as a problem separate from humans to an issue directly connected to human well-being.[4]

Responsibility and Blame: Contamination as a Blameless Phenomenon. Given that news articles must reflect journalistic objectivity, it was expected that responsibility for contamination—where actors and organizations are identified as being responsible for creating chemical hazards—would be framed in subtle ways, for example through general references to industrial pollution or regulatory problems and delays. However, even subtle references to responsibility are not articulated in a considerable proportion of the articles from the first decade (48%) and in the majority of articles (66%) from the second decade. More specifically, these articles express concern over contamination and describe how contamination occurs, but do not give the reader any sense of the role of government or industry in creating these problems. In one article, for example, we learn that "the air inside your new home could be 5–20 times more polluted than the outdoor air in the most industrialized cities," owing to chemical off-gassing from paint, furniture, and floor coverings (Gillespie 2005). Despite citing multiple household goods as sources of potentially harmful chemicals, the article does not point to the industrial producers of these household products, nor does it question the efficacy of government safety assessments that allowed these products to go on the market. . . . By failing to identify responsible parties, these articles paint a picture of body burdens as a "blameless" phenomenon.

Of the proportion of articles that *do* identify a responsible party, most of the articles in the earlier period identify industry as responsible, while most of the articles in the later period identify government as responsible. There is a significant decrease in articles identifying industry responsibility between the two decades (p < .01). While industry is blamed for having "saturated" the environment with chemicals (McInnes 1988), government is blamed for being slow "to respond to environmental threats" (Bergman 1989), and is said to have "deliberately ignored and tried to explain away powerful evidence . . . of extremely high toxic doses" (young 1990).

Solutions: Collective Action versus Precautionary Consumption. Frames around solutions to body burdens fall into two categories: collective action and what I call "precautionary consumption." Forms of collective action include the implementation of stronger regulations, a reduction in chemical production and improvements in risk assessments. One article, for example, suggests that "clearing the deadly toxins from the animals and fish . . . requires an international ban on the chemicals that contaminate them" (Canadian Press 1989). Another article states that "with an eye on the 1,000 new synthetic chemicals introduced worldwide every year and the 100,000 already on the market, the International Joint Commission is saying that if a chemical belongs to a problem family, manufacturers, importers, and users should have to prove it is not persistent and toxic before it can be introduced or remain on the market" (Smith 1996). Another puts the onus on government, suggesting that we need a "higher level of concern and commitment by the government to find out about the relationship of these chemicals to human disease" (Cone 1998). Articles identifying forms of collective action as a solution decline dramatically ($p < .01$): from 92 percent in 1986–1996 to 69 percent in 1997–2006.

Corresponding with this drop is a significant increase ($p < .001$) in the number of articles using a "precautionary consumption" frame (13% in 1986–1996 and 47% in 1997–2006). Many of these articles employ a language related to the precautionary principle—a "better safe than sorry" policy mandate to act despite a lack full scientific certainty to prevent harm to human health or the environment (O'Riordan, Cameron, and Jordan 2001). This principle represents a "public paradigm" (Brown 2007), however, in the case of this frame, precaution is decidedly self-interested and enacted at the individual consumer level. A store owner in one article, for example, is quoted as saying "our store is all about the precautionary principle" (Morphet 2005), while another article quotes a "healthy" shopper who claims to be "very cautious in what I use. I am very good at reading a label" (Wigod 1998). Most importantly, this frame suggests that specific changes to one's behavior as a consumer can reduce exposure to potentially harmful chemicals. Readers are advised to refrain from heating leftovers in plastic, to reduce their use of cosmetics and perfumes, and to buy furniture and electronics that have not been treated with brominated flame retardants. One article entitled "How pure can you get? Worried about how many harmful chemicals are in the food you eat and the air you breathe? There are ways to keep the toxic levels down" notes the universal and involuntary nature of contamination, but concludes with a section advising readers to eat organic food and to avoid many cosmetic products and conventional cleaners (Morphet 2005). Another observes that avoidance is impossible—"don't eat, don't drink and don't breathe"—but includes tips on buying goods without contaminants (Mittelstaedt 2006). The precautionary consumption frame implies that involuntary chemical exposures can be avoided through informed consumer choices or through changes to everyday behaviors.

To summarize, over a period of two decades between 1986 and 2006, we see a significant shift in the framing of body burdens. There is a drop in the proportion of articles whose problem frame discusses forms of collective action as a solution to chemical bioaccumulation, such as the implementation of stronger regulations or placing restrictions on chemical production. The precautionary consumption frame becomes more prominent, while frames articulating institutional or industrial responsibility for contamination fade away. In other words, while news frames in the earlier period reference the broader institutional arrangements involved in the production of chemical hazards, more recently, chemical risks are framed as an individual consumer and health problem, even when the efficacy of individual protective actions are unclear.

Discussion and Conclusions

When we track variations in the framing of body burdens over a 20-year period, we see two important trends. First, the framing of responsibility for contamination becomes increasingly vague. Second, the notion of precautionary consumption, a form of privatized self-protection, emerges as a key frame. Its emergence coincides with the appearance of the "consumption fallacy" (Altman et al. 2008) and "personal commodity bubble" (Szasz 2007) observed in other studies. Self-protection, therefore, is a dominant frame in the discourse of environmental and technological risks, even though it undermines the importance of structural reform.

Precautionary Consumption and the Individualization of Risk

The emergence of the precautionary consumption frame contributes to the individualization of risk and responsibility, and the market for environmentally friendly goods is presented as an alternative whereby individuals can enact their own standards of precaution. When, for example, an article expresses concern about chemicals present in ordinary brands of dish soap (those approved by federal regulators), and readers are advised to avoid these brands in favor of "ecological" alternatives, the article is encouraging the enactment of a precautionary—but individualized—risk assessment. By conceptualizing protection as a consumption problem, a precautionary consumption frame defines responsibility at the individual level, thereby eliminating the discursive space in which one might contemplate the role of the state. The meaning of precaution in this frame contrasts sharply with the precautionary principle, which is inherently supportive of collective mechanisms of universal protection.

The increasing focus on self-protection found in the media articles is at least partly explained by the increasing interest in the human health effects of environmental degradation. As Brown et al. (2001) similarly argue, a focus on the human health dimensions of environmental pollution turns the discussion away from themes of collective responsibility and institutional complicity toward a "personal health" model, prioritizing healthy behaviors and the individual's responsibility in the maintenance of well-being. The individualizing effect is bolstered by the articles' focus on consumer goods as pathways of exposure to chemicals. At the same time, there is a sharp decline in the number of articles linking environmental contamination to the presence of contaminants in food; food is no longer conceptualized as part of the environment, but is seen as a commodity. Unlike universal environmental exposures—such as exposure to polluted air—contact with contaminants in everyday commodities is more personal than collective, as it relates to a specific consumer *choice*. . . .

Precautionary consumption hints at the possibility of controlling a fundamentally uncontrollable phenomenon. It offers self-protection through consumption—a tangible and immediate solution to chemical exposure—thereby suggesting that individuals are empowered to manage this risk. Rather than look to (and wait for) the state to regulate chemical production and prevent pollution at the outset, responsibility is transferred to the individual—in this case, the reader—who gains immediate control over his or her exposure through specific measures of chemical avoidance and "green" consumption. The precautionary consumption frame provides a false sense of empowerment and comfort, given that self-protection through consumption offers a tangible, immediate and artificial solution to the systemic and structural problem of chemical exposure. The tendency of news articles to draw on this frame may furthermore reflect journalistic norms, as articles strive to address the underlying question of "what can I do?" with even the most uncontrollable risk. Ironically, the more inherently unmanageable a universal risk is, the more vulnerable it becomes to this sort of individualization, owing to the struggle to provide the reader with a sense of control.

Theoretical Implications

The growing attention to consumer products highlights how *multiple* sources of risk complicate the framing of responsibility for risk production. Chronic chemical bioaccumulation is an emergent risk. It is created by multiple and integrated sources, making it difficult to identify a single perpetrator and ultimately leading to what Beck calls organized irresponsibility. Although Beck gives brief theoretical consideration to organized irresponsibility, this study of body burdens illustrates how this notion exists empirically as a dominant news frame and is associated with "emergence" as an objective characteristic of risk. In the mid-to late-1980s, for example, when the focus on chemical bioaccumulation centered on industrial activity in the Great Lakes region, articles were more likely to highlight the institutional dimensions of responsibility, especially the failure of government to control pollution. After the mid-1990s, a greater number of articles concentrated on universal exposure through the combination of environment, food, and consumer good path-ways—coinciding with a much weaker framing of blame and responsibility.

Finally, the ideological congruence of precautionary consumption with neo-liberalism cannot be ignored. The precautionary consumption frame draws on the language of choice and individual autonomy, and presents chemical bioaccumulation as a lifestyle and consumer problem. Here, the risk-as-governance perspective is appropriate for understanding shifts in framing. An emphasis on the consumer's role in mitigating environmental and technological risks resonates with a neo-liberal model of governance, where the market, rather than the state, becomes the primary institution for social change. As this literature demonstrates, neo-liberalism is reflected in the discursive transfer of responsibility away from the state to the market, and away from individuals as citizens to individuals as consumers (Ilcan 2009). Owing to their role in producing "safer" consumer alternatives, the green consumer product and organic food industries stand to gain from such a shift, as they are the only organizations that can enable self-protection. The sheer range of green alternatives available to consumers is astounding. Major supermarkets now carry a range of non-toxic water bottles, children's toys, computers, and cosmetics, as well as a full range of organic food-stuffs. As this market expands, the frame of precautionary consumption becomes increasingly accessible to the news media, particularly when such products are marketed as "non-toxic" and "natural."

There are significant social and political implications associated with these shifts in framing. By encouraging individualized responses to a universal risk, news articles divert attention away from the role of state institutions, industry, science and technology in creating and managing the hazards associated with industrialization and chemical production. Individuals may be able to reduce their exposure to some contaminants by changing their shopping habits, but the frame of precautionary consumption overstates this ability, particularly when contaminants are present in air, water, and soil. This frame also leaves unspoken the social and economic constraints on access to "green" products. In fact, universal protection will require collective measures that motivate greater regulatory stringency, better chemical safety assessments, and substantial decreases in chemical production.

Notes

MacKendrick, Norah A. 2010. "Media Framing of Body Burdens: Precautionary Consumption and the Individualization of Risk." *Sociological Inquiry* 80 (1): 126–49.

[1] In this article, "body burden" and "chemical bioaccumulation'" are used interchangeably. . . .

[2] In 1998, Canada hosted the first round of negotiations to draft a global United Nations-sponsored treaty to deal with these pollutants.

[3] Farmed fish can contain higher levels of toxins than wild-caught fish, and most retailers use labels to distinguish wild-caught from farmed fish. Farmed fish are a specific commodity with a higher risk of contamination and do not represent a ubiquitous or universal risk.

[4] This is especially the case with the discourse of environmental justice. By drawing attention to social and racial stratification in proximity to industrial pollution and toxic waste, this discourse similarly connects environmental degradation to human health (Taylor 2000).

References

Allan, S. 2002. *Media, Risk, and Science.* Philadelphia, PA: Open University Press.

Altheide, D. L. 1997. "The News Media, the Problem Frame, and the Production of Fear." *The Sociological Quarterly* 38: 647–68.

Altman, R. G., R. Morello-Frosch, J. G. Brody, R. Rudel, P. Brown, and M. Averick. 2008. "Pollution Comes Home and Gets Personal: Women's Experience of Household Chemical Exposure." *Journal of Health and Social Behavior* 49: 417–35.

Antilla, L. 2005. "Climate of Scepticism: US Newspaper Coverage of the Science of Climate Change." *Global Environmental Change-Human and Policy Dimensions* 15: 338–52.

Beck, U. 1992. *Risk Society: Towards a New Modernity.* London: Sage Publications.

———. 1995. *Ecological Politics in an Age of Risk.* Cambridge: Polity Press.

———. 1999. *World Risk Society.* Malden, MA: Polity Press.

———. 2006. "Living in the World Risk Society." *Economy and Society* 35: 329–45.

Beck, U., and E. Beck-Gernsheim. 2002. *Individualization: Institutionalized Individualism and Its Social and Political Consequences.* Thousand Oaks, CA: SAGE.

Bergman, B. 1989. "Toxins Place Arctic Way of Life in Jeopardy." *Toronto Star* 15: A22.

Binder, A. 1993. "Constructing Racial Rhetoric: Media Depictions of Harm in Heavy Metal and Rap Music." *American Sociological Review* 58: 753–67.

Brown, P. 2007. *Toxic Exposures: Contested Illnesses and the Environmental Health Movement.* New York: Columbia University Press.

Brown, P., and S. Zavestoski. 2004. "Social Movements in Health: An Introduction." *Sociology of Health & Illness* 26: 679–94.

Brown, P., S. M. Zavestoski, S. McCormick, J. Mandelbaum, and T. Luebke. 2001. "Print Media Coverage of Environmental Causation of Breast Cancer." *Sociology of Health & Illness* 23: 747–75.

Brown, P., S. Zavestoski, S. McCormick, B. Mayer, R. Morello-Frosch, and R. G. Altman. 2004. "Embodied Health Movements: New Approaches to Social Movements in Health." *Sociology of Health & Illness* 26: 50–80.

Bueckert, D. 1991. "Pollution, Infertility Studied." *The Globe and Mail* 20: E13.

Cable, S., T. E. Shriver, and T. L. Mix. 2008. "Risk Society and Contested Illness: The Case of Nuclear Weapons Workers." *American Sociological Review* 73: 380–400.

Carvalho, A., and J. Burgess. 2005. "Cultural Circuits of Climate Change in UK Broadsheet Newspapers, 1985–2003." *Risk Analysis* 25: 1457–69.

Centers for Disease Control and Prevention (CDC). 2005. *Third National Report on Human Exposure to Environmental Chemicals.* Atlanta, GA: Centers for Disease Control and Prevention, National Center for Environmental Health.

Canadian Environmental Law Association (CELA). 2006. *Confidentiality and Burden of Proof under the Canadian Environmental Protection Act (CEPA). Submission to the House of Commons Standing Committee on Environment and Sustainable Development.* Toronto, ON: Canadian Environmental Law Association.

Clarke, L. 1993. "The Disqualification Heuristic." *Research in Social Problems and Public Policy* 5: 298–312.

Clarke, L., and J. F. Short. 1993. "Social-Organization and Risk: Some Current Controversies." *Annual Review of Sociology* 19: 375–99.

Colborn, T., D. Dumanoski, and J. P. Myers. 1996. *Our Stolen Future: Are We Threatening Our Fertility, Intelligence, and Survival? A Scientific Detective Story.* New York: Dutton.

Coleman, C.-L., J. Hartley, and D. Kennamer. 2006. "Examining Claimsmakers' Frames in News Coverage of Direct-to-Consumer Advertising." *Journalism and Mass Communication Quarterly* 83: 547–63.

Cone, M. 1998. "Pollution Is Feminizing World's Wildlife, Study Indicates." *The Vancouver Sun* September 24: A16.

Corburn, J. 2005. *Street Science: Community Knowledge and Environmental Health Justice.* Cambridge, MA: The MIT Press.

D'Angelo, P. 2002. "News Framing as a Multiparadigmatic Research Program: A Response to Entman." *Journal of Communication* 52: 870–88.

Doyle, Aaron. 2007. "Introduction: Trust, Citizenship and Exclusion in the Risk Society." In *Risk and Trust: Including or Excluding Citizens?*, edited by Law Society of Canada, 7–22. Black Point, NS: Fernwood Pub.

Driedger, S. M. 2007. "Risk and the Media: A Comparison of Print and Televised News Stories of a Canadian Drinking Water Risk Event." *Risk Analysis* 27: 775–86.

Eldridge, J., J. Reilly, N. Pidgeon, R. E. Kasperson, and P. Slovic. 2005. "Risk and Relativity: BSE and the British Media." In *The Social Amplification of Risk*, edited by N. Pidgeon, R. E. Kasperson, and P. Slovic, 138–54. Sterling, VA: Earthscan.

Entman, R. M. 1993. "Framing: Toward Clarification of a Fractured Paradigm." *Journal of Communication* 43: 51–8.

Ericson, R. V., P. M. Baranek, and J. B. L. Chan. 1989. *Negotiating Control: A Study of News Sources*. Toronto, ON: University of Toronto Press.

Ericson, R. V., A. Doyle, and D. Barry. 2003. *Insurance as Governance*. Toronto, ON: University of Toronto Press.

Freudenburg, W. R. 1993. "Risk and Recreancy: Weber, the Division of Labor, and the Rationality of Risk Perceptions." *Social Forces* 71: 909–32.

Freudenburg, W. R., C.-L. Coleman, J. Gonzales, and C. Helgeland. 1996. "Media Coverage of Hazard Events: Analyzing the Assumptions." *Risk Analysis* 16: 31–42.

Gamson, W. A., and A. Modigliani. 1989. "Media Discourse and Public Opinion on Nuclear Power: A Constructionist Approach." *American Journal of Sociology* 95: 1–37.

Gillespie, C. 2005. "Guarding your Home against Pollutants." *Toronto Star*, August 6: N12.

Grandjean, P., and P. J. Landrigan. 2006. "Developmental Neurotoxicity of Industrial Chemicals." *The Lancet* 368: 2167–78.

Griffin, R. J., S. Dunwoody, and C. Gehrmann, 1995. "The Effects of Community Pluralism on Press Coverage of Health Risks from Local Environmental Contamination." *Risk Analysis* 15: 449–58.

Hilgartner, S., and C. L. Bosk. 1988. "The Rise and Fall of Social-Problems—A Public Arenas Model." *American Journal of Sociology* 94: 53–78.

Ilcan, S. 2009. "Privatizing Responsibility: Public Sector Reform under Neoliberal Government." *The Canadian Review of Sociology and Anthropology* 46: 207–34.

Iles, A. 2007. "Identifying Environmental Health Risks in Consumer Products: Nongovernmental Organizations and Civic Epistemologies." *Public Understanding of Science* 16: 371–91.

Kasperson, R. E., O. Renn, P. Slovic, H. S. Brown, J. Emel, R. Goble, J. X. Kasperson, S. Ratick, and N. Pidgeon. 2005. "The Social Amplification of Risk: A Conceptual Framework." In *The Social Amplification of Risk*, edited by N. Pidgeon, R. E. Kasperson, and P. Slovic, 13–46. Sterling, VA: Earthscan.

Kitsuse, J. L., and M. Spector. 1973. "Toward a Sociology of Social Problems: Social Conditions, Value Judgments, and Social Problems." *Social Problems* 20: 407–19.

Kroll-Smith, J. S., P. Brown, and V. J. Gunter. 2000. *Illness and the Environment: A Reader in Contested Medicine*. New York: New York University Press.

Lupton, D. 1999. *Risk*. New York: Routledge.

MacKendrick, N. A. 2009. "Protecting Ourselves from Chemicals: A Study of Gender and Precautionary Consumption." Consuming Chemicals Paper Series. Toronto, ON: National Network on Environments and Women's Health.

McAndrew, B. 1998. "Cleansers Cause Fears over Cancer. Group Thinks It's Possible There's a Link." *Toronto Star* 24: A8.

McCormick, S., P. Brown, and S. Zavestoski. 2003. "The Personal Is Scientific, the Scientific Is Political: The Public Paradigm of the Environmental Breast Cancer Movement." *Sociological Forum* 18: 545–76.

McInnes, C 1988. "Chemical Pollution a Motherhood Issue." *The Globe and Mail* 23: D4.

Miller, M. M., and B. P. Reichert. 2000. "Interest Group Strategies and Journalistic Norms." In *Environmental Risks and the Media*, edited by S. Allan, B. Adam, and C. Carter, 45–54. New York: Routledge.

Mittelstaedt, M. 2006. "Want a Full-Time Job? Live Chemical-Free; Consumers Bombarded by Toxins Are Fighting Back. But There's Really Only One Way to Win—'Don't Eat, Don't Drink and Don't Breathe'." *The Globe and Mail* 1: A10.

Moneo, M. 2006. "Study Examining Safety of Seafood Eaten by Natives." *The Globe and Mail* 17: S1.

Morello-Frosch, R., J. G. Brody, P. Brown, R. G. Altman, R. A. Rudel, and C. Perez. 2009. "Toxic Ignorance and Right-to-Know in Biomonitoring Results Communication: A Survey of Scientists and Study Participants." *Environmental Health* 8: 6.

Morphet, S. 2005. "How Pure Can You Get? Worried about How Many Harmful Chemicals Are in the Food You Eat and the Air You Breathe? There Are Ways to Keep the Toxic Levels Down." *The Globe and Mail* 26: F4.

Nelkin, D. 1987. *Selling Science: How the Press Covers Science and Technology.* New York: W.H. Freeman.

O'Malley, P. 2004. *Risk, Uncertainty and Government.* London: GlassHouse.

O'Riordan, T., J. Cameron, and A. Jordan. 2001. *Reinterpreting the Precautionary Principle.* London: Cameron May.

Oliver, P., and H. Johnston. 1999. "What a Good Idea! Ideology and Frames in Social Movement Research." *Mobilization* 5: 37–54.

Perrow, C. 1984. *Normal Accidents: Living with High-Risk Technologies.* New York: Basic Books.

Press, C. 1989. "Inuit Told Food Supply Shows Safe PCB Levels." *The Vancouver Sun* 8: A.9.

Roberts, J. A., and N. Langston. 2008. "Toxic Bodies/Toxic Environments: An Interdisciplinary Forum." *Environmental History* 13: 629–35.

Robinson, E. E. 2002. "Community Frame Analysis in Love Canal: Understanding Messages in a Contaminated Community." *Sociological Spectrum* 22: 139–69.

Rose, N. 1999. *Powers of Freedom: Reframing Political Thought.* Cambridge: Cambridge University Press.

Scheufele, D. A. 1999. "Framing as a Theory of Media Effects." *Journal of Communication* 49: 103–22.

Scheufele, D. A., and D. Tewksbury. 2007. "Framing, Agenda Setting, and Priming: The Evolution of Three Media Effects Models." *Journal of Communication* 57: 9–20.

Scott, D. N. 2007. "Risk as a Technique of Governance in an Era of Biotechnological Innovation: Implications for Democratic Citizenship and Strategies of Resistance." In *Risk and Trust: Including or Excluding Citizens?*, edited by Law Society of Canada, 23–56. Black Point, NS: Fernwood Pub.

Seale, C. 2003. "Health and Media: An Overview." *Sociology of Health & Illness* 25: 513–31.

Sexton, K., L. L. Needham, and J. L. Pirkle. 2004. "Human Biomonitoring of Environmental Chemicals." *American Scientist* 92: 38–45.

Simpson, S. 2001. "Flushing Is Start of Toxic Journey: No Escaping Daily Load of Chemicals Entering Strait of Georgia, Scientist Discovers." *The Vancouver Sun*, October 25: B7.

Smith, C. 1996. "Of Estrogen Mimics and Breast Cancer." *Toronto Star*, August 3: E6.

Stallings, R. A. 1990. "Media Discourse and the Social Construction of Risk." *Social Problems* 37: 80–95.

Steinemann, A. 2004. "Human Exposure, Health Hazards, and Environmental Regulations." *Environmental Impact Assessment Review* 24: 695–710.

Szasz, A. 1994. *Ecopopulism: Toxic Waste and the Movement for Environmental Justice.* Minneapolis: University of Minnesota Press.

———. 2007. *Shopping Our Way to Safety: How We Changed from Protecting the Environment to Protecting Ourselves.* Minneapolis: University of Minnesota Press.

Taylor, D. E. 2000. "The Rise of the Environmental Justice Paradigm." *American Behavioral Scientist* 43: 508–80.

Tuchman, G. 1978. *Making News: A Study in the Construction of Reality.* London: Collier Macmillan.

Ubelacker, S. 2005. "Reheating Leftovers? Make Sure Plastic Containers Are Microwave-Safe." *The Calgary Herald*, December 29: C1.

Wigod, R. 1998. "Cosmetics Crusader Urges Users to Think Critically about Products." *The Vancouver Sun*, October 26: B11.

Young, M. L. 1990. "High Dioxin Found in Mothers' Breast Milk." *The Vancouver Sun* 29: B2.

Zavestoski, S., K. Agnello, F. Mignano, and F. Darroch. 2004. Issue Framing and Citizen Apathy toward Local Environmental Contamination. *Sociological Forum* 19: 255–83.

Legitimating the Environmental Injustices of War
12
Toxic Exposures and Media Silence in Iraq and Afghanistan

Eric Bonds

War causes major environmental harms, but because there tends to be a high level of secrecy surrounding the military, it can be difficult to obtain data to show exactly how environmental degradation results from military practices. In addition, in wartime, our media often fail to question governmental decisions. This reading shows how the news media contribute to the legitimation of military activities and thus help to maintain a system that fails to question war and its attendant injustices. Bonds uses the case of toxic exposures resulting from open burn pits in Iraq and Afghanistan to show that the U.S. media have remained largely silent about the impacts on civilians and thus the larger humanitarian costs of war.

Introduction

During the Iraq and Afghanistan wars, the US Department of Defense (DoD) burned the majority of its solid waste in open-air pits or trenches. For many years, in fact, this was the only form of waste disposal at most bases (GAO 2010, IoM 2011). It is well known that the uncontrolled burning of plastics, Styrofoam, electronics, unexploded weapons, and other manufactured and highly processed materials releases harmful toxins and particulate matter into the air. US journalists have paid increasing amounts of attention to the suspected impacts of this pollution. Their attention, however, has focused almost exclusively on the potential harm to US soldiers and veterans, while it ignores potential civilian impacts. For instance, news stories such as the following frequently give accounts of veterans who link their sicknesses to pollution from the burning refuse.

> When Wendy McBreairty got back from Iraq in 2004, she desperately tried to understand what was causing her medical symptoms, including shortness of breath, muscle fatigue, muscle spasms, fatigue and dry eyes. The 32-year-old Air National Guard staff sergeant found that others had similar, often equally puzzling, problems. Among the 40 people in her shop alone, five have neurological

or respiratory issues. One thing they had in common was that they all lived . . . a mile southeast of an open-air burn pit at Joint Base Balad, Iraq. (Kennedy 2010)

Within a few months of arriving at Bagram Air Field, Afghanistan, in April 2010, [Capt. Rebecca] Selby, an Air Force logistician, could no longer explain away a growing list of ailments, including digestive problems, rashes and pain throughout her body that made it hard to walk. The only culprit she or her doctors could suggest was the acrid smoke that filled her lungs most days as she drove past Bagram's burn pit, where tons of trash was burned daily, releasing a toxic bouquet of chemicals. (Carrol 2013)

While US journalists covering the burn-pit controversy tell such stories, it is never mentioned that if military base pollution made soldiers sick, it must have made civilians living nearby sick as well. And when journalists describe the pollution itself, how it billowed over military bases and covered living quarters with ash and soot, such accounts never mention that this pollution would not have stopped at the cement barricades and concertina wire at base boundaries, but must have also settled over civilians' homes and the surrounding landscapes.

In what follows, I make a contribution to our collective understanding of environmental inequality by providing descriptions of satellite imagery to show that US bases (including those that burned waste in the open air for years) were not located in the uninhabited deserts or wastelands that many Americans might imagine Iraq and Afghanistan to be. On the contrary, I show that many bases were located next to farmsteads, townships, cities, cropland, orchards, and rivers. By so doing, I establish that US burn pits are an important, albeit neglected, environmental justice issue. I also provide a content analysis of major US newspaper coverage of the burn-pit controversy to show how it has almost entirely ignored potential civilian impacts. These findings raise a vital question: why is it that US news accounts attend to the potential toxic effects caused by military pollution on US service members, while completely ignoring the fact that civilians living next to bases would be impacted too? I argue that this has much to do with the legitimation of military violence in America, which involves both US governmental attempts to shape media coverage of war, along with a complicit news media, that together produce a style of reporting that sociologists Altheide and Grimes (2005) have called "war programming." . . .

Legitimating the Environmental Injustices of War

Preparing for and fighting wars is inherently damaging to the environment. Making weapons is highly resource intensive (NRC 2008). Training for wars produces a great deal of pollution, and can severely damage whole landscapes (Seager 2003, Schmidt 2004), while deploying troops overseas and maintaining navies across the world's oceans requires a tremendous amount of energy (Yergin 1991). War itself often means the intentional destruction of environments, and unexploded ordnance (bombs, landmines, and cluster munitions) can continue to kill long after wars are officially declared over. Because the US government maintains the world's largest military, with bases and a navy fleet deployed across the globe, and because it has regularly fought wars since World War II, it has an especially large environmental footprint (Gould 2007, Clark and Jorgenson 2012).

The treadmill of destruction model calls attention to the increasingly severe environmental impacts of war and militarism (Hooks and Smith 2004, Clark and Jorgenson 2012). According to this model, governments are continuously building up their capacities for violence as they vie with one another for territory and power. As they do so, they develop and utilize increasingly complex technologies that produce greater environmental footprints. Studies on the treadmill of destruction model also stress that the environmental impacts of militarism are not borne equally. On the contrary, militarism is an important domestic cause of environmental injustice, as the DoD places a disproportionate toxic burden on poor communities and communities of color for weapons development and testing in the United States (Hooks and Smith 2004, Sbicca 2012). The US government also creates global environmental inequalities as a result of the wars it fights, both through the pollution it creates as part of its daily operations and through its embrace of high-tech weapons that can cause long-lasting and widespread environmental impacts, such as the use of Agent Orange in the Vietnam War (Frey 2013), other herbicides used to destroy drug crops in Colombia (Smith et al. 2014), the use of cluster munitions in Laos, and the use of depleted uranium in Iraq (Nixon 2013).

The treadmill of destruction model calls attention to the relationship between militarism and environmental degradation, while also pointing to the US military as an important global driver of environmental injustice. The model, however, has not yet been extended to explain how these outcomes are legitimated. This is an issue of major concern because the often-devastating environmental impacts of militarism are at odds with widely upheld human rights norms and ideals of the "just war," which requires that militaries restrain themselves to prevent unnecessary widespread or severe civilian harm. Without legitimation work to reconcile these impacts with widely shared norms, then the US government might see reduced domestic support for its war-fighting, and may have to deal with increased antiwar sentiment (Bonds 2013).

The environmental injustices of war are legitimated in several ways, but two key methods have to do with the denial of environmental risks and, perhaps more importantly, the ability to limit public awareness of the civilian impacts cause by conflict. First, when the US government is accused of exposing soldiers and civilians to toxins, its response—like the response of most actors vested in the use of potentially harmful chemicals—is denial. According to Nancy Langston (2010, p. viii) "since World War II, synthetic chemicals in plastics, pharmaceuticals, and pesticides have permeated bodies and ecosystems throughout the United States, often with profound health and ecological effects." This regulatory failure is often due to inadequate scientific models that cannot initially track the ways that exposures to even small amounts of some chemicals can produce negative health impacts, or the ways that potential toxins can interact with one another and the surrounding environment to produce unanticipated illnesses (Nash 2006, Langston 2010). And toxins may escape initial regulatory control because their effects are slow to develop and may not be known for years, which is especially the case for chemicals that impact fetal development and therefore cause health impacts across generations (Langston 2010). But companies and government agencies that have vested interests in the use of potential toxins also play a role here, first by insisting that the absence of scientific proof linking chemicals to ill-health effects means that they are safe, and then by working to raise doubts about the scientific evidence that does emerge (Nash 2006, Langston 2010). The US DoD, which has a long history of producing toxic contaminants both internationally and within the nation's border, has regularly used such tactics in an effort to avoid liability and cleanup expenses (Bonds 2011).

But these methods are necessary only when toxic contamination is brought to light and contested. Another important way that the environmental impacts of war are legitimated in America is that they are so difficult for Americans to see. This is bound up with the more general legitimation of militarism and conflict through a complicit news media, something David Altheide and Jennifer Grimes (2005, p. 618) call "war programming" to refer to "the organization and structure of the discourse of recent reportage about wars, and not mere content. It encompasses content as well as thematic emphases and dominant frames." War programming is part of the ongoing legitimation of America's regular war-fighting. It begins with the demonization of leaders in certain foreign countries that have been targeted for military action, along with coverage of military experts anticipating and planning for an impending war (see also Solomon 2005). War programming continues with coverage of the war itself, largely from a US perspective and reliant upon US government sources. . . .

A fundamental aspect of war programming is that the humanitarian costs of war are largely hidden from view. Numerous analyses have shown that the US news media regularly fails to report on the civilian impacts of US wars (Herman and Chomsky 1988, Tumber and Palmer 2004, Ravi 2005, Lindner 2009). This outcome is not, of course, the result of direct censorship, but is due to a more subtle coincidence of interests between large news media companies and the US government that produces a similar effect (Herman and Chomsky 1988). For one, these media companies need to keep reporting costs down and are wary about the risks of sending reporters out into war zones, so they become highly reliant on official sources of information. The US government, however, focuses attention away from the death and suffering experienced by civilians in its wars—General Tommy Franks famously summed up this position in the Iraq War by stating, "We don't do body counts" (quoted in Tumber and Palmer 2004).

While the US government withholds some kinds of uncomfortable or troublesome information, it also actively works to encourage reporters to cover wars from a military perspective. It provides opportunities for journalists to "embed" with US military forces. This might be attractive to reporters, as it is much safer than going into a war zone without the protection of US troops, but it also means that they will be more likely to take the perspective of soldiers, and not that of civilians, when covering the war (Lindner 2009). The US government also provides video footage, satellite imagery, and high-tech graphics of weapons to news organizations, which gives them ample opportunity to devote extensive coverage to the technical marvels of the US military, but again does not present much of an opportunity to consider civilian impacts (Thussu 2004). Finally, news companies, at least at the start of US wars, may avoid reporting on civilian deaths and suffering due to fear of criticism for not being patriotic or supportive of US troops and the war effort (Herman and Chomsky 1988, Hallin 2013), though Altheide and Grimes (2005) note that this dynamic might change when wars drag on and become increasingly unpopular. Taken together, these factors help produce a kind of war reporting in the US news that largely ignores civilian death and suffering. As war programming, it is a kind of framework for perception and discourse that is taken for granted and a matter of convention. The following case study on the news coverage of US military burn pits in Iraq and Afghanistan will show that, like the impacts of war more generally, the environmental consequences of conflict are legitimated when they are left out of news reports, and so become impossible for Americans to see.

The Wastes of War in Iraq and Afghanistan

While the DoD used open-pit burning as its primary means of waste disposal in Iraq and Afghanistan, it did not keep records of how much waste it burned. It has, however, estimated that every US service member in each theater produced on average 8–10 pounds of waste per day (IoM 2011). This means that, at the peak of the two wars' respective "surges," roughly 900,000 pounds of waste was produced a day by the US military in Afghanistan in 2011 and approximately 1,700,000 pounds of waste was produced per day in Iraq in 2008, the majority of which was burned without any pollution controls.[1] To grasp how much waste was put into burn pits, it is helpful to know that an estimated 90,000 empty plastic water bottles were burned each day at the Balad air base in Iraq alone (Kennedy 2008a). Overall, the waste burned at US bases in Iraq and Afghanistan included paper, plastic, Styrofoam, rubber, petroleum products such as oil and lubricants, chemicals such as solvents and paint, metals, unexploded weapons, electronics, batteries, and medical waste (Senate Hearing 2009, GAO 2010, IoM 2011).

. . . While the US government has known for several decades that uncontrolled burning of waste produces negative health impacts, this has been the typical method of waste disposal by the US DoD during times of war (IoM 2011). While it has always posed risks, a panel from the US National Academy Institute of Medicine (2011, p. 3) warns that, "technologic advances in recent military conflicts mean that new items are being burned—plastic bottles and electronics, for example—and the burning of such items presents new health risks." Many large US bases in Iraq and Afghanistan were not, after all, rugged and sparsely furnished outposts, but highly developed areas with movie theatres, fast-food restaurants, swimming pools, and "big box"-like PX retail stores (Senate Hearing 2009). They had, in other words, many of the furnishings of contemporary American life without contemporary forms of waste disposal. This development is consistent with the treadmill of destruction model, which anticipates that militaries will have increasingly large environmental footprints both as more resources are put into militarism and as militaries become more high tech (Hooks and Smith 2004).

Thousands of US soldiers and other coalition troops were exposed to burn-pit emissions, which have been linked with toxins known to cause negative health outcomes. These chemicals include dioxins, "acetaldehyde, acrolein, arsenic, benzene . . . ethylbenzene, formaldehyde, hydrogen cyanide, hydrogen chloride, hydrogen fluoride, various metals, nitrogen dioxide, phosgene, sulfuric acid, sulfur dioxide, toluene, trichloroethane, trichloropropane, and xylene" (Curtis 2006). And just as important as the actual chemical constituents, burning trash at US bases without controls produced pollution in the form of very small particulate matter, which can get lodged in human lungs and is associated with a number of chronic illnesses (IoM 2011).

The acrid smell of burn-pit emissions, sometimes billowing plumes reaching upwards high into the air or hanging as a low haze, was a much remarked upon part of daily life at some bases. Much less discussed, however, is the fact that civilians in Iraq and Afghanistan were exposed to this pollution too, making burn pits an under-recognized environmental justice issue. Because Americans oftentimes imagine war zones as uninhabited wildernesses or as unpopulated "sacrifice zones" (Seager 1993, Kuletz 1998), it is important to take a closer look.

. . . In 2009, for instance, burn pits were being operated in 30 out of 45 medium-sized bases and in 19 out of 45 large bases in Iraq (GAO 2010).

. . . Google satellite imagery gives a sense of the features, such as the presence of farmland and rivers. . . . A discussion of these specific examples underscores the more general point that bases were often located next to farmland, agricultural communities, rivers, and urban neighborhoods. As such, civilians were widely exposed to burn-pit pollution from US military bases during the Iraq (2003–2011) and Afghanistan (2001–present) wars.

Balad Air Base, Iraq

The burn-pit operation at Balad air base was the largest in Iraq and Afghanistan, operating from 2003 to late 2009. At its peak, it burned up to 200 tons of waste per day (GAO 2010, IoM 2011). The continuously burning pit produced a column of smoke that was "such an invariable part of the horizon that software engineers writing a program to help fighter pilots navigate their way onto the base made it a central part of the digitally simulated skyline" (LaPlante 2008a). The former US base is surrounded on all sides by cropland. Numerous farmsteads are within one mile of the base, and at least six small communities lie within two miles. The Tigris River, an important source of drinking and irrigation water in Iraq, runs parallel to the base within a distance of one mile.

Forward Operating Base Marez, Iraq

This former US base complex at Mosul Airport sits adjacent to cropland on its east side. The Tigris River runs parallel to the base, within two miles of its boundary. The city of Mosul, with a population of more than one million, begins on the immediate north side of the former base. The base was established in

2004 and continued to use open-pit burning as late as 2009 as its main form of solid-waste disposal, though it also had an unlined landfill (GAO 2010). US government inspectors found that the base was not in compliance with DoD directives on waste disposal, and continued to burn plastics, batteries, aerosol cans, and electronics (GAO 2010).

Forward Operating Base Warhorse, Iraq

Like Marez, this US military base was without a controlled incinerator, and continued to burn plastics and other potentially toxic materials in the open air well into 2009 (GAO 2010). The base was located roughly three miles from the city of Baqubah, and was immediately surrounded by farmland. Two small townships and fruit orchards are located within a mile of the former facility. The Diyala River runs about a mile from the base.

Shindad Air Base, Afghanistan

This base, which was first established in 2004, housed an estimated 4000 US service members and Afghan troops in 2013. It continued to burn its solid waste—including plastics and batteries—as late as 2013, according to US government inspectors (SIGAR 2014). Agricultural fields and small townships are located directly south of the Shindad base.

Forward Operating Base Salerno, Afghanistan

This base was first developed in 2002 and continued to burn its waste in the open air until at least 2013 (SIGAR 2013a). While the DoD paid to have two trash incinerators built at this base in 2010, US government inspectors found that they had not been used. Instead, the base was burning 14 tons of trash per day in the open air (SIGAR 2013b). The Salerno base is located within a mile of the town of Khost and is surrounded by agricultural lands and small townships on all sides.

These specific examples show that US open-pit burning created toxic air pollution that did not stay on US bases, but also spread into Iraqi and Afghan homes, cropland, and waterways. The examples also indicate that civilian exposures to this pollution from US bases across Afghanistan and Iraq—as a whole—were widespread, even if we do not know the specific years and the specific locations where this pollution occurred. The DoD, after all, reported that in 2010 it was operating 184 burn pits in Afghanistan and 52 burn pits in Iraq (GAO 2010). This initial analysis demonstrates that open-pit burning is an important, even if unrecognized environmental justice issue. The next section considers how this inequality has been legitimated in the US news media.

Denial and Disappearing Civilians from View

Up until 2008, there was little public regard for the DoD's preferred method of waste disposal in Iraq and Afghanistan, and the issue had been given practically no major press coverage. This changed in 2008 when a reporter from the Army Times named Kelly Kennedy ran the first of several stories on the issue that year. Her first article was based on a newly uncovered memo from Air Force Lt. Col. Darrin Curtis (2006)—a PhD in environmental engineering—who wrote to his supervisors stating that, "in my professional opinion, there is an acute health hazard for individuals. There is also the possibility for chronic health hazards associated with the smoke."

The 2008 articles by Kennedy galvanized US veteran activism. Sick veterans who were largely unable to explain their illnesses recognized that exposure to burn-pit emissions was a possible cause of their suffering, and the Disabled Veterans of America began raising awareness about the potential health problems that toxic emissions from burn pits could cause (Kennedy 2008b, 2008c). Members of Congress soon began raising inquiries into the DoD's waste-disposal policies in Iraq and Afghanistan (Kennedy 2008d, Senate Hearing 2009), and it became the subject of media attention, to which the DoD had to respond.

At first, the DoD steadfastly denied that the emissions from its open-pit burn sites could have made soldiers and veterans sick. Military officials told reporters, for instance, that there are "no short- or long-term health risks, and no elevated cancer risks are likely among personnel deployed to Balad [air base]" (quoted in LaPlante 2008a). Officials also stated that, despite some initial monitoring, "we have not identified anything, where there are troops, where it would have been hazardous to their health" (quoted in LaPlante 2008b). The DoD, of course, had a vested interest in making such claims in order to avoid healthcare expenses and compensation payments to sick veterans that could easily cost billions of dollars. The DoD also had an interest in denying these risks in order to protect its own legitimacy, which would be put at risk if it appeared as though it had unnecessarily damaged its soldiers' health.

Faced with the continued advocacy of veterans groups and ongoing press coverage of sick soldiers, the DoD soon undertook a more nuanced approach to denial. The same officials that earlier denied potential health impacts told the press, "we feel at this point in time that it's quite plausible—in fact likely—that there are a small number of people that have been affected with longer-term health problems" (quoted in LaPlante 2009). While the official DoD position now acknowledges that exposure to burn pits could combine with other risk factors to make some troops sick (Levine 2009), it also maintains that there is insufficient evidence to prove that any single illness was caused by burn-pit emissions exposure, rather than having other genetic, individual behavioral, or other environmental causes (USAPHC 2014). Like other controversies regarding potential or actual toxins, the DoD is denying responsibility for illnesses in the absence of medical certainty, which often takes years to produce due to limitations in scientific understanding, because contaminants can create different kinds of health impacts in different environmental conditions, and because these impacts themselves are not always immediate but can take years—or even a generation—to develop fully (Nash 2006, Langston 2010). But due to public and Congressional pressure, the DoD and the Department of Veterans' Affairs is continuing to study this issue. Congress, in fact, mandated that it do so in a 2013 law that establishes a "burn-pit registry" in order to track the health outcomes of service members who believe they were exposed to burn-pit emissions (Kennedy 2014). There has been no attempt from the US government, however, to understand how the health of Iraqi and Afghan civilians has been impacted by the military's open-air waste incineration, though they certainly were subjected to the same pollution and were just as vulnerable, if not more so. Civilians, in fact, have been almost entirely excluded from the news discourse regarding the burn-pit controversy.

In order to understand how civilian impacts of burn pits were—or were not—discussed in the US news media, I conducted a survey of major domestic newspapers from 2007 through 2014.[2]. Newspaper stories continue to be an important way that Americans learn about the world each day, and while their importance may be waning compared with TV news and other online news formats, newspaper articles are easily searchable and are typically representative of other forms of mainstream news. The search produced 49 distinct stories. While five of these stories made passing reference to civilian impacts, and one story mentioned potential impacts to civilians on par with impacts to soldiers, the vast majority of news stories made no mention that Iraqi and Afghan civilians might also have been harmed by the US military's burning of waste . . .

The omission of Iraqi and Afghan civilians from the burn-pit controversy in the US news creates quite a puzzle. After all, they would have been just as much impacted by exposure to toxins in burn-pit emissions as the US soldiers who were featured in news accounts. Toxic pollution, after all, does not harm all people equally (Langston 2010). While babies and young children, pregnant women, and sick individuals are the most vulnerable to toxins, healthy adult men are—as a whole—the least susceptible. And, of course, US soldiers are disproportionately men who are screened before deployment to ensure that they are physically fit, whereas the Iraqi and Afghan communities living near bases included a multitude of more vulnerable persons. Additionally, while deployments for US soldiers could be as long as 15 months, and some soldiers were deployed multiple times, there was no reprieve for Iraqis and Afghans living near the bases, who also were potentially further exposed to toxins from burn-pit pollution through contaminated water used both for drinking and for irrigating food crops. To sum things up, if US soldiers were endangered by open-pit burning at US bases, then so too were Iraqi and Afghan civilians. Any mention of these potential impacts, however, is almost entirely absent from mainstream US press coverage. These findings beg the question, why?

The answer likely has two parts. First, in this era of limited funding for investigative work, reporters are often relegated to covering the "actions" of important players relevant to topics thought to be of interest to readers and media audiences (McChesney 2003). Consequently, the "facts" regarding an issue tend to be of secondary importance compared with what important players—which may be politicians, government agencies, celebrities, or advocacy groups—think, say, or do (McChesney 2003). This means that such actors have tremendous leverage—whether they intend it or not—to frame how issues are perceived (Best 2013). In this particular case, those who were framing the issue of burn-pit pollution were not doing so in a way that included civilians. This is borne out in my content analysis of newspaper coverage regarding the controversy, in which I assessed and categorized the "action" or story that each article covered . . .

Taken as a whole, veterans' actions were the most common stories told in newspaper accounts of burn pits in Iraq and Afghanistan. News stories ran about individuals or groups of veterans demanding adequate healthcare and compensation for sicknesses they linked with burn-pit exposure. A number of stories were written to update readers about developments regarding a class-action lawsuit filed by veterans seeking compensation from KBR, a private contractor that operated many burn pits on behalf of the US military at bases in Iraq and Afghanistan. . . .

Another common story type focused on US government agency actions in relation to burn pits, for instance an agency's decision to fund a new study, the unveiling of new waste-disposal policies, or internal government investigations, namely those undertaken by the Government Accountability Office (GAO) or the Special Inspector General for Afghanistan Reconstruction (SIGAR). . . .

. . . News reports on health studies and on the scientific controversy over the extent of harm caused by burn pits also failed to register that civilians in these wars might have been exposed too. Finally, when veterans' advocates in the US Congress took up the issue, they predictably did so in a way that excluded Iraqi and Afghan civilians. For instance, the Senate Democratic Policy Committee was the first group to hold a hearing on the issue in 2009, entitling it "Are Burn Pits in Iraq and Afghanistan Making Our Veterans Sick?" (Senate Hearing 2009). Framing the event this way precluded an acknowledgment that these are countries populated by civilians who would be similarly endangered. . . .

Taken together, the results of my content analysis show that each of the most common story types cover the "actions" of players who are very likely to frame burn-pit pollution in terms of impacts on US soldiers and to exclude impacts on Iraqi and Afghan civilians. While this goes some way in helping to explain why potential civilian impacts were mostly excluded and hidden from view in US news stories, it does not provide a complete account. After all, even in an era of media cutbacks and declines in investigative reporting, there is nothing stopping reporters from at least mentioning that if soldiers are being made sick from pollution at US bases, then civilians would likely be impacted too. For this reason, part of the answer must also be that reporters were simply following a more general convention in US news coverage of the nation's wars, which typically fails to consider civilian impacts across the board.

Instead, mainstream news media sources mostly give a US-focused account of fighting (Altheide and Grimes 2005, Solomon 2005), and while this style of coverage sometimes draws attention to the costs of war, the deaths and injuries of soldiers are given far more weight than civilian suffering and casualties (Solomon 2005). Needless to say, such coverage has a legitimating effect in that it prevents many Americans from developing a full appreciation of the destruction and anguish caused by war. The analysis at hand indicates that the environmental injustices caused by conflict and overseas militarism are similarly legitimated in the US through the coincidence of interests between the US military and major news companies that together produce war programming.

Conclusion: A New Generation's Agent Orange?

Politicians and news reporters have called open-pit burning "this generation's Agent Orange."[3] Both the use of Agent Orange in the Vietnam War and open-air trash incineration in Iraq and Afghanistan produced large amounts of pollution. US soldiers coming home from both of these conflicts linked their otherwise unexplained illnesses with wartime toxic exposures. In the cases of both Agent Orange and open-pit burning, the US government at first denied that toxic exposures caused sicknesses, and refused to compensate soldiers for their losses. Eventually, the US government changed course for Vietnam Veterans, even if only after a long political fight that left many sick veterans without compensation or adequate healthcare (Wilcox 2011a). Sick veterans and their advocates hope that the US government, after creating a congressionally mandated burn-pit registry, will similarly make the decision to compensate US veterans who have fallen ill from exposure to emissions from open-air waste incineration. But it is important to keep in mind that most of the people harmed by Agent Orange were not US soldiers—they were Vietnamese civilians who were exposed to defoliants but were never recompensed for the loss of their livelihoods or health (Wilcox 2011b, Frey 2013).

Journalists' and politicians' comparisons between Agent Orange contamination and burn-pit emissions is very telling. For one, it bolsters Altheide and Grime's (2005) model on war programming as an ongoing legitimating process for US militarism and conflict. In the model, the mainstream news media largely paves the way for new wars by demonizing foreign leaders, portraying looming wars as inevitable, and failing to fully consider the possible costs of—or alternatives to—conflict. In the early days of war, the press mostly reports on the war uncritically and from a US military perspective, largely ignoring civilian impacts. At the war's conclusion, or after it has dragged on and become increasingly unpopular, members of the media may engage in some reflection and limited self-criticism, which will be superficially incorporated into the next round of war coverage.

These dynamics are illustrated in the analogies journalists make between burn pits and Agent Orange, as they self-consciously state that they hope that the US government and public will not ignore sick Iraq and Afghanistan War veterans, as happened after the Vietnam War. With this cautionary lesson in mind, and a desire to avoid the journalistic failings of the past, they are willing to tell the tragic stories of sick soldiers who believe their illnesses were caused by exposure to burn-pit emissions. At the same time, as Altheide and Grimes (2005) anticipate, even as war programming was altered in response to this critique of Vietnam-era reporting, it also remained largely the same in crucial ways. Most importantly here, media accounts continued to ignore or exclude civilians from coverage, creating the impression that only US soldiers were impacted by toxins from the US military. Taken together, these reports seem to portray a war zone in which civilians do not exist.

But of course they do exist, and they were also exposed to such toxins . . .

Notes

Bonds, Eric. 2016. "Legitimating the Environmental Injustices of War: Toxic Exposures and Media Silence in Iraq and Afghanistan." *Environmental Politics* 23 (3): 395–413.

[1] These estimates were made using the lower eight pounds of waste per day average.
[2] This search was conducted in LexisNexis Academic, using such search terms as: "Burn pit and (Iraq or Afghanistan)" and "waste and burn and (Iraq or Afghanistan)."
[3] For examples, see Amin and MacVicar (2014), LaPlante (2010), Senate Hearing (2009), and Simonich (2011).

References

Altheide, D. L., and J. N. Grimes. 2005. "War Programming: The Propaganda Project and the Iraq War." *The Sociological Quarterly* 46 (4): 617–43. doi:10.1111/j.1533–8525.2005.00029.x.

Amin, S., and S. MacVicar. 2014. "Former VA Official: Burn Pits Could Be the New Agent Orange." *Aljazeera News*, December 3. http://america.aljazeera.com/watch/shows/america-tonight/articles/2014/12/3/burn-pits.html (accessed December 30, 2014).

Best, J. 2013. *Social Problems*. New York: W.W. Norton & Co.

Bonds, E. 2011. "The Knowledge Shaping Process: Elite Mobilization and Environmental Policy." *Critical Sociology* 37 (4): 429–46.

——— 2013. "Hegemony and Humanitarian Norms: The U.S. Legitimation of Toxic Violence." *Journal of World-Systems Research* 19 (1): 82–106.

Carrol, C. 2013. "Legislation Would Fund Research into Effects of Burn Pits on Troops." *Stars and Stripes*, June 27. LexisNexis Academic (accessed December 16, 2014).

Clark, B., and A.K. Jorgenson. 2012. "The Treadmill of Destruction and the Environmental Impacts of Militaries." *Sociology Compass* 6 (7): 557–69. doi:10.1111/j.1751–9020.2012.00474.x.

Curtis, D. L. 2006. "Burn Pit Health Hazards." *Department of the Air Force Memorandum*, December 20. https://www.dpc.senate.gov/hearings/hearing50/memo_burn_pits_122006.pdf (accessed December 16, 2014).

Frey, R.S. 2013. "Agent Orange and America at War in Vietnam and Southeastern Asia." *Human Ecology Review* 20 (1): 1–10.

GAO. 2010. "DoD Should Increase Its Adherence to Guidance on Open Pit Burning and Solid Waste Management." *Government Accountability Office Report* #11–63. http://gao.gov/assets/320/311365.pdf (accessed January 6, 2015).

Gould, K.A. 2007. "The Ecological Costs of Militarization." *Peace Review* 19 (3): 331–34. doi:10.1080/10402650701524873.

Hallin, D.C. 2013. "Between Reporting and Propaganda: Power, Culture and War Reporting." In *Selling War: The Role of the Mass Media in Hostile Conflicts from World War I to the "War on Terror,"* edited by J. Seethaler, M. Karmasin, G. Melischek, and R. Wöhlert, 93–106 Bristol: Intellect Books.

Herman, E.S., and N. Chomsky. 1988. *Manufacturing Consent: The Political Economy of Mass Media*. New York: Pantheon Books.

Hooks, G., and C.L. Smith. 2004. "The Treadmill of Destruction: National Sacrifice Areas and Native Americans." *American Sociological Review* 69 (4): 558–75. doi:10.1177/000312240406900405.

IoM (Institute of Medicine). 2011. *Long-Term Health Consequences of Exposure to Burn Pits in Iraq and Afghanistan* [online]. Washington, DC: The National Academies Press. http://www.nap.edu/catalog.php?record_id=13209.

Kennedy, K. 2008a. "Burn Pit at Balad Raises Health Concerns: Troops Say Chemicals and Medical Waste Burned at Base Are Making Them Sick, but Officials Deny Risk." *Army Times,* October 27. http://archive.armytimes.com/news/2008/10/military_burnpit_102708w/ (accessed December 17, 2014).

———. 2008b. "Burn Pit Fallout." *Army Times,* November 13. http://archive.army times.com/article/20081113/NEWS/811130332/Burn-pit-fallout (accessed December 17, 2014).

———. 2008c. "Troops Complain about Burn Pits." *Army Times,* November 26. http://archive.airforcetimes.com/article/20081126/NEWS/811260330/Troops-air-com plaints-about-burn-pits (accessed December 17, 2014).

———. 2008d. "Senator Wants Answers on Dangers of Burn Pits." *Army Times,* November 7. http://www.armytimes.com/article/20081107/NEWS/811070317/Sena tor-wants-answers-dangers-burn-pits (accessed December 16, 2014).

———. 2010. "Balad Burn Pit Harmed Troops Living 1 Mile Away." *Navy Times,* January 18. http://archive.navytimes.com/article/20100118/NEWS/1180316/Balad-burn-pit-harmed-troops-living-1-mile-away (accessed December 17, 2014).

———. 2014. "Vets May Sign PP for Registry after Dust, Smoke Exposure." *USA Today,* June 23. http://www.usatoday.com/story/nation/2014/06/23/burn-pit-registry-opens/11273885/(accessed January 6, 2015).

Kuletz, V. 1998. *The Tainted Desert: Environmental Ruin in the American West.* New York: Routledge.

Langston, N. 2010. *Toxic Bodies: Hormone Disruptors and the Legacy of DES.* New Haven, CT: Yale University Press.

LaPlante, M. 2008a. "Hill AFB Officer Worries That Iraqi Burn Pit Threatens Troops' Health." *Salt Lake Tribune,* October 29. LexisNexis Academic (accessed December 17, 2014).

———. 2008b. "Military Mum on Dirty Air in Iraq." *Salt Lake Tribune,* November 23. LexisNexis Academic (accessed December 17, 2014).

———. 2009. "Military: Burn Pits Caused Illnesses." *Salt Lake Tribune,* December 16. LexisNexis Academic (accessed December 17, 2014).

———. 2010. "Vets: Burn Pits Are Killing Us." *The Salt Lake Tribune,* January 15. Lexis-Nexis Academic (accessed December 17, 2014).

Levine, A. 2009. "Military: Burn Pits Could Cause Long-Term Damage to Troops." *CNN,* December 18. http://www.cnn.com/2009/POLITICS/12/18/military. burn.pits/ (accessed January 6, 2014).

Lindner, A.M. 2009. "Among the Troops: Seeing the Iraq War through Three Journalistic Vantage Points." *Social Problems* 56 (1): 21–48. doi:10.1525/sp.2009.56.1.21.

McChesney, R. 2003. "The Problem of Journalism: A Political Economic Contribution to an Explanation of the Crisis in Contemporary US Journalism." *Journalism Studies* 4 (3): 299–329. doi:10.1080/14616700306492.

Nash, L. 2006. *Inescapable Ecologies: A History of Environment, Disease, and Knowledge.* Berkley: University of California Press.

Nixon, R. 2013. *Slow Violence and the Environmentalism of the Poor.* Cambridge, MA: Harvard University Press.

NRC. 2008. *Minerals, Critical Minerals, and the U.S. Economy.* Washington, DC: National Academies Press.

Ravi, N. 2005. "Looking beyond Flawed Journalism: How National Interests, Patriotism, and Cultural Values Shaped the Coverage of the Iraq War." *The International Journal of Press/Politics* 10 (1): 45–62. doi:10.1177/1081180X05275765.

Sbicca, J. 2012. "Elite and Marginalised Actors in Toxic Treadmills: Challenging the Power of the State, Military, and Economy." *Environmental Politics* 21 (3): 467–85. doi:10.1080/09644016.2012.671575.

Schmidt, C.W. 2004. "The Price of Preparing for War." *Environmental Health Perspectives* 112: a1004–a1005. doi:10.1289/ehp.112-a1004.

Seager, J. 1993. *Earth Follies: Coming to Feminist Terms with the Global Environmental Crisis.* New York: Routledge.

Senate Hearing. 2009. "Are Burn Pits in Iraq and Afghanistan Making Our Soldiers Sick." *Senate Democratic Policy Hearing,* November 6. http://www.dpc.senate.gov/hearings/hearing50/transcript.pdf (accessed January 6, 2015).

SIGAR. 2013a. "Forward Operating Base Salerno: Inadequate Planning Resulted in $5 Million Spent for Unused Incinerators and the Continued Use of Potentially Hazardous Open-Air Burn Pit Operations." *Special Investigator for Afghanistan Report* #13–8. http://www.sigar.mil/pdf/inspections/2013-04-25-inspection-13-08.pdf.

———. 2013b. "Memorandum on Camp Leatherneck Waste Disposal, July 11." *Special Investigator for Afghanistan Alert,* 13–4. http://www.sigar.mil/pdf/alerts/SIGAR%20Alert%2013–4_Camp%20Leatherneck%20Solid%20Waste%20Disposal.pdf.

———. 2014. "Shindad Airbase: Use of Open-Air Burn Pit Violated DoD Regulations." *Special Investigator for Afghanistan Report* #14–81. http://www. sigar.mil/pdf/inspections/SIGAR-14–81-IP.pdf.

Simonich, M. 2011. "Vet Speaks Out against Open-Air Burn Pits." *Alamogordo Daily News* (New Mexico), November 10. LexisNexis Academic (accessed December 17, 2014).

Smith, C.L., G. Hooks, and M. Lengefeld. 2014. "The War on Drugs in Colombia: The Environment, Treadmill of Destruction and Risk-Transfer Militarism." *Journal of World-Systems Research* 20 (2): 185–206.

Smith, J. 2014. "'Environmental Poisoning' of Iraq Is Claimed." *The New York Times,* March 27, 8.

Solomon, N. 2005. *War Made Easy: How Presidents and Pundits Keep Spinning Us to Death.* Hoboken, NJ: Wiley.

Thussu, D.K. 2004. "Live TV and Bloodless Death: War, Infotainment, and 24/7 News." In *The media and war,* edited by D.K. Thussu and D. Freeman, 117–32. London: Sage.

Tumber, H., and J. Palmer. 2004. *Media at War: The Iraq Crisis.* London: Sage.

USAPHC. 2014. "Joint Base Balad Burn Pit." *US Army Public Health Command Fact Sheet* 47–002–0214. http://phc.amedd.army.mil/PHC%20Resource%20Library/Balad_Burn_Pit_JBB_FS_47–002–0214.pdf.

Wilcox, F.A. 2011a. *Waiting for an Army to Die: The tragedy of Agent Orange.* 2nd ed. New York: Seven Stories Press.

———. 2011b. *Scorched Earth: Legacies of Chemical Warfare in Vietnam.* New York: Seven Stories Press.

Yergin, D. 1991. *The Prize: The Epic Quest for Oil, Money, and Power.* New York: Simon & Schuster.

PART V DISASTER

The BP Disaster as an *Exxon Valdez* Rerun

13

Liesel Ashley Ritchie, Duane A. Gill, and J. Steven Picou

The three readings in this section are sociological explorations into the causes and consequences of disasters. While the first two readings are about oil spills, it is important to note that environmental sociologists investigate the social causes and consequences of all types of disasters—from major storms (like Hurricane Katrina, which is discussed in the third reading of this section) to nuclear meltdowns. Also, while some sociological studies focus on the social organization of how a disaster occurs, others examine the consequences for human societies. Using lessons learned from the BP and the Exxon Valdez *disasters, Liesel Ashley Ritchie, Duane A. Gill, and J. Steven Picou reflect on the consequences of human-made disasters for communities; for example, the authors show how communities such as those along the Gulf Coast affected by the BP disaster and those in Alaska affected by the* Exxon Valdez *spill often suffer economically, as they are dependent on resources and a clean environment for income-generating activities such as fishing or tourism.*

On April 20, 2010, eleven people were killed when the Deepwater Horizon drilling rig exploded. Owned by Transocean Ltd. and contracted to BP, the rig burned in the Northern Gulf of Mexico until it eventually collapsed, leaving a breached wellhead gushing an estimated 55,000 barrels of oil per day. Some 185 to 205 million gallons of crude oil were released before the wellhead could be capped in July and permanently sealed in September, nearly five months after the initial explosion.

As this environmental disaster unfolded, the nation witnessed the riveting reporting of oil penetrating the marshes and washing ashore on white sandy beaches. At the same time, individuals and communities who depend on the Gulf of Mexico worried about the economic and socio-cultural effects as well as how to get compensated for damages.

Over a year later, there is no more 24-hour CNN coverage. If one were to rely solely on media accounts, it might seem that the Deepwater Horizon disaster is over. One *New York Times* headline from February 1, 2011 read: "Report

Foresees Quick Gulf of Mexico Recovery"—yet the article goes on to say, "The 39-page report acknowledges that any definitive assessment at this point is impossible, and that fully understanding the spill's ecological effects will take years." Not surprisingly, more pressing national and international events have caught the media's attention. Most Gulf waters have been reopened for fishing, seafood is being tested and deemed safe for human consumption, and beaches are being declared clean and, in fact, inviting. Still, economic and social upheaval will likely leave the communities along the Northern Gulf of Mexico (the Gulf) in disarray for the foreseeable future.

Indeed, lessons learned from 21 years of social science research on community impacts of the 1989 *Exxon Valdez* oil spill (EVOS) suggest this recent environmental disaster is far from over. If the BP disaster follows the EVOS script, we can anticipate some of the problems in store for residents of Gulf communities. The severity of the disaster depends on how the natural and social environments interrelate and the amount of damage to the ecosystem. We know that social and psychological stress are heightened by the uncertainty that comes from toxic contamination. The complexities of local and regional economies and the upheaval created by a temporary economic boom created by cleanup operations also contribute to the disaster. Finally, as injured parties seek compensation for damages from this environmental disaster, many are further traumatized by impersonal bureaucratic structures and protracted litigation processes.

It is especially difficult for people who have not experienced an environmental or "technological" disaster to fully understand what people and communities along the Gulf are going through. Among both researchers and in the general public, there is a broad tendency to distinguish between "natural" disasters and "technological" or "environmental" disasters. It is perhaps most useful to think about disasters as being on a continuum, with "natural" disasters at one end of the continuum and human-caused disasters at the other. While these events' qualities and characteristics overlap in terms of their social impacts, what we typically refer to as "natural" disasters such as tornadoes, earthquakes, and hurricanes, are considered "Acts of God." There is no one to blame and, in most cases, communities are able to come together to rebuild and move on. At the other end of the continuum, technological disasters such as Chernobyl, Bhopal, and the more recent BP oil spill are triggered by human error. These tend to result in long-term community disruption and chronic psychological stress.

As Michael Edelstein has noted, "outsiders" don't understand why environmental disaster survivors can't just move on. And if outsiders can't understand, they are less likely to offer support, and that further impedes community recovery. It is our hope that the past can offer some lessons in how best to approach long-term recovery in the Gulf.

Community Ties to the Environment

Sociologists Steve Kroll-Smith and Steve Couch's ecological-symbolic approach to disasters asserts that community recovery and interpretive processes are influenced by the type of environment damaged and the community's relationship to that environment. Renewable resource communities (RRCs) like those affected by the EVOS in Alaska are particularly vulnerable to the effects of an environmental disaster. They depend on renewable natural resources for their social, cultural, and economic existence. As noted by Steve Picou and Duane Gill, sociologists who have studied the EVOS for more than 20 years, understanding environment-social relationships between these aspects of life in RRCs is essential to understanding the often ignored human side of environmental disasters.

The fishing community of Cordova, Alaska—located on Prince William Sound and considered "ground zero" for the EVOS—is a prime example of how the fate of an RRC is tied to its natural environment. Although more than 20 years have passed since the supertanker *Exxon Valdez* ran aground, only 10 of 26 resources/species are classified by the EVOS Trustee Council as having recovered from the contamination of the oil spill. The herring biomass has yet to return to a level where harvesting is viable—a critical problem for the fishing fleet and community that once depended heavily on the revenues associated with the short, but highly profitable, herring season each spring. Despite Exxon's claims that the herring population's crash following the spill had nothing to do with its oil, a substantial body of recent research indicates otherwise. In short, academics have documented long-term impacts of the toxic crude oil for the Prince William Sound ecosystem. Among these is the persistence of volatile levels of *Exxon Valdez* oil in intertidal regions, on beaches, and in salmon streams, as well as significant declines in local fisheries.

Particularly for commercial fishermen and Alaska Natives in Cordova and other oiled communities, the consequences of EVOS-related environmental degradation have been significant. These include empirically documented elevated levels of collective trauma, social disruption, economic uncertainty, community strain, and psychological stress. Notably, the chronic community impacts of the EVOS are directly tied not only to the actual loss of various types of resources, but also to the threat of the *future* loss of resources.

It is in this context that we consider the situation for communities along the Gulf in the aftermath of the Deepwater Horizon blowout. This disaster damaged marine ecosystems and resources—the extent of which has yet to be fully realized. Additionally, the dispersants used to mitigate the impacts of the spill threaten several "at-risk" industries along the northern Gulf. Commercial and recreational fishing, tourism, and other enterprises tied to natural resources have already suffered severe economic losses and the threat of loss continues. In particular, there remains uncertainty regarding the recovery of shrimp, oysters, crab, and other fish. As in Alaska, there are Native American groups such as the Houma in Louisiana with strong cultural ties to the natural environment that was threatened or damaged by the oil and dispersants.

Data collected in communities along the Gulf of Mexico in the months since the BP spill reveal high levels of psychological stress among groups that rely on renewable natural resources for their livelihoods, as well as concerns related to health as a result of exposure to the oil and dispersants. In Alabama, for example, we found that the strongest predictors of stress were family health concerns, commercial ties to renewable resources, and concern about economic future, economic loss, and exposure to the oil. Unfortunately, the experiences of Gulf communities resonate with the ongoing narratives of Alaskans over the two decades following the EVOS.

Invisible Trauma

In the late 1980s, anthropologist and physician Henry Vyner wrote about the psychological effects of environmental contamination—"invisible trauma." Similar issues are addressed by Michael Edelstein in *Contaminated Communities: The Social and Psychological Impacts of Residential Toxic Exposure*. One of the primary concerns regarding environmental impacts and how these impacts ultimately affect communities and groups has to do with ecological damage that is difficult to detect. Even more challenging is assessing damage that may not emerge until years after initial contamination. A January 2011 report by the Harte Research Institute for Gulf of Mexico Studies at Texas A&M University-Corpus Christi concluded: "Realistically, the true loss to the ecosystem and fisheries may not be accurately known for years, or even decades." According to a 2005 National Research Council study, determining ecosystem recovery in the aftermath of an event such as the BP spill is more challenging than assessing initial impacts.

In a milieu of general uncertainty regarding the range and scope of environmental contamination, there is also confusion about the nature and extent of health-related impacts, economic impacts, and social impacts. Social impacts are particularly detrimental when they are the result of the kind of long-term disruption often associated with technological disasters.

The Money Spill

The complexity of the coastal economy along the Northern Gulf of Mexico far exceeds that of the region affected by the EVOS in Alaska. To date, the diverse yet intertwined oil and gas extraction economies, tourism or "beach" economies, and commercial fishing economies have all experienced both direct and indirect impacts of the BP oil spill. The temporary Federal moratorium on deepwater drilling hurt Louisiana communities dependent on the oil and gas industry for their livelihoods. Many areas in Alabama and Florida were hit by declines in tourism because potential visitors were concerned about oiled beaches and toxic waters unsuitable for swimming, as well as restrictions on recreational marine boating and sport fishing. And similar concerns related to the safety of seafood from the Gulf have been far-reaching, shaking consumer confidence and causing disruptions to the commercial fishing industry, including seafood harvesting, processing, and distributing. This hurts markets that were already mired in the process of rebounding from the effects of Hurricanes Katrina and Rita in 2005. From this perspective, the economic recovery of coastal communities is strongly related to perceptions of seafood consumers and tourists around the country—which further confounds community recovery.

Conversely, in the weeks and months after the Deepwater Horizon blowout there was considerable social and economic disruption associated with the "money spill"—large amounts of money that were spent on oil mitigation and clean-up efforts. As was the case in the *Exxon Valdez* disaster, in which those who made money on clean-up efforts were referred to as "spillionaires" and "*Exxon* whores," the money spill and ensuing economic disruption along the Gulf fostered problems in many communities. In media accounts and in our own interviews, some residents expressed frustration with perceived inequalities in chances to work on the Vessels of Opportunity program. For example, some locals felt shut out by the contracting process, growing particularly agitated in cases where they saw people from outside the area having their boats hired. Unequal access to resources associated with the money spill pitted family, friends, and neighbors against each other.

As in the wake of the EVOS, this money spill contributed to social upheaval and the emergence of a "corrosive community." This term, coined by environmental sociologists Bill Freudenburg and Timothy Jones in the 1990s, refers to post-disaster social environments in which social relationships are altered, social support breaks down, and civil order is disrupted. Notably, corrosive communities typically emerge following human-caused disasters, rather than in the context of natural disasters. If experience with the EVOS is any indication, this damage to social networks and the long-term potential for diminished social capital— trust and social ties—is very real. Loss of social capital may further increase stress levels, affecting overall community well-being. In addition, there is a strong possibility for outmigration from coastal communities over the long term as a result of declining economic conditions related, at least in part, to the BP spill. In Alabama, one-third of our survey sample indicated a desire to move from their community. Should this occur, communities will lose not only social capital, but also human capital.

Impersonality and Collective Stress

Although separated by more than 20 years, both the EVOS and the BP oil disasters highlight critical issues of recreancy (blame), responsibility, and loss of trust in corporations and government. These are characteristic of human-caused disasters. In the months following the Deepwater Horizon blowout, frustration and anger over accountability, lack of transparency, and finger-pointing over who was responsible escalated in coastal states and the halls of Washington. Following the Exxon playbook, BP continues to downplay the amount of oil that was released into the Gulf and tries to spread the blame to Halliburton for a faulty cement job and Transocean for operating an unsafe rig. This corporate posturing sets the stage for BP to minimize financial responsibility for resulting environmental, economic, and social damages.

Relative to the grounding of the *Exxon Valdez*, public responses to the rig explosion suggest higher levels of perceived complicity and corruption between BP and the former Minerals Management Service (now the Bureau of Ocean Energy Management, Regulation, and Enforcement). More than two decades of data regarding attitudes toward government, big business, and the U.S. legal system in the context of the EVOS show that that beliefs about trust and blame are related to frustration, anger, alienation, and stress. Given this empirical evidence, we can expect these or similar outcomes to escalate over time among Gulf Coast residents.

Corporations like Exxon and BP aren't designed to react to the demands of being responsible for a disaster. When it comes to dealing with claimants and the public, their corporate cultures and bureaucratic structures seem impersonal and insensitive. Government bureaucracies and our justice system are not much better in responding to survivors of environmental disasters. Interacting with the impersonality of corporate and government bureaucracies and engaging with the legal system not only exacerbates existing stress, it creates new stress for victims.

In the months since the BP disaster, the financial claims process has, indeed, become a bureaucratic and legal obstacle, as well as a source of contention and stress. Despite the unprecedented establishment of a $20 billion escrow fund to compensate those affected by the spill, the Governor of Alabama has character-ized the process as "extortion." In March 2011, eleven months into the cleanup, the Gulf Coast Claims Facility reported that of 155,000 claims applications sub-mitted for full or interim payments, just 25 percent had been processed. Many people filing claims have expressed frustration and anger with the futility of the process. With its pronouncements to pay "legitimate" claims and to make survi-vors "whole," BP's public relations campaign since shortly after the explosion on the Deepwater Horizon rig has also mirrored that of Exxon's in 1989. Similarly, President Obama's remark that the compensation funds represent "an important step towards making the people of the Gulf Coast whole again" hasn't instilled confidence among those who lived through the EVOS. As one Alaskan recently put it: "That's lawyer speak for 'we're going to pay out as little as possible.'"

Litigation as "Secondary Trauma"

Minimal financial support from the BP claims process has caused many who believe that they've incurred unpaid losses to resort to litigation. This, too, is reminiscent of the predicament in which many commercial fishermen in Alaska found themselves after the EVOS. Following the EVOS, and now in the wake of the BP disaster, it is apparent that there are some things that are not valued or even considered in the calculation of damages. How does one calculate the value of the base of a food chain—plankton, larval fish, and many bottom dwell-ing organisms—species that are difficult to see with the naked eye? How do we assign value to a way of life that is lost to human-caused environmental contami-nation? In the case of the EVOS, Alaska Native claims of damage to their cultural heritage and subsistence practices were dismissed in Federal Court. Moreover, actual damage claims did not include losses incurred after the case went to trial in 1994. Shortly after going to trial, the multi-million dollar herring fishery col-lapsed and, to this date, has not recovered.

BP spill-related litigation will be delayed until 2013 and is likely to result in protracted legal processes for literally thousands of Gulf Coast residents. Research tells us that this will prolong the social impacts of the disaster, leaving most plaintiffs with a lack of closure. With the precedent set by *Baker v. Exxon*, which took almost two decades to wind through the U.S. judicial system, we can anticipate prolonged litigation-related stress for those involved. If it follows the pattern documented in Alaska, this "secondary trauma" could prove to be as stressful as the initial disaster itself. Coupled with other long-term, chronic social impacts previously discussed we can expect high levels of stress among individuals, groups, and communities to continue, just as with the EVOS. If damage awards through the BP claims process remain contentious and redress through the courts is delayed, mental health problems and community disruption will persist for decades along the Gulf Coast.

Disasters like the BP oil spill are what Kai Erikson, a sociologist who studies social consequences of catastrophic events, describes as a "new species of trouble" that "scare human beings in new and special ways, . . . [and] . . . elicit an uncanny fear in us." And as German sociologist Ulrich Beck suggests, these events also represent risks that are a major feature of contemporary society. The cumulative impacts of hurricanes Katrina, Rita, and Wilma in 2005, Ivan in 2004, and now the BP oil spill disaster are intensified by the current global economic crisis and our reliance on fossil fuels. Together, these stressors take a toll on various forms of community capital, including financial, human, social, built, political, natural, and cultural capital.

Like the EVOS, and environmental disasters in general, the BP oil spill will continue to reveal "contested" scientific evidence concerning ecological damages, emerging secondary traumas such as the claims process and litigation, and serious community conflict and mental health problems. Given what we know about the trajectory of economic and social recovery for survivors of the EVOS more than 20 years ago, the situation in Gulf communities warrants close monitoring and attention by researchers, mental health care providers, policy-makers, and perhaps most importantly, the public at large.

Note

Ritchie, Liesel Ashley, Duane A. Gill, and J. Steven Picou. 2011. "The BP Disaster as an *Exxon Valdez* Rerun." *Contexts* 10: 30.

References

Beck, Ulrich. 1992. *Risk Society: Towards a New Modernity.* London: SAGE Publications, Inc.

Edelstein, Michael. [1988] 2004. *Contaminated Communities: The Social and Psychological Impacts of Residential Toxic Exposure.* Boulder, CO: Westview Press.

Edelstein, Michael. 2000. "Outsiders Just Don't Understand." In *Risk in the Modern Age: Social Theory, Science and Environmental Decision-Making*, edited by M. J. Cohen, 123–42. New York: St. Martin's Press, Inc.

Exxon Valdez Oil Spill Trustee Council (EVOSTC). 2010. *Update on Injured Resources and Services, 2010.* Anchorage, AK: *Exxon Valdez* Oil Spill Trustee Council [November]. http://www.evostc.state.ak.us/Universal/Documents/Publications/2010 IRSUpdate.pdf.

Exxon Valdez Revisited: Rights and Remedies. 2009. *University of St. Thomas Law Journal* 7: 1.

Freudenburg, William R., and Robert Gramling. 2011. *Blowout in the Gulf: The BP Oil Spill Disaster and the Future of Energy in America.* Cambridge, MA: The MIT Press.

Freudenburg, William R., and Timothy Jones. 1991. "Attitudes and Stress in the Presence of Technological Risk: A Test of the Supreme Court Hypothesis." *Social Forces* 69 (4): 1143–68.

Gill, Duane A., J. Steven Picou, and Liesel A. Ritchie. 2012. Forthcoming. "The *Exxon Valdez* and BP Oil Spills: A Comparison of Initial Social and Psychological Impacts." *American Behavioral Scientist* 56 (1): 3–23.

Kroll-Smith, J. Stephen, and Stephen R. Couch. 1991. "What is a Disaster?" An Ecological Symbolic Approach to Resolving the Definitional Debate." *International Journal of Mass Emergencies and Disasters* 9: 355–66.

Picou, J. Steven, and Duane A. Gill. 1997. "Commercial Fishers and Stress: Psychological Impacts of the *Exxon Valdez* Oil Spill." In *The* Exxon Valdez *Disaster: Readings on a Modern Social Problem*, edited by J.S. Picou, D.A. Gill, and M. Cohen, 211–36. Dubuque, IA: Kendall-Hunt.

Picou, J. Steven, Cecelia Formichella, Brent K. Marshall, and Catalina Arata. 2009. "Community Impacts of the *Exxon Valdez* Oil Spill: A Synthesis and Elaboration of Social Science Research." In *Synthesis: Three Decades of Social Science Research on Socioeconomic Effects Related to Offshore Petroleum Development in Coastal Alaska*, edited by Stephen R. Braund and Jack Kruse, 279–307. MMS OCS Study Number 2009–006. Minerals Management Service, OCS Region, Anchorage, Alaska.

Ritchie, Liesel Ashley. 2004. *Voices of Cordova: Social Capital in the Wake of the* Exxon Valdez *Oil Spill*. PhD diss, Department of Sociology, Anthropology, and Social Work. Mississippi State University. http://www.colorado.edu/hazards/research/voicesof cordova_ritchiedissertation.pdf.

Ritchie, Liesel Ashley, and Duane A. Gill, D.A. 2007. "Social Capital Theory as an Integrating Framework for Technological Disaster Research." *Sociological Spectrum* 27: 1–26.

Vyner, Henry. 1988. *Invisible Trauma: The Psychological Effects of Invisible Environmental Contaminants*. Lexington, MA: Heath.

Silent Spill 14
The Organization of an
Industrial Crisis

Thomas D. Beamish

In this reading, Thomas Beamish examines one of the biggest oil spills in U.S. history—the spill at Guadalupe Dunes in California. Unlike the fast-moving BP and Exxon Valdez disasters, discussed in the previous chapter, the Guadalupe Dunes disaster happened slowly over many years. As Beamish shows, the cleanup at Guadalupe Dunes came late; the original problem had been allowed to mushroom over many years of inaction. Beamish, in his investigation of the systemic and institutional underpinnings of this long-term event, argues that the extent of this disaster was exacerbated by both the slowness in which it occurred and the style of decision making that takes place in particular kinds of organizations.

> There's a strange phenomenon that biologists refer to as "the boiled frog syndrome." Put a frog in a pot of water and increase the temperature of the water gradually from 20°C to 30°C to 40°C . . . to 90°C and the frog just sits there. But suddenly, at 100°C . . ., something happens: The water boils and the frog dies. . . . Like the simmering frog, we face a future without precedent, and our senses are not attuned to warnings of imminent danger. The threats we face as the crisis builds—global warming, acid rain, the ozone hole and increasing ultraviolet radiation, chemical toxins such as pesticides, dioxins, and polychlorinated biphenyls (PCBs) in our food and water—are undetectable by the sensory system we have evolved.
> —Gordon and Suzuki 1990

Underneath the Guadalupe Dunes—a windswept piece of wilderness 170 miles north of Los Angeles and 250 miles south of San Francisco—sits the largest petroleum spill in US history. The spill emerged as a local issue in February 1990. Though not acknowledged, it was not unknown to oil workers at the field where it originated, to regulators that often visited the dunes, or to locals who frequented the beach. Until the mid-1980s, neither the oily sheen that often appeared on the beach, on the ocean, and the nearby Santa Maria River nor the

215

strong petroleum odors that regularly emanated from the Unocal Corporation's oil-field operations raised much concern. Recognition, as in the frog parable, was slow to manifest. The result of leaks and spills that accumulated slowly and chronically over 38 years, the Guadalupe Dunes spill became troubling when local residents, government regulators, and a whistleblower who worked the field no longer viewed the periodic sight and smell of petroleum as normal. . . .

I first heard of the Guadalupe spill on local television news in August 1995. (My home was 65 miles from the spill site.) The scene included a sandy beach, enormous earth-moving machinery, a hard-hatted Unocal official, and a reporter, microphone in hand, asking the official how things were proceeding. The interplay of the news coverage and Unocal's official response caught my attention more than anything else. The representative asserted that Unocal had extracted 500,000 gallons of petroleum from a large excavated pit on the beach just in view of the camera. The newscaster ended the segment by saying (I paraphrase), "It's nice that Unocal is taking responsibility to get things under control." This offhand remark about responsibility set me to thinking about the long-term nature of the spill and about why it had not been stopped sooner, either by Unocal managers or by regulators.

A few months later, a colleague and I drove to the beach. My colleague, a geologist who was familiar with the area, had suggested that we visit the Guadalupe Dunes for their scenic beauty. We walked the beach and the dunes that border the oil field, alert for signs of the massive spill. The pit that Unocal had recently excavated had been filled in. The only hint of the project that remained was a small crew that was driving pilings into the sand to support a steel wall intended to stop hydrocarbon drift (movement of oil on top of groundwater) and the advancing Santa Maria River, which threatened to cut into an underground petroleum plume and send millions more gallons into the ocean.

Unocal security personnel followed along the beach, watching suspiciously as we took pictures. In fact, the spill was so difficult to perceive (only periodically does the beach smell of petroleum and the ocean have rainbow oil stains) that my impressions wavered. Was this really a calamitous event? The whole visit was imbued with the paradox of beauty and travesty.

Under my feet was the largest oil spill in California, and most likely the largest in US history. . . . Yet the "total amount spilled" continue to be, as one local resident noted in an interview, a matter of "political science." There is still controversy over just how big this spill really is. The smaller of the two estimates . . . (8.5 million gallons) comes from Unocal's consultants. State and local regulatory agencies do not endorse it (Arthur D. Little et al. 1997). The estimates quoted most often by government personnel put the spill at 20 million gallons or more, which would make it the largest petroleum spill ever recorded in the United States.

At first glance, it seems strange that so many individuals and organizations missed the spillage for so many years; "passivity" seems to be the word that best characterizes the personal and institutional mechanisms of identification and amelioration. It is also clear that the Guadalupe spill is very different from the image of petroleum spills that dominates media and policy prescriptions and the public mind: the iconographic spill of crude oil, complete with oiled birds and dying sea creatures.

The Guadalupe Dunes spill is only the largest *discovered spill*. Representing an inestimable number of similar cases, it exemplifies a genre of environmental catastrophe that portends ecological collapse.

Describing his impression of the spill in a 1996 interview, a resident of Orcutt, California, explained why he remained unsurprised by frequent diluent seeps: "When you grow up around it—the smell, the burning eyes while surfing, the slicks on the water—I didn't realize it could be a risk. It was normal to us." In a 1997 interview, a local fish and game warden—one of those initially responsible for the spill's investigation—responded this way to the question "Why did it take so long for the spill to be noticed?": "It is out of sight, it's out of mind. I can't see it from my back yard. It is down there in Guadalupe, I never go to Guadalupe. You know, I may have walked the beach one time, but I never saw anything. It smelled down there. What do you expect when there is an oil field? You know, you drive by an oil production site; you are bound to smell something. You are bound to."

In the days and weeks after my initial visit to the dunes, I wondered why the spill had gained so little notoriety. Beginning my research in earnest, I visited important players, attended meetings, took official tours of the site, and followed the accounts in the media.

What makes the Guadalupe spill so relevant is that it represents a genre—indeed a pandemic—of environmental crises (Glantz 1999). Collectively, problems of this sort—both environmental and non-environmental—exemplify what I term *crescive troubles*. According to the *Oxford English Dictionary*, "crescive" literally means "in the growing stage" and comes from the Latin root "crescere," meaning to "to grow." "Crescive" is used in the applied sciences to denote phenomena that accumulate gradually, becoming well established over time. In cases of such incremental and cumulative phenomena (particularly contamination events), identifying the "cause" of injuries sustained is often difficult if not impossible because of their long duration and the high number of intervening factors. Applied to a more inclusive set of social problems, the idea of crescive troubles also conveys the human tendency to avoid dealing with problems as they accumulate. We often overlook slow-onset, long-term problems until they manifest as acute traumas and/or accidents (Hewitt 1983; Turner 1978).

There are also important political dimensions to the conception of crescive troubles. Molotch (1970), in his analysis of an earlier and more infamous oil spill on the central coast of California (the 1969 Santa Barbara spill), relates a set of points that resonate with my discussion. In that article, Molotch examines how the big oil companies and the Nixon administration "mobilized bias" to diffuse local opposition, disorient dissenters, and limit the political ramifications of the Santa Barbara spill. Two of his ideas have special relevance: that of the *creeping event* and that of the *routinization of evil*. A creeping event is one "arraigned to occur at an inconspicuously gradual and piecemeal pace" that in so doing diffuses consequences that would otherwise "follow from the event if it were to be perceived all at once" (ibid., p. 139). . . .

Our preoccupation with immediate cause and effect works against recognizing and remedying problems in many ways. It is mirrored in the way society addresses the origin of a problem and in the way powerful institutional actors seek to nullify resistance and diffuse responsibility. The courts and the news media, for instance, often disregard the underlying circumstances that led to many current industrial and environmental predicaments, focusing instead on individual operators who have erred and pinning the blame for accidents on their negligence (Perrow 1984; Vaughan 1996; Calhoun and Hiller 1988). Yet this ignores the systemic reasons why such problems emerge. In short, most if not all of our society's pressing social problems have long histories that predate their acknowledgment but are left to fester because they provide few of the signs that would predict response—for example, the drama associated with social disruption and immiseration. . . .

The inability of our current remedial systems, policy prescriptions, and personal orientations to address a host of pressing long-term environmental threats is frightening. There are, however, numerous examples of disconnected events—seemingly unrelated individual crises recognized after the fact—that have received widespread public attention. Through national media coverage, images of ruptured and rusting barrels of hazardous waste bearing the skull and crossbones have become icons that fill many Americans with dread (Szasz 1994; Erikson 1990, 1994). But these are only the end results of ongoing trends that have been repeated across the country with less dramatic consequences. In view of the startling deterioration of the biosphere, much of which is due to slow and cumulative processes, more attention should be devoted to how such scenarios unfold. . . .

My specific intent is to uncover how and why the Guadalupe spill went unrecognized and was not responded to even though it occurred under unexceptional circumstances. The industrial conditions were quite normal, and the regulatory oversight was typical. It would seem that there was nothing out of the ordinary, other than millions of gallons of spilled petroleum. This is, in part, why the spill is so instructive. It represents a perceptual lacuna—a blank spot in our organizational and personal attentions. . . .

Why didn't local managers report the seepage, as the law requires? How did field personnel understand their role? How could pollution of such an enormous magnitude be left so long before receiving official recognition and action? Why did the surrounding community take so long to react? . . .

The reality that surrounds crescive circumstances is characterized by polluters who are unlikely to report the pollution they cause, authorities who are unlikely to recognize that there is a problem to be remedied, uninterested media, and researchers who take interest only if (or when) an event holds dramatic consequence. In short, all those who are in positions to address crescive circumstances are disinclined to do so. Forms of degradation that lack direct and immediate impact on humans, dramatic images of dying wildlife, or other archetypal images of disaster tend to be downplayed, overlooked, and even ignored.

The national print media certainly mirrored the propensity to ignore the Guadalupe spill (Hart 1995). Over the period 1990–1996, the national press devoted 504 stories to the *Exxon Valdez* accident and only nine to the Guadalupe spill.

In a 1996 interview, a reporter for the *Santa Barbara News Press* offered his opinion as to why the Guadalupe spill had received little public attention until 1993. His view resonates with three of the four social factors articulated above (social disruption, stakeholders, and media fit):

We didn't see black oily crude in the water and waves turning a churning brown. We didn't see dead fish and dead birds washing up. We didn't see boats in the harbor with disgusting black grimy hulls. This is largely an invisible spill. It took place underground. . . . Because it was not so visual, especially before Unocal began excavation for cleanup, I think that it just didn't capture the public. . . . But after Unocal began excavations, driving sheet pilings into the beach, scooping out massive quantities of sand, setting up bacteria eating machines, burning the sand. It began to dawn on people the magnitude of this thing, but again it wasn't in their back yards, Guadalupe is fairly remote. . . . And it's not a well-to-do city [the city of Guadalupe]—comparatively, anyway, with the rest of our area. . . . So I don't think it really sparked the public interest as much as it could have or would have if it was . . . a surface spill. . . .

Central to my research were field interviews with members of the local oil industry, government regulators, community members, and environmental activists. These interviews were tape recorded, transcribed, and systematically analyzed. In addition to the interviews, there were many spontaneous conversations—in hallways, in office waiting rooms, in the homes of those that were the intended interviewees—with individuals I had not originally contacted or planned to meet. Though not recorded, these conversations should not be seen as any less important than the others. I also pursued ethnographic context, recording scores of informal conversations concerning the spill. I accumulated and analyzed a substantial collection of archival materials, and I have followed media portrayals of the spill closely since 1989. . . .

In its early stages (from 1953 until 1978 or 1979), the leakage at the Guadalupe field was not troubling, nor was there anyone to whom to report it. Because it was part of routine fieldwork, it received little attention. According to those who read the meters that tracked the coming and going of the diluent, "many times there were little leaks; that was just normal" (field worker, telephone interview, 1996). A worker quoted in a local newspaper went so far as to say that "diluent loss was a way of life at the Guadalupe oil field" (Friesen 1993). Dumping hundreds of gallons of diluent into the dunes, as long as it was done a gallon at a time, was an ordinary part of production. This is not a great leap of reason; oil work obviously involves oil. Until the 1970s, Unocal sprayed the dunes with crude oil to keep them from shifting and thus to make field maintenance and transportation easier. If spraying crude oil over the dunes was unproblematic, why would diluent leaks, which were largely invisible as soon as they hit the sand, be unsettling? Although workers mention that they became alarmed in the 1980s when puddling diluent periodically appeared as small ponds on the surface of the dunes, the chronic leaks themselves evoked little attention. In brief, at Guadalupe the normalcy of spilling oil of all kinds (crude oil, lubricants, and diluent) worked to blunt perceptions of the leaks as problematic. The leaks were an expected part of a day in the life of an oil worker. According to the *Telegram-Tribune* (Greene 1993b): "A backhoe [operator] at the field . . . for 12 years . . . cited 'an apparent lack of concern about the immediate repair of leaks or the detection of leaks.' Diluent lines would not be replaced unless they had leaked a number of times or were a 'serious maintenance problem. . . .' Although workers checked meters on the pipelines and looked for leaks if there was a discrepancy, often a problem wasn't detected until the stuff flowed to the surface' said . . . a field mechanic."

By both historical and contemporary accounts, oil spills have long been a common occurrence in oil-held operations. This seems to have been especially the case at fields operated by Union Oil. But this does not help us understand why, once field personnel recognized the spillage as a significant problem, they denied it and failed to report it (as specified by state and federal law) for 10 years or more. A first step in understanding why workers failed to report their spill to the authorities once it had "tipped" toward becoming a grievous problem requires us to attend to the vocabulary, the structure, and the enactment of work and how these factors not only molded workers' perceptions of the leaks but also kept them from reporting outside their local work group.

The "Company Line," 1978–1993

Organizationally, oil work at the Guadalupe field was arranged, like work in many traditional industrial settings, around a hierarchical seniority system. Recruitment and promotion were internally derived, meaning the field workers relied on their immediate foremen and supervisors for instruction, guidance, and ultimately, future chances at success (promotion, salary increase, choice of shifts, and so forth). . . . This organizational structure helps to explain Unocal employees' silence about the Guadalupe spill after it was recognized as a threat. Even when the leaks began to look more like a bona fide spill, the rank-and-file workers were insulated from reporting it themselves by their position within the field's hierarchy and their immediate responsibilities. Reporting outside the work group was management's domain.

A Unocal field worker I interviewed in 1997 articulated his experience of the change from a normal to a problematic spill as follows: "You come up and you see a clamp [on a] diluent [pipe]line. It is leaking. You tighten it up, you change it, you . . . fix it and it . . . has made a puddle. That is not something you would turn in. When it went into the ocean . . . and you see the waves break and they weren't breaking white [but] brown water, there is a problem. [That happened] sometime in the 1980s. . . . We all knew right then . . . we had some kind of problem. Well, we all kind of estimated it could be rather large considering that this field had been here so long before we ever got there." Corroborating this worker's impressions, another worker quoted in the *Telegram-Tribune* (Greene 1993b) remembered finding large concentrations of diluent that were no longer the "leaks" that had created them but looked more like a typical "oil spill": "In 1980 a large puddle of diluent that had saturated the sand and bubbled up to fill a spot 5 to 10 feet wide. . . ." He told investigators that he and his co-workers realized at the time there were problems with the diluent system "even though management seemed to ignore the problems."

By this time, the problems brought on by "normal operating procedures" were obvious and destructive. This became especially apparent in the mid-1980s, when accelerated spillage periodically slowed oil production at the field (Greene 1992b, 1993a; Rice 1994). Yet, instead of self-reporting the spillage, the field workers turned to denial and secrecy. . . .

In view of the hierarchy in the field, they were not responsible; their managers were. The hierarchical insulation from responsibility thus helped to keep workers who watched diluent spill into the dunes from feeling obligated to do something about it. When relieved of making decisions, people tend to cede their personal responsibility to those who are in control (Milgram 1974; Asch 1951). . . .

The field's hierarchy had five major levels. A new worker began as a utility man, then worked his way up to pumper and then to field mechanic. If able, with long enough tenure at the field he could become a foreman. Over the foremen were the field supervisors, who headed operations at specific fields; over them was a superintendent who oversaw Guadalupe and another oil operation in the area. . . .

The culpability of all those at the field, but especially the superintendent, supervisors, and field foremen, coupled with the field's organizational characteristics, meant that explicit knowledge concerning the scope and scale of the leaks stayed inside the local operation. "Each field is its separate own little field," said a field worker interviewed in 1997. "We were kind of out in the middle of nowhere. So once we reported to our superior [a field foreman] then he has to report it to the field supervisor, who has to report to the regional superintendent, who then reports it to Los Angeles. Somewhere along the line I think it stopped. I think that it stopped with the field supervisor." This field worker was describing a loosely *coupled* organizational arrangement—one with organizational units that are "somehow attached, but [whose] attachment may be circumscribed, infrequent, weak in its mutual affects, unimportant, and/or slow to respond" (Weick 1976, p. 3). In this case, the slack that existed at the local field between workers and between workers and managers and the loose organizational coupling that existed between the local field and corporate offices (including environmental divisions) were reflected in the technical division of responsibilities, in the authorities of office, and in the expectations placed on each. A great deal of flexibility existed between these units as long as certain goals were met. In this case, petroleum continued to be produced and sent out at an acceptable rate. In view of the local field's autonomy and field personnel's collective interest in remaining a viable production unit, not telling outsiders about the spill made a great deal of sense. . . .

The long-term nature of the Guadalupe spill made it especially problematic for all those who worked the field for any length of time. Liability for it was diffuse—indeed, organization wide. For those in the lower echelons, going outside the proscribed line of command to report the spill created triple jeopardy: Not only would they risk being personally associated with an organizational offense; they also would have been informing on co-workers and endangering their careers by implicating their superiors. One does not succeed within a vertically organized work setting by "ratting out" one's superiors or co-workers. Fear of social and organizational reprisal was evident in my discussions with field personnel, in California wardens' accounts of their interactions with subpoenaed field personnel, and in local newspapers' stories such as Greene 1993b: "Current employees contacted for this story were surprised and dismayed their names would become public because of what they told the state investigators. They worried about their superiors and co-workers at Unocal finding out."

Workers at the Guadalupe field did not want to go over the head of their field foreman, their supervisor, or their superintendent. A field worker, interview by telephone in 1997, said: "There is somebody above you and someone above them and someone above them. One thing that you don't want to do is break the chain of command . . . that causes friction." Informing might have affected how many hours of work one received, one's chances of promotion, and ultimately whether or not one would keep a well-paying job. . . .

A Culture of Silence

Local managerial power and organizational routines did not wholly determine behavior at the field. The normative framework that prevailed there was also attributable to the subculture of oil-field work and to individual workers' agency. . . .

To understand more fully why workers kept quiet about a spill they knew was patently illegal while field managers covered it up and lied to authorities about its origins, we must look beyond matters of hierarchy and seniority. We must look at individual motivations and at the social glue that bound workers to their work group. In short, we must look at the dominant social milieu at the oil field in order to see how social relations between workers played into the initial normalization of the spill and how they reinforced the intra-organizational conditions that discouraged self-reporting. Taken separately, both structural and cultural explanations would predict that self-reporting was unlikely; together, they make self-reporting appear a dubious regulatory strategy. . . .

Social Ties and Field Secrets

Workers at Guadalupe inherited and developed a set of norms and beliefs about what were and were not appropriate in-group behaviors. This is a normal part of group unity. Moreover, that this unity led to the coverup of an ongoing petroleum spill becomes more understandable (even if socially inexcusable) when we address the threat it posed to each individual at the field and to the local outfit as a whole. . . .

. . . [P]ressure to keep the spill a secret, based in a de facto culture of silence was observable in how field workers reacted when they found out that one of them had called the authorities. (. . . the first admission came in an anonymous telephone call to state officials in February of 1990.) When interviewed in 1997, the field worker who initially blew the whistle related being overheard by the field office's secretary and described the secretary's reaction to his phone call as follows:

> I got on the phone in the office. I say [to the health department official], "Okay, I'll talk to you later," and I hear his click, and I'm still on the phone, and I hear another click. The secretary eavesdropped and heard my conversation. She came in, and she started yelling at me, "What are you doing! We will all lose our jobs!" And I said, "Not if we didn't do anything! If it isn't ours, why would we lose our jobs? We are not going to lose our jobs!" We knew [about the spill]. But I never thought it would come to the point where they would shut everything down. What I thought would happen is they would isolate the problem and go on producing.

. . . Individuals, in protecting themselves from association with the spill, also collectively shielded the organization from harm, at least in the short term. The threats of a shutdown of the field and a loss of jobs and the social pressure to remain silent kept workers from reporting the spill. (Once the spill was "discovered" by regulators, Unocal's corporate headquarters did shut the field down, and all the workers were either transferred or laid off.)

Moreover, breaking with one's peers and eliciting an out-of-group admission about what was (initially at least) a "normal" part of production was also unlikely for a set of more socially relevant reasons. Even once the spill had accumulated and became noticeable, reporting it would mean informing on co-workers and facing their opinions. Once his identity became known at the site, the whistleblower was ostracized by many of his fellow workers. . . .

Inter-Organizational Location as Amplification

In conjunction with the organizational location of workers relative to one another and to management and with the culture of silence that characterized the field, the Guadalupe field's structural isolation from outside interference (both physically and organizationally from regulatory authorities and Unocal's head offices) and the corporate incentives worked against self-reporting. Like many other corporations, Unocal was not a monolithic undifferentiated body with a single objective or universally shared knowledge. In organizational form, Unocal consisted of loosely coupled upstream corporate offices, production units, and downstream refinery and vending segments. Insulation from outside interference amplified the power that field routines and the local production culture had over individual perceptions and over field workers' choices.

Because the Guadalupe field was largely autonomous from its head offices, its day-to-day domestic affairs were largely internal. A report of an incident had to go to the top before making its way to outside authorities. Because the information stopped in the field's chain of command, it never made it out of the field, where action could be taken to stem it. This is not a claim that Unocal headquarters could not have known about the spill if they had wanted to investigate it. The argument forwarded here is more passive: Headquarters was interested only in specific information from Unocal's extraction divisions, and this information tended to consist of production quotas rather than of information as to whether environmental matters were being addressed. Again, Unocal, as a corporation, seemed to care little about how local operations performed their production as long as the fields continued to produce profits. . . .

Had efficiency included not wasting diluent, a case could be made that the loss of diluent into the dunes would have been a sign to those on the outside that something was amiss. In this instance, Unocal's head offices may have taken a more active interest if dollars were being lost. Had hundreds of thousands of gallons of refined petroleum product been purchased from an external source and subsequently lost, it would seem expensive and hard to cover up. But spilling was considered a part of producing oil at Unocal's operations, and it also was rather normal for others in the industry. Furthermore, it was considered largely an internal affair. The diluent used at the site beginning in the early 1950s originated at Unocal's refinery situated at the edge of the Guadalupe Dunes, literally a part of the Guadalupe field's production infrastructure. Oil extracted from the Guadalupe field was piped to the Nipomo refinery for initial separation. Diluent, as a by-product of this refining process, was then pumped back to Guadalupe for use. At Guadalupe, diluent was stored at a number of tank farms; from there it was transferred via pipeline to individual extraction wells. If production was consistent, lost diluent would not be missed, especially in view of the normality of spilling and the shoddy records that were being kept (because the price of refining was internal). Losing diluent cost the local operator little (at least, relative to getting caught or facing the prospects of personally reporting it), as long as crude oil was being produced at the expected rate. On the other hand, if the field supervisor reported the spills (which had "tipped" toward the obvious in the 1980s) he would have known that he had a big monetary and criminal problem on his hands. It would have tarnished his personal record, reflected badly on Unocal's image as a whole, potentially shut down the local operation, and presented the possibility of criminal prosecution. What is more, the potential fines for having not reported the spill are significant. . . .

Two examples of the penalties associated with pollution of this sort illustrate the predicament that field managers confronted when deciding whether to report the spill. The federal Clean Water Act specifies that violators can be fined between $5000 and $50,000 a day per violation for being "knowingly" negligent. Estimating the potential fines involved for this single act would require starting with the date of the amendment's passage (in 1973) and calculating daily fines up to 1990 (when Unocal ceased using diluent at the field). The estimate ranges from $31,025,000 to $310,250,000. Likewise, under California's Proposition 65 (a citizen-sponsored "right to know" act passed in 1987) Unocal was also liable for not reporting its release of petroleum into local river and ocean waters frequented by recreationalists. Proposition 65 caps fines against violators at $2,500 per person per exposure day. These are but two examples.

Moreover, the field supervisor and superintendent personally stood to lose thousands of dollars in potential bonuses that were paid for meeting corporate expectations. Field supervisors received incentives in the form of commendations, quick advancement, and end-of-the-year bonuses for keeping production costs down and petroleum yields high. High production costs would have resulted from capital outlays for such items as Guadalupe's pipeline infrastructure. Much as in the system that prevailed in the Soviet Union into the 1980s, costs were "hidden" by a reward system that recognized only production goals and the accompanying steady income stream. Thus, the primary goal was keeping production high, not worrying about diluent costs that (at least on paper) were trivial, being locally internalized. According to a Unocal supervisor interviewed in 1997 for this research, why it took 38 years for the spill to be reported by field managers was rather easy to understand: "Unocal [did not report the spill] to the public because local managers received financial incentives to keep costs low. The corporate culture of the production outfits saw spills as a normal part of their routine." Although this was not the only reason that local Unocal managers would continue to spill, it certainly provided a strong incentive not to report it or stop the leakage at the field once it had become organizationally ominous. Only negative personal and organizational repercussions would have resulted if local managers reported the spill. As a latent product of the pressures articulated thus far, spilling and not reporting makes a great deal of sense from the production side of the equation. . . .

In brief, organization-sponsored complicity, the culture of silence, and the inter-organizational isolation of the field combined to make reporting of the Guadalupe spill improbable until the accumulation of diluent had gotten so bad that neither insiders (field personnel) nor outsiders (regulators) could fail to recognize it, a society of environmentalists was there to be concerned about it, and the insistent local media were eager to report on it. These are all factors that society can ill afford to either count on or wait for. . . .

. . . The predominantly social, cultural, and structural explanations I have put forth are powerful in part because of the nature of industrial regulation in the United States, where it has been left to corporate actors to report their own excesses. There is little interdiction, investigation, or active following up of problems by government authorities until a situation is so dire that a coverup is impossible to sustain. Thus, outside of personal motivation on the part of a worker, a foreman, or a supervisor to report a leak, there is little (aside from morals) that would impel anyone in a company to do so. And that is a slippery slope that takes us back inside the social dynamics that characterized the normative and cognitive institutions that characterized Unocal's local field operations. . . .

Note

Beamish, Thomas D. 2002. *Silent Spill: The Organization of an Industrial Crisis.* Cambridge, MA: The MIT Press.

References

Arthur D. Little Inc. in association with Furgro West, Headley and Associates, Marine Research Specialists, and Science Applications International Corp. 1997. Guadalupe Oilfield Remediation and Abandonment Project.

Asch, S. 1951. "Effects of Group Pressure upon the Modification and Distortion of Judgments." In *Groups, Leadership, and Men*, edited by H. Guetzkow. Pittsburgh, PA: Carnegie Press.

Bensman, J., and I. Gerver. 1963. "Crime and Punishment in the Factory: The Function of Deviancy in Maintaining the Social System." *American Journal of Sociology* 28: 588–98.

Brown, P., and E. Mikkelsen. 1990. *No Safe Place: Toxic Waste, Leukemia, and Community Action.* Berkeley: University of California Press.

Calhoun, G., and H. Hiller. 1988. "Coping with Insidious Injuries: The Case of Johns-Manville Corporation and Asbestos Exposure." *Social Problems* 35 (2): 162–81.

Dinno, R. 1999. *Protecting California's Drinking Water from Inland Oil Spills.* Sacramento: Planning and Conservation League.

Elliston, F., J. Keenan, P. Lockhart, and J. Van Schaick. 1985. *Whistleblowing: Managing Dissent in the Workplace.* Santa Barbara, CA: Praeger.

Erikson, K. 1990. "Toxic Reckoning: Business Faces a New Kind of Fear." *Harvard Business Review* 90: 118–26.

———. 1994. *A New Species of Trouble: The Human Experience of Modern Disasters.* New York: W.W. Norton.

Friesen, T. 1993. "Criminal Charges May Be Eliminated against Unocal." *Five Cities Times-Press-Recorder*, December 7.

Garfinkel, H. 1956. "Conditions of Successful Degradation Ceremonies." *American Journal of Sociology* 61: 420–24.

Glantz, M., ed. 1999. *Creeping Environmental Problems and Sustainable Development in the Aral Sea Basin.* Cambridge: Cambridge University Press.

Glazer, M. 1987. "Ten Whistleblowers: What They Did and How They Fared." In *Corporate and Governmental Deviance*, 3rd ed., edited by M. Ermann and R. Lundman, 229–249. Oxford: Oxford University Press.

Gordon, A., and D. Suzuki. 1990. *It's a Matter of Survival.* Cambridge, MA: Harvard University Press.

Greene, J. 1992b. "Unocal: A Leaky Environmental Record." *Telegram-Tribune*, August 5.

———. 1993a. "Unocal Spills May Have Gone Unreported." *Telegram-Tribune*, July 1.

———. 1993b. "Unocal Workers Confirm Leaks." *Telegram-Tribune*, June 17.

Hart, G. 1995. "How Unocal Covered Up a Record-Breaking California Oil Spill." In *The News That Didn't Make the News and Why*, edited by C. Jensen. Four Walls, Eight Windows.

Hawkins, K. 1983. "Bargain and Bluff: Compliance Strategy and Deterrence in the Enforcement of Regulation." *Law and Policy Quarterly* 5 (1): 35–73.

Hewitt, K., ed. 1983. *Interpretations of Calamity: From the Viewpoint of Human Ecology.* Australia: Allen & Unwin.

Milgram, S. 1974. *Obedience to Authority.* San Francisco: Harper & Row.

Molotch, H. 1970. "Oil in Santa Barbara and Power in America." *Sociological Inquiry* 40 (Winter): 131–44.

Perrow, C. 1984. *Normal Accidents: Living with High Risk Technologies.* New York: Basic Books.

Pratt, J. 1978. "Growth or a Clean Environment? Responses to Petroleum-related Pollution in the Gulf Coast Refining Region." *Business History Review* 52 (1): 1–29.

———. 1980. "Letting the Grandchildren Do It: Environmental Planning during the Ascent of Oil as a Major Energy Source." *Public Historian* 2 (4): 28–61.

Rice, A. 1994. "Endless Bummer." *Santa Barbara Independent*, March 17.

Skillern, F. 1981. *Environmental Protection: The Legal Framework.* New York: McGraw-Hill.

Szasz, A. 1994. *Ecopopulism: Toxic Waste and the Movement for Environmental Justice.* Minneapolis: University of Minnesota Press.

Turner, B. 1978. *Man-Made Disasters.* London: Wykeham.

Vaughan, D. 1996. *The* Challenger *Launch Decision: Risky Technology, Culture, and Deviance at NASA.* Chicago: University of Chicago Press.

Weick, K. 1976. "Educational Organizations as Loosely Coupled Systems." *Administrative Science Quarterly* 21 (March): 1–19.

———. 1995. *Sensemaking in Organizations: The Mann Gulch Disaster.* London: Sage.

Wolf, S. 1988. *Pollution Law Handbook: A Guide to Federal Environmental Law.* Connecticut: Quorum Books.

Yeager, P. 1991. *The Limits of the Law: The Public Regulation of Private Pollution.* Cambridge: Cambridge University Press.

Left to Chance 15
Hurricane Katrina and the Story of Two New Orleans Neighborhoods

Steve Kroll-Smith, Vern Baxter, and Pam Jenkins

Poor people, people of color, and the elderly are typically the most vulnerable in times of disaster. It's hard to overstate the injustices that were revealed by Hurricane Katrina and its aftermath in 2005. In particular, those without cars and without a useable credit card had a particularly difficult time evacuating and relocating. In their book Left to Chance: Hurricane Katrina and the Story of Two New Orleans Neighborhoods, *Steve Kroll-Smith and colleagues explore the personal stories of Katrina's evacuees and how they dealt with the unpredictability created by disaster. Through in-depth interviews with residents of two different African American neighborhoods, this reading illustrates how peoples' experiences, while varied and contingent on a variety of factors, including luck, were also strongly shaped by social class.*

In the classic tune "When the Levee Breaks," recorded in 1929, Joe McCoy and Memphis Minnie foretell the journey many people in the city would take seventy-six years later to escape Hurricane Katrina's historic floodwaters:[1]

> Mean old levee taught me to weep and moan
> Mean old levee taught me to weep and moan
> Got what it takes to make a Mountain Man leave his home.

In 2000, five years before Katrina, Walter Maestri, the emergency manager for Jefferson Parish, made public his assessment that a powerful storm could fill up the bowl that is the Crescent City, topping the roofs of houses and rising as high as three to four stories. In Maestri's scenario the city could be under water for up to ten weeks. While recognizing that getting everyone out of harm's way is more pipe dream than practical policy, Maestri was certain that evacuating the city was the only reasonable option in the event of a calamity of this magnitude.[2] Walter Maestri, as we now know, was all too prophetic in both his warning and his counsel.

On August 28, the day before Katrina made landfall, an alarmed mayor of New Orleans ordered the first mandatory evacuation in the history of the city. With gravitas fitting the moment, Mayor Ray Nagin warned, "This is a threat we've never faced before." His simple command, to get out of town, created for many a mess of complications.[3] For one thing, Katrina's timing could not have been worse for tens of thousands of New Orleans families. The twenty-eighth day of the month is a date on the calendar when the paycheck is spent, or nearly so. The prospect of an emergency trip is likely to stretch if not snap an already thin budget, a problem all too common among Hollygrove's working class.

Hollygrove resident Denise Anders begins her story of evacuating New Orleans with her son a day before the storm:

> It's just him and I. Money is tight. What can I bring? So I get the ice chest. We have orange juice, water. . . . We had ham. I made two sandwiches, cut them in half and we had cereal. . . . Let's just pack clothes for five days. . . . Well, we're in Shreveport and I'm crying and panicking. I'm crying out, "Lord, I'm going to have to make sure my savings is right." . . . All night driving. My son was napping in between and the blessing was he would get up. "Mama, wake up!" I was swerving. "Mama get up!"

In her clipped, spur-of-the-moment telling, Denise decants the life of the evacuee in knowing, wrought-iron sentences. Imagine having to move abruptly from the worrisome but more or less predictable life of a single mother to the unmoored life of an evacuee. From the predictable to the wholly unforeseen, she is fleeing ahead of danger on roads leading somewhere, if traffic permits, if money allows, if the car is able. Her only goal is to get on the road, to put distance between her and danger.

The words "emergency" and "evacuation" have a certain affinity for one another. Together they conjure images of unusual events and circumstances, experiences that defy our center of gravity, throwing us off balance if not knocking us down. But there is at least one noteworthy difference in the meaning of these two words. If an emergency is a sudden, unexpected occurrence requiring immediate action, evacuation is the act of protecting oneself and others by withdrawing from a place in an orderly fashion. The definitional emphasis on "orderly" is echoed in the literature on the sociology of disaster. A foundational paper on the topic concludes that

> to the extent that there are research observations, they show that the withdrawal movement itself usually proceeds relatively well. The flight tends to be orderly, reasonable from the perspective of the evacuees, and generally effective in removing people from danger.[4]

Our quarrel with the sentence that flight from danger appears orderly is not that at times it can appear so to others; we suspect it does. Our squabble is with the subject of the sentence. Chances are quite good that "from the perspective of the evacuees," words like "orderly" and "organized" would not capture their immediate and felt experiences of fleeing from harm's reach. Now, had the passage read, "from the perspective of those sociologists who examine evacuation from what Geertz might call an 'experience-distant' vantage point," the abstraction "orderly" might well make sense.[5] But for Denise Anders—who recounts, "All night driving. My son was napping in between and the blessing was he would get up. 'Mama, wake up!' I was swerving. 'Mama get up!'"—fleeing Katrina's wind and water was anything but methodical.

If we move closer to the lived moments of evacuation, if we get inside the car by listening to the accounts of those on the road, the relation of the unforeseen and the flight from danger is plain to see. Denise continues:

> So my sister and her husband, my little niece and her boyfriend, they was at the [New Orleans] convention center. There was a cop that told them he didn't know when the bus was coming. But he had a truck. . . . He was a police officer working but he said he was leaving because he couldn't take it anymore. . . . He said, "I can't take care of y'all but if y'all hide away in the back of my truck, I have to go to Baton Rouge. I'm leaving. I can drop y'all off."

In words that spill out one after another as if they themselves were running from danger, Denise invites us to find some alternative vocabulary for making sense of the lived experiences of evacuation. Far from orderly in the academic or conventional meaning of the word, her sister and her sister's husband and her niece and the niece's boyfriend find themselves on an aleatory journey, a miscellaneous collection of fate-filled moments.

Our task in this chapter is to call to life a few of the interanimating contingencies that make up the moments of evacuation and the experience of being evacuees in Pontchartrain Park and Hollygrove. In doing so we reveal a little of how people cope with a chance-filled world by at least in part not recognizing it for what it is. But there is more afoot in the intersection of evacuation, contingency, and our two neighborhoods. The quantity and quality of the haphazard are not completely random, though they are likely to appear so to those who find themselves on the road. Contingency is typically shaped in part by life chances tethered to the wheel of material well-being. "Material well-being" sounds a bit like an experience-distant abstraction. We can bring it closer to the experiences of the people from the neighborhoods by calling it "cheese."

"Cheese"

"There's a word heard from time to time in the black community. Some people got it; some people don't. The word is 'cheese,'" explains Cheryl Haden, a college professor who lives in Pontchartrain Park. "Cheese" refers to a variety of things, all of them related to material resources of one sort or another. Cheese might be a credit card, a small square of hard plastic that opens the world of commodities and services to those who own one and keep current with their monthly payments. Or it might be a paycheck for those with jobs, or a dependable car. Cheryl describes the uses of "cheese" in her experience of evacuation:

> Okay, where to begin. Well, the difference between me and swimming in floodwater was my American Express card. I'm a middle-class person, I'm a professor at a university. I live within my means; I certainly don't live beyond them. But

I don't live anywhere below my means either. I sometimes think of myself as an alligator in the lagoon. I have exactly expanded to the size of the territory that will sustain me.

It was the 29th, which is a day or two before the first payday of the fall semester for me. I'd been living off my summer salary and the little bit of savings that I had. And when we got the notification about the size and the danger of Hurricane Katrina, I used my American Express card and paid for reservations to reserve my room at the Best Western. I happen to have had a car, the car I still have. It was just three years old at the time and in good working order. It had four good tires on it; it didn't overheat. There are so many people who have city cars, just cars that are great for knocking around town and getting where you need to go so that you're not on public transportation or on a bicycle but are not going to survive the drive someplace else. If you don't have a credit card, you can't reserve a hotel room. If you don't have a credit card, you cannot rent a car.

If social class, to borrow from David Harvey, is not some abstract "force that operates outside the web of life, the day-to-day struggles, routines, or the ordinary," its immediacy in human lives will be dramatically enhanced during a flight from impending danger that sends one from home to road.[6] Denise Anders, who lives in Hollygrove, and Cheryl Haden, who lives in Pontchartrain Park, are both homeowners. They are both single moms, each with one child. Both are African American, separated in age by less than ten years. These similarities are striking, but a fundamental difference between them sends these two women on quite different journeys down the evacuation highway. Both women were drawn into a world governed by chance, but they did so with different resources at their disposal. While "chance" has a different look and feel for those with cheese, both stories emphasize the place of fluke in the kismet-filled life on the road.

Cheryl continues:

I wasn't completely hip to the whole contraflow explanation, and I had the book and the map and it didn't make a damn bit of sense to me, so I just tried to leave a day earlier, you know, avoid it entirely. So my sister and I had taken our own cars in the past, but this time, we all rode in my little baby SUV, with my mom and my sister and my daughter and her fish.

Some other people followed us in their cars. Professor Lawrence Jenkins, who's the chair of our department, and Christopher Saucedo, who's a sculpture professor, made up a caravan. A new professor from Canada, Marge, who's since left, also came along, and Andy Arden and his wife, Jen.

We all stayed at the same Best Western on Main Street in Houston because it's . . . about six blocks from the Houston Museum of Fine Arts, really close to the original Ninfa's. Our plan was to have some decent Mexican food and drink martinis. And I was responsible for the tequila, and Marge brought the bourbon, and Christopher had the Irish whiskey. So, we had our plan, we had wives and children and pets, and we were all in the same place at the same time. And we watched the Weather Channel with bated breath.

And we watched this monster grow. We started getting nervous and praying for it to go to Mississippi, and it went to Mississippi, and we were all so relieved. So we all went to Cheesecake Bistro in the Galleria and bought things we shouldn't have and ate too much and drank too much and came back to our hotel and went to sleep. Next morning Lawrence Jenkins was knocking on our hotel room door saying that the city's flooded. And I turned on the television and I saw Wolf Blitzer on CNN, who said that in all likelihood it may be as late as Christmastime, December, before residents were allowed back in their homes.

And I thought, "Oh my God, how are you going deal with this?" So I spoke to my sister about it, and we decided that the first thing we needed to do was get my mom someplace where she would be safe and comfortable while we dealt with her house and my house and my sister's house, because at this point we didn't know.

And I called my cousin Alice because being a black person from New Orleans, people stay here, people live here, people die here, and they're only as limited as their family members who actually move away and are prepared to shelter them from those things that push us out of the city. My cousin Alice's son and wife and daughter had recently bought a home just outside of Atlanta, in Stone Mountain.

Cheryl, her daughter, and her sister left their friends in Houston and made the twelve-hour road trip to Georgia. There they shared a house with dozens of extended family members who were also escaping Katrina's reach.

At one time there were about thirty family members in this one house, and seventeen of them were under the age of ten. No one shut their eyes for ninety-six hours because we had two family members who stubbornly refused to leave New Orleans. And we were watching CNN, and we were hoping to get a glimpse of them but also hoping not to get a glimpse of them. They were both safe, they got put on the bus and put on a plane, and one ended up in Utah, and one ended up in Arizona. Apparently you didn't know where you were going until the plane had taken off.

Okay, we were in Georgia, Stone Mountain, that's where we were. We were in Stone Mountain in Georgia, our New Orleans home is flooded, and all I had was two pair of underwear and one pair of shoes, and you know, two changes of clothes, and that's all my child had. . . . So we're going to Walmart to at least get something. I didn't know how much money I really had anymore. Whitney Banks were sort of shut down. I lived on my American Express card. They were excellent; I will always be an American Express card customer. They said, "We're so sorry this happened to you, buy whatever you want to, you don't have a credit limit." They didn't charge us interest for six months. They asked if we needed doctors, if we needed to buy a car.

One of my former students who I lived with during the renovation of the house, Bonnie Slaughter, and her husband, Scott, she called and she asked me, "Do you have a car? I'm going send you $5,000." And I said, "Well, I actually don't need any money right now." A guy I dated, I can't even remember, called to see if we were okay or if we needed anything, and he had two cars and he wanted to send one to us. And I said, "We have a car," but you know, these are just some of the people who popped up in my life. . . .

So, you know, I got a job at Walmart because I didn't know if my university would still be in business when I returned.

Cheryl recounts a story that begins with what she anticipated would be a short, pleasant trip out of town, an unscheduled vacation to avoid a hurricane that would most likely miss the city, as most do. Some people in New Orleans call these unexpected trips an "e-vacation," a play on "evacuation." In the initial hours of her evacuation, Cheryl, accompanied by family and friends, acted on the basis of a well-thought-out plan to move a caravan of kin and colleagues to a comfortable location in Houston, one well-suited for those who enjoy the arts, an array of spirits, and good food. From there they would wait it out in reasonable comfort the passage of the hurricane.

Calamity, however, in all its guises, from the random accident to breached seawalls weakened by years of neglect, hangs suspended over our best-laid plans, ready to change the course of our lives in the most abrupt and unexpected manner. Freud would have us "treat chance as worthy of determining our fate" in our most quotidian of moments.[7] Life in disaster leaves us no choice in the matter.

Once it was clear that Hurricane Katrina had no intention of avoiding an opportunity to demonstrate New Orleans' poorly built and maintained levee system, her capricious winds shoved the errant waters of the lakes and canals into the city, turning Cheryl's planned evacuation into a protean, shape-shifting journey, a trek that takes her from a spontaneous holiday in Houston to Stone Mountain, Georgia, life lived among thirty-plus relatives, and finally to a job at a local Walmart. While on the road Cheryl encounters a former student who offers to send her $5,000. She hears from the people at American Express that her credit is limitless and no interest charges will accrue to her balance for six months. People from her past pop up to offer her material assistance.

It takes no great leap of the imagination to connect Cheryl's relative good fortune in this turn of events to her social and material resources. Cheese, we might say, has far more than an accidental connection to daily life. But all is not shaped or predicted by material resources. For a middle-class, single mom on the road fleeing a hurricane, cheese intersects with the mutable force of the accidental.

There is little that is predictable about her evacuation experience; it is more a cascade of colliding coincidences than a prudent, well-conceived journey contrived from a first-this-then-that plan of action. If we stop at this juncture and conjure up an orderly abstraction of this personal account of bedlam bordering on entropy, we risk losing what is most forceful about Cheryl's story: to wit, its blow-on-blow account of fate and fortuity. It is plain, however, that the fragments of credit, money, friends, and family from which she cobbled together her journey are not themselves wholly coincidental; they are, rather, connected in some fashion to the supple, elastic materials of class that reach deep into the tissue of human life lived in both its banal and its unnerving moments.

Consider now Denise Anders's account, beginning with her decision to get out of town. Denise left the city in haste, her son in tow, with a few sandwiches, some cereal, and a little money; she drove madly out of harm's way. No e-vacation for her. By the time she left, it was all too clear that much of New Orleans was destined to be under water:

> Well, Saturday morning, watching the news, it was just in my spirit. I was scared. It was like, we need to leave. So me being a single parent, I'm like, who can I get

to leave with me because gas and a hotel room can be expensive. So I called up friends and family members. "The storm is coming our way. We need to go." They like, "We're doing this, we're staying, we have a party." . . . So I'm panicking. I'm at the gas station getting gas. I call a girlfriend of mine; she says her sister has some rooms in Tyler. I said, "I can't get anybody from my family to go. I may just stay." And she said, "No, you can come with us." I had never heard of Tyler, Texas, let alone been to Tyler, Texas.

We get to Tyler. It took us sixteen hours. We get to Tyler at six in the morning. We drove all night. My girlfriend, her husband, and kids, drove in one car. Us in another. They didn't have a cell phone, so it's not like I can call them. I'm just trying to keep up with them 'cause I didn't know where I was going. So when we get to Tyler the next morning we go get breakfast. We . . . see what's happening to New Orleans.

Denise evacuated with her son, but like Cheryl and so many people in New Orleans, her kin, people she carries in her heart and mind, would fast become a part of her life on the road:

I could not get in contact with my family in New Orleans for a couple of days. The plan was my sister who worked at Days Inn on Read Boulevard would take everybody with her. So everybody went to the Days Inn, and the last I heard from them the hotel was flooding and they had to go onto the roof. There were police officers and their families on the roof too. Boats came to get them. They said the boats would come back for everyone else. But they didn't. The police got saved but not my family.

I spoke with someone at the shelter that was doing rescue, and I told them about my family. So a boat finally came. It got my mom, two of my sisters, two of my brother-in-laws, and my little niece and her boyfriend. But then the boat capsized twice. They said my mom almost drowned flipping over. Then they had to go to the convention center. They were at the convention center for, like, three days.

Finally my mom and all get picked up by a helicopter. They fly them to Louis Armstrong Airport. They took a jet from Louis Armstrong to Arlington, Texas. So I had to drive from Tyler to Arlington, Texas, to pick them up. So a guy at the hotel who knew one of my friends said, "I know how to get to Arlington, I'll ride with you." So on the way to Arlington my truck started smoking. I never had a problem with my truck before. So, we can't drive past seventy or it starts smoking, something with the fuel pump, I don't know.

So it takes us like five hours to get to Arlington. And there are all these people at the airport, we go from check-out place to check-out place trying to find them. So we finally get them and we drive back to Tyler. We were able to get another room at the hotel. It was like five in my room and five in another room. So we get together, then we get the money split up.

Meanwhile, Denise's niece and her two children, who had somehow ended up in Houston, were told they had to leave the shelter that was housing them. She told her niece to come to Tyler, where she would be near her family and could get care for her asthmatic son:

I told her they'll give you the nebulizer; they'll give you medicine. So my niece came to Tyler. That was a relief because she was crying every day. So I took her to the shelter. She got the nebulizer and all the medicines. So it was her and I together. She said, "Well, at least I have you."

Denise was able to pay for a third hotel room to house her niece and her niece's children. At last, though far from New Orleans, more than a dozen members of the Anders clan were together. The night her niece arrived, some local folks they had met in Tyler chipped in some cash that Denise distributed among family members.

But even with the help of strangers, money was running low. The motel was only a stop along the way. Denise found a new apartment complex nearby that let people move in without a deposit and offered one month's free rent. Several family members moved into the new apartments, while she moved to a nearby duplex and took her son and mother along with her. Good fortune struck again in the form of a $500 gift card and donations of "blankets, comforters, china, utensils, toilet tissue, everything" from a local church.

In her animated voice, Denise continues:

> When everybody had a cot or bed, a place to stay, we started a routine. Every morning we got up, we piled up as many as we could into the truck, dropped them at the shelter, come back, filled up again and back to the shelter. The shelter fed us. So it took two or three trips to get everybody at the shelter, then to bring everybody back.
>
> On that first week of driving back and forth, on my last trip of the day, it was me, my son, my sister, and my brother-in-law in the truck. I'm in the middle lane and there's an F-150 and a Yukon on the side of me. Coming in the opposite direction, this lady runs a red light. She hit the truck and the SUV and smashed us in the middle. I was okay initially. My brother-in-law, he was in the back, he was hurt. Maybe a week later, my back begins to hurt. I started getting pain from my neck all the way down. So maybe for three weeks I was going to therapy.

Fleeing harm's way, Denise ran headlong into a haphazard world with its own troubles. Her place in that world was complicated by the trials and tribulations of at least fifteen other displaced family members all short on money. There is little in Denise's story that foreshadows what might happen next.

Cheese and Kin

The importance of family is woven into the evacuation stories we collected in Hollygrove and Pontchartrain Park. Not surprisingly, many of these stories sort themselves out by the relative access to cheese. Contingency and its sometimes obvious, sometimes nuanced connection to class is underlined in the following two stories about evacuation and kin. In Gwen Rigby's almost effortless effort to reach her kin and Jesse's angst-filled search for his brother, we meet once again at the intersection of material well-being and life in disaster.

Gwen Rigby from Hollygrove works for Amtrak. A practical person, Miss Rigby planned her evacuation around access to train routes that would take her to friends scattered about the country. Gwen knew that several relatives were stranded on the causeway bridge and at the convention center in the first few days after the storm. She desperately needed to know her kin were okay. A text message would do:

> Once I texted everybody and found out that everybody was okay, it was alright. But then I couldn't find my grandmother. Come to find out my grandmother was evacuated to Houston. So I stayed there long enough to get her situated, and I went on. When I was hopping from train to train at that particular time, I was thinking about my family. "Where are they? Are they okay?" Because even though I could text them, it was a long time before they texted me back. I really didn't get in contact with them until I was in Chicago, which is the reason why I nixed my plans to go from Chicago to Atlanta and to go from Chicago to Houston. So that's all I was thinking about. I wasn't thinking about the house, what it looked like. I was just wondering where was everyone?

Jesse Gray, who calls Hollygrove home, might be speaking for most people in both neighborhoods when he notes with a raw plainness, "Oh, yes, it's important to know where your family is when you're in a disaster. Because you worry if you don't know where your people are." Jesse evacuated and lost touch with his brother Bo, who rode out the storm in his home just a couple blocks from Jesse's house:

> We lost contact with Bo. We didn't know where he was. It was about three weeks before we heard. We thought he had drowned. We made contact with one of my first cousins, and he says Bo was in a boat rescuing people throughout the neighborhood. You know, but we still, we hadn't heard nothing from Bo, and we lost contact with my first cousin 'cause he was still in New Orleans too. Then Bo called, finally, and I remember crying because I thought he was dead. I remember crying on the phone, I was so glad to hear his voice.

Whither the Cheese

While we've no scale or measure affirming this, these varied accounts of evacuation allow us to discern the subtle connections between the agile reach of cheese into the moment-by-moment events, circumstances, and coincidences of evacuation. It might be said we learn far more about the reach of market forces into the interstices of ordinary human life when all hell breaks loose than we do in the quotidian world of the mundane.

As with Cheryl, Denise, and countless other evacuees with their varying resources—or well-practiced resourcefulness—so too for Gwen and Jesse: the trouble to locate kin along the road appeared in one case to be relatively easy while in another to be a source of prolonged worry and concern. In their cases, "cheese" explains a good deal of the variance.

No one would relish being forced to flee home in the face of impending disaster. But a close-up look at the life of an evacuee suggests that once on the road, the relative access to material resources plays a part in sorting out who is likely to move about in space and time with a measure of confidence and who will find the unforeseen just around every turn. If uncertainty is a moment when it is impossible to contrive a reliable probability distribution for outcomes, those whom disasters push to the road are likely to face more or less of it based, at least in part, on their relative access to cheese.

For many if not most residents of Pontchartrain Park and Hollygrove, the long, strange trip of evacuation would prove to be only the beginning. With nowhere to live in a city brought to its knees by turbid waters freed from the shoddy levees meant to contain them, returning to the city would set them on often meandering journeys from evacuee to exile before finding a way home.

Notes

Kroll-Smith, Steve, Vern Baxter, and Pam Jenkins. 2015. Left to Chance: Hurricane Katrina and the Story of Two New Orleans Neighborhoods. Austin: University of Texas Press.

[1] Kansas Joe McCoy and Memphis Minnie, 1929, "When the Levee Breaks," audio recording, June 18 (New York: Columbia Records).

[2] In James West and Chris Vaccaro, 2000, " 'Big Easy' a Bowl of Trouble in Hurricanes," *USA Today*, July (republished August 28, 2005), http://www.usatoday.com/weather/news/2000/wnoflood.htm.

[3] Gordon Russell, 2005, "Nagin Orders First-Ever Mandatory Evacuation of New Orleans," *Times-Picayune*, August 28, 1.

[4] E. L. Quarantelli, 1980, "Evacuation Behavior and Problems," miscellaneous report 27 (Columbus: Disaster Research Center, Ohio State University), 4. See also Thomas Drabek, 1969, "Social Processes in Disaster: Family Evacuation," *Social Problems* 16 (3): 336–49.

[5] Geertz, 1983, 57.

[6] Harvey, 2006, 80.

[7] Sigmund Freud, quoted in Richard Rorty, 1989, *Contingency, Irony, and Solidarity* (Cambridge, England: Cambridge University Press), 22.

PART **VI** SOCIAL MOVEMENTS

People Want to Protect Themselves a Little Bit

16

Emotions, Denial, and Social Movement Nonparticipation

Kari Marie Norgaard

Citizen action is an important way to bring about social and environmental change. But such action depends not just on the perceived existence of a problem but also on people's understandings of that problem and their emotional responses to it. Kari Marie Norgaard looks at the failure of a community in Norway to take action on global climate change. Given that Norway's weather has already been measurably affected by climate change and given Norwegians' relatively high level of knowledge about this issue, Norgaard asks why many Norwegians avoid participation in social action. Norgaard argues that part of the answer lies in the negative emotions—feelings of fear, guilt, helplessness, and so forth—that this issue elicits. Her case study documents how community members engage in collective avoiding of the topic because it makes them unhappy or uncomfortable. Collectively, they seek to distance themselves from the problem. While Norgaard focuses on climate change, her argument may be useful in thinking about many different environment-related problems.

. . . Global climate change is arguably the single most significant environmental issue of our time. Scientific reports indicate that global warming will have widespread ecological consequences over the coming decades including changes in ecosystems, weather patterns, and sea level rise (IPCC 2001). Impacts on human society are predicted to be widespread and potentially catastrophic as water shortages, decreased agricultural productivity, extreme weather events, and the spread of diseases take their toll. Potential outcomes for Norway include increased seasonal flooding, decreased winter snows and the loss of the gulf stream that currently maintains moderate winter temperatures, thereby providing both fish and a livable climate to the northern region. In Norway public support for the environmental movement as well as public awareness of, and belief in, the phenomenon of global warming have been relatively high. In Bygdaby the weather was noticeably warmer and drier than in the past. Yet, in spite of the fact that people were clearly aware of global warming as a phenomenon, everyday life went on as though global warming, and its associated risks, was not a possibility. Despite the apparent heaviness and seriousness of the issue, it was not discussed in the local newspaper, or the strategy meetings of local political, volunteer or environmental meetings I attended. Aside from casual comments about the weather, everyday life went on as though global warming, and its associated risks—did not exist. Instead, global warming was an abstract concept, that was not integrated into everyday life. Mothers listened to news of unusual flooding as they drove their children to school. Families watched evening news coverage of failing Hague climate talks followed by American made sitcoms. Few people even seemed to spend much time thinking about it. It did not appear to be a common topic of either political or private conversation. How did people manage to outwardly ignore such significant risks? Why did such seemingly serious problems draw so little response?

The people of Bygdaby are not unique. Despite the extreme seriousness of global warming, the pattern of meager public response in the way of social movement activity, behavioral changes or public pressure on governments exists worldwide. Public apathy on global warming has been identified as a significant concern by environmental sociologists (e.g., Kempton et al. 1995, Dunlap 1998, Rosa 2001, Brechin 2003). Existing literature emphasizes the notion (either explicitly or implicitly) that information is the limiting factor in public non-response. Yet the people I met were generally well informed about global warming. They expressed concern frequently, yet this concern did not translate into action.

Over time I noticed that conversations about global warming were emotionally charged and punctuated with awkward pauses. The people I spent time with and interviewed raised a number of emotional concerns including fear of the future and guilt over their own actions. During an interview, Eirik described the complexity of the issue:

> We go on vacation and we go shopping, and my partner drives to work every day. And I drive often up here to my office myself. We feel that we must do it to make things work on a practical level, but we have a guilty conscience, a bit of a guilty conscience.

. . . It became further apparent that community members had a variety of tactics for normalizing these awkward moments and uncomfortable feelings—what Arlie Hochschild calls practices of emotion management (1979, 1983, 1990). This paper describes how the presence and management of unpleasant and troubling emotions associated with global warming worked to prevent social movement participation in this rural Norwegian community.

In Bygdaby the possibility of global warming was both deeply disturbing and almost completely invisible, simultaneously unimaginable and common knowledge. The people I spoke with *did* believe global warming was happening, expressed concern about it, yet lived their lives as though they did not know. . . .

Methods: An Ethnography of the Invisible

The observations in this paper are based on one year of field research including 46 interviews, media analysis, and eight months of participant observation. The people I spent time with lived in a rural community of about 14,000 inhabitants in western Norway. Because my research question concerned why people were *not* more actively engaged with the issue of global warming, gathering information required a number of strategies to minimize the tendency for people to begin talking about global warming because it was a topic they knew I found interesting. I kept the specific focus of my research vague, telling people that my work was on issues such as "political participation" and "how people think about global issues." . . . As a participant observer I attended to the kinds of things people talked about, how issues were framed, and especially noted topics that were not discussed. I watched regional television news and read the local and national newspapers. I paid particular attention to beliefs, emotions and cultures of talk with respect to global warming, i.e., whether it was discussed, if so how it came up, how people seemed to feel talking about it. . . . I interviewed as wide a variety of people as I could find. Those I interviewed ranged in age from 19 to early 70s. Respondents were from a variety of occupations, and from six of the nine active local political parties. . . .

. . . I attempted to minimize the degree to which respondents felt a moral pressure to provide a particular answer by first listening to see whether global warming was volunteered as an issue. If global warming was not raised (as it often was not), I asked what people thought about the recent weather (which was widely described as abnormal), and followed with more specific questions such as when they first began thinking about global warming and whether they spoke about global warming with family or friends.

Why Norway?

Despite the salience of my questions to the situation in all Western nations, a case study set in Norway is particularly useful. Anyone who begins to talk about movement non-participation, denial and political action in the U.S. immediately encounters a host of relevant questions: "Do people really know the information?" "Is global warming really happening? I thought it was still controversial." "Do people really have enough time and money to spare that we can consider it denial that they are not acting?" "People in the U.S. are apathetic in general, why would it be any different on this issue?" Each of these are valid questions that complicate an analysis like mine. Yet each of these factors is either absent or minimized in Norway: Norway has one of the highest levels of GDP of any nation and a fifty year history of welfare state policies that has redistributed this wealth among the people (UNDP 2004). In terms of political activity, Norwegians again are exemplary. High percentages of Norwegians vote and are active in local politics. When it comes to information and knowledge, Norwegians are a highly educated public. Norway and Japan are tied for the highest level of newspaper readership in the world. Furthermore, in contrast to the situation in the U.S. (Gelbspan 2004), Norwegian media did raise the issue of global warming in their coverage of the unusual weather, and described potential future weather scenarios and impacts. Although there were certainly skeptics about global warming in Bygdaby, such skepticism is much less than in the United States where large counter campaigns have been waged by industry (McCright and Dunlap 2000, 2003). Finally, Norwegians have been proud of their relationships to nature, environmentalism and leadership on global environmental issues including global warming (Eriksen 1993, 1996). If any nation can find the ability to respond to this problem, it must be in such a place as this, where the population is educated, cared for, politicized and environmentally engaged.

Research in Norway is also unique due to the particularly strong contradiction between professed values and the nation's political economy. In Norway there is strong identification with humanitarian values and a heightened concern for the environment (Reed and Rothenberg 1993). Yet as the world's sixth largest oil producer, Norway is one of the nations of the world that has benefited most from oil production. The presence of high levels of wealth, political activity, education, idealism and environmental values together with a petroleum based economy made the contradiction between knowledge and action particularly visible in Norway.

"Bygdaby" was selected because its size allowed me access to a wide cross section of the community, and the fact that residents spoke a dialect I was able to understand. The presence of a nearby lake (that failed to freeze) and ski area (that opened late) were not conditions I selected for, but ones that nonetheless added to the visibility and salience of global warming for community members.

The Winter of 2000–2001 in Bygdaby

Global warming was clearly salient for Norwegians on both the local and national levels during the time period of my research. A number of unusual weather events took place in the fall and winter that year. Most tangible was the very late snowfall and warmer winter temperatures in Bygdaby. Temperatures for the community as reported by the local newspaper showed that average temperature in the Bygdaby region on the whole was warmer than in the past. In fact, as of January 2001 the winter of 2000 was recorded as the second warmest in the past 130 years. Additionally, snowfalls arrived some two months late (midto late January as opposed to November). As a result of these conditions, the ski area opened late, with both recreational and economic effects on the community, and the ice on the lake failed to freeze sufficiently to allow ice fishing, once a frequent activity. In fact, not only did the local ski season start late, the downhill ski area opened with 100% artificial snow—a completely unprecedented event. A woman who was walking on the lake drowned when the ice cracked and she fell through, although this sort of accident could have happened in the spring of any year when the ice normally broke up.

The topic of global warming was also very visible in the media. In addition to weather events, a number of national and international political events brought global warming to Bygdabyingar's minds. In November, several thousand miles to the South, the nations of the world held climate meetings at the Hague. Both the King and Prime Minister mentioned global warming in their New Year's Day speeches. Three weeks later on January 22, 2001, the United Nations Intergovernmental Panel on Climate Change released a new report on climate. In March U.S. President George Bush declared that following the Kyoto Protocol was not in the economic interests of the United States and flatly rejected it. Each of these events received significant attention in the regional and national press. . . . Indeed I was continually impressed with the level of up to date information that people had regarding global warming. Here is an excerpt from a focus group that I conducted with five female students in their late teens the week after the climate talks at the Hague failed. Note that these young women are aware of the failed talks, are familiar with (and critical of) the fact that Norway is required to decrease carbon dioxide emissions 5% by the year 2008, and that they feel that global warming is a real issue, observable in their immediate surroundings:

KARI: What have you heard about global warming?
SIRI: I have heard about the conference, I became a bit afraid when they didn't reach agreement . . .
TRUDI: Our Minister of Environment! In 2008 we will decrease our emissions by five percent (General laugher)
METTE: That will help!
KARI: And is it something that you feel is really happening, or . . . (Several speaking at once)
METTE: Now it is incredible, five degrees Celsius is, you know, really strange . . . mmm, Ja-
SIRI: (Interrupting) There should be snow [now].
TRUDI: It comes in much closer for us. It is here . . . you notice it. you know, it's getting worse and worse . . . Last year there was snow at this time of year. And actually that is the way it should have been for quite some time now.

A Series of Troubling Emotions

Although the sense that people fail to respond to global warming because they are too poorly informed (Read et al. 1994, Kempton et al. 1995, Dunlap 1998, Bord et al. 1998, Brechin 2003), too greedy or too individualistic, suffer from incorrect mental models (Bostrom et al. 1994) or faulty decision making processes (Halford and Sheehan 1991), underlies much of the research in environmental sociology, the people I spoke with expressed feelings of deep concern and caring and a significant degree of ambivalence about the issue of global warming. People in Bygdaby told me many reasons why it was difficult to think about this issue. In the words of one man, who held his hands in front of his eyes as he spoke, "people want to protect themselves a bit." Community members described fears associated with loss of ontological security, feelings of helplessness, guilt and the associated emotion of fear of "being a bad person.". . .

Not only were these emotions unpleasant in themselves, the feelings that thinking about global warming raised went against local emotion norms. . . . Emotion norms in Bygdaby (and Norway generally) emphasized the importance of maintaining control (beholde kontroll) and toughness (å være tøff), and for young people, being cool (kult)—especially in public spaces. Adults, especially men and public figures, faced pressures to be knowledgeable and intelligent. In some settings, especially for educators, there was an emotion norm of maintaining optimism. Educators described balancing personal doubts and deep feelings of powerlessness with the task of sending a hopeful message to students. When I spoke with Arne, a teacher at the local agricultural school:

> I am unfortunately pessimistic. I just have to say it. But I'm not like that towards the students. *You know, I must be optimistic when I speak with the students.*

Note that Arne's use of the phrase "you know" highlights the sense that this reality, the need to be optimistic with students, is taken for granted, incontestable. . . .

Risk, Modern Life, and Fears Regarding Ontological Security

> **Automobile and plane crashes, toxic chemical spills and explosions, nuclear accidents, food contamination, genetic manipulation, the spread of AIDS, global climate change, ozone depletion, species extinction and the persistence of nuclear weapons arsenals: the list goes on. Risks abound and people are increasingly aware that no one is entirely safe from the hazards of modern living. Risk reminds us of our dependency, interdependency and vulnerability (Jaeger et al. 2001, 13).**

One day in mid-December my husband and I, disappointed with the lack of snow in Bygdaby, decided to take the train a few hours away to a neighboring community. The temperature was about minus five and the sun was shining brightly on the bare fields surrounding our house as we loaded our skis into the taxi and headed down the road to the train. "Do you like to ski?" I asked our driver? "Oh yes, but I don't do much of that anymore," he replied. "When I was a kid we would have skis on from the first thing in the morning to the end of the day. There was so much more snow back then. When you think of how much has changed in my fifty years it is very scary."

Global warming threatens biological conditions, economic prospects and social structure (IPCC 2001). At the deepest level, large scale environmental problems such as global warming threaten people's sense of the continuity of life, what Anthony Giddens calls ontological security (1984, 1991). What will Norwegian winters be like without snow? What will happen to farms in the community in the next generation? . . .

Feelings of Helplessness—"You Have to Focus on Something You Can Do"

> I think that there are a lot of people who feel that no matter what I do I can't do anything about that anyway.

As Hege Marie, a student in her late teens described, a second emotion that the topic of global warming evoked was helplessness. The problem seemed so large and involved the cooperation of people in so many different countries. Governments were unable to reach agreement. Perhaps entire economic structures would have to change. Thus it is not surprising that rather than feeling that there was much that could be done, Liv, a woman in her late sixties, pronounced that, "we must take it as it comes," and Gurid told me, "you have to focus on something you can do or else you become completely hopeless."

Fear of Guilt

Thinking about global warming was also difficult because it raised feelings of guilt. Members of the community told me they were aware of how their actions contributed to the problem and they felt guilty about it.

> So many times I have a guilty conscience because I know that I should do something, or do it less. But at the same time there is the social pressure. And I want for my children and for my wife to be able to experience the same positive things that are normal in their community of friends and in this society. It is very . . . I think it is a bit problematic. I feel that I could do more, but it would be at the expense of, it would create a more difficult relationship between me and my children or my partner. It really isn't easy.

Guilt was also connected to the sense of global warming as an issue of global inequity: Norwegian wealth and high standard of living are intimately tied to the production of oil. Given their high newspaper readership and level of knowledge about the rest of the world, community members were well aware of these circumstances. This understanding contrasted sharply with the deeply ingrained Norwegian values of equality and egalitarianism (Jonassen 1983, Kiel 1993), thus raising feelings of guilt.

Fear of "Being a Bad Person"—Identity: Self and National Images

Another source of concern that comes with awareness of global warming is the threat it implies for individual and national self-concepts. . . .

Norwegian public self-image includes a strong self-identification of being environmentally aware and humanitarian (Eriksen 1993, 1996). Norwegians have been proud of their past international leadership on a number of environmental issues including global warming. Stereotypical characterization of Norwegians describes a simple, nature loving people who are concerned with equality and human rights (Eriksen 1993, 1996). Yet Norway has increased production of oil and gas threefold in the last ten years. Expansion of oil production in the 1990s contributed significantly to the already high standard of living, making Norway one of the countries in the world that has most benefited from fossil fuels. In 2001 Norway was the world's sixth largest oil producer and the world's second largest oil exporter after Saudi Arabia (MoPE 2002). Information about global warming—such as Norway's inability to reach Kyoto reduction quotas, increasing carbon dioxide emissions and government expansions of oil development—makes for an acute contradiction between the traditional Norwegian values and self-image and the present day economic situation in which high electricity use, increasing consumption and wealth from North Sea oil make Norway one of the larger per capita contributors to the problem of global warming. . . . For Norwegians, information on global warming not only contradicts their sense of being environmentally responsible. As a problem generated by wealthy nations for which people in poor nations disproportionately suffer, knowledge of global warming also challenges Bygdabyingar's sense of themselves as egalitarian and socially just. . . .

A Cultural Tool Kit of Emotion Management Strategies

If the emotions of fear, guilt, hopelessness or "fear of being a bad person" worked against social change in Bygdaby, how might this have happened? In Bygdaby there were active, observable moments which, although fleeting, pointed to the role of emotions in the generation of non-participation as an active process—what Nina Eliasoph calls the *production* of apathy (1998). If what a person feels is different from what they want to or are supposed to feel they may engage in some level of *emotional management* (Hochschild 1979, 1983, 1990, Thoits 1996). While the act of modifying, suppressing or emphasizing an emotion is carried out by individuals, emotions are being managed to fit social expectations, which in turn often reproduce larger political and economic conditions. . . . In the case of global warming in Bygdaby, emotions that were uncomfortable to individuals were also uncomfortable because they violated norms of social interaction in the community. And at least some of these emotion norms in turn normalized Norway's economic position as a significant producer of oil.

. . . In Bygdaby people managed the unpleasant emotions described in the previous section by avoiding thinking about them, by shifting attention to positive self-representations, and—especially in terms of the emotion of guilt—by framing them in ways that minimized their potency. When it came to the strategy of framing and of shifting attention to positive self-representations, community members had available a set of "stock" social narratives upon which to draw, many of which were generated by the national government and conveyed to the public through the media. . . .

Ann Swidler uses the metaphor of "tools in a tool kit" to describe the set of resources available to people in a given culture for solving problems (1986). Using this metaphor I will briefly describe how these culturally available strategies served as tools that were used to achieve selective attention and perspectival selectivity—and thereby to manage thinking in such a way as to manage emotions.

Selective Attention

Selective attention can be used to decide what to think about or not to think about, screening out for example painful information about problems for which one does not have solutions (e.g., "I don't really know what to do, so I just don't think about that"). Strategies of emotion management in the form of selective attention were primarily aimed at managing the emotions of fear and helplessness. Here I describe the techniques of controlling exposure to information, focusing on something you can do and not thinking too far into the future.

"We Can't Dig Ourselves into Depression, Right?": Controlling Exposure to Information

. . . Community members described feelings of uncertainty as being easily evoked by too much information, thus adhering to the emotion norms of maintaining optimism and control required managing exposure to information on global warming. Educators and activists in particular had to be careful not to become overwhelmed in order to continue their work:

> No, but you can't—you know I feel that in a way the philosophy of all this is happening so fast. *I do as much as I can,* and *we can't focus on what's painful.* We don't go in and have meetings and talk about how gruesome everything is. We talk about how it is and *can't dig ourselves down into depression, right?*

Another activist described how she reads very little of the details, that it is in fact "better not to know everything." People were aware that there was the potential that global warming would radically alter life within the next decades, and when they thought about it they felt worried, yet they did not go about their days wondering what things would be like for their children, whether these could be the last years farming could take place in Bygdaby, or whether their grandchildren would be able to ski on real snow. They spent their days thinking about more local, manageable topics. Mari described how, "you have the knowledge, but you live in a completely different world." . . .

"I Don't Allow Myself to Think so far Ahead"

There is a lot of unrest in the country. There is a lot that is negative. Then I become like—yeah, pfff! But when someone has something that they are working on, in relation to that you are trying to influence—then it's like, okay to be optimistic after all. But I think that this can just explode around us, and so it is well that *I don't allow myself to think so far ahead.*

The most effective way to manage unpleasant emotions was to turn your attention to something else, as Lise, a young mother describes in the above passage, or by focusing attention onto something positive, as she also describes. . . .

Focusing on Something You Can Do

Similar to the strategies of controlling exposure to unpleasant information and not thinking too far ahead was the strategy of focusing on something that you could do. . . . Peter, a local politician, describes how global warming is a theme that "everyone is interested in" but which does not receive attention on the local level because there "isn't so much that you can do."

> Yes, it is of course a theme that everyone is interested in, but locally it isn't discussed much because . . . well climate change, you know there isn't so much you can do with it on a local level, but of course everyone sees that something must be done . . .
>
> Peter's comments are similar to the earlier passage with Lise who describes both the need for optimism and the underlying hopelessness that global warming raised.

Perspectival Selectivity

Unpleasant emotions of guilt and those associated with a "spoiled identity" could be managed through the cognitive strategy of perspectival selectivity. Perspectival selectivity, "refers to the angle of vision that one brings to bear on certain events" (Rosenberg 1991, 134). Euphemisms, technical jargon and word changing are used to dispute the meanings of events such as when military generals speak of "collateral damage" rather than the killing of citizens. Stanley Cohen writes, "Officials do not claim that 'nothing happened' but what happened is not what you think it is, not what it looks like, not what you call it" (2001: 7). Here I describe two "stock" social narratives that were frequently used to change the angle of vision one might bring to the facts about Norway's role in the problem of global warming.

"Amerika" as a Tension Point

Bygdabyingar knew an amazing number of facts about the U.S. References to the U.S. appeared in numerous conversations I participated in and overheard while in Bygdaby. I use the Norwegian spelling of the word to indicate that I am talking about a stereotypical Norwegian view of the U.S., what Steinar Bryn calls Mythic America (Bryn 1994). There are many stereotypical images of the U.S. in Norway, but to me what is most interesting is not the images themselves but how they were used.

Stories about "Amerika" were often told in strategic moments to deflect Norwegian responsibility and shortcomings and to support notions of Norwegian exceptionalism (we may not be the best, but we aren't anything as bad as they are). For example, in late April of 2001 U.S. President Bush made the infamous statement that he would not sign the Kyoto Protocol on the grounds that it was, "not in the U.S. economic interests." Many Bygdabyingar took the opportunity to tell me of their criticism of this position. Bush's comment was widely repeated and discussed in the Norwegian press and in public commentary. Here the statement was used in a motivational speech by a local young woman on May 1, 2001:

The Kyoto agreement is about cutting carbon dioxide emissions by 5 percent. And even that ridiculous pace was too much for the climatehooligan George W. Bush in the United States. The head of the USA's environmental protection department said that "We have no interest in meeting the conditions of the agreement." Well, that may be so. But it is other countries that will be hit the hardest from climate change . . .

Yet despite widespread criticism of the United States for taking such a position, this is essentially the same move that the Norwegian government made in dropping national emissions targets, increasing oil development, taking a leading role in the development of the carbon trading schemes known as the Kyoto and Clean Development Mechanism and shifting the focus from a national to an international agenda (Hovden and Lindseth 2002). In this context, criticizing the poor climate record of the U.S. directs attention away from Norway's shifting behavior, sending the message that at home things are not that bad. . . .

"Norway Is a Little Land"

A second narrative, "Norway is a little land" deflected troubling information and emotions connected to Norway's role in global warming with the subtext that, "we are so few, it doesn't really matter what we do anyway." While it often conveyed a genuine sense of powerlessness, this discourse also worked to let people off the hook, creating the sense of "why bother." During a conversation about his opposition to Norway joining the European Union, Joar, a Bygdabyingar in his early 50s explained how this emphasis on Norway's size, while in some sense true, is also a strategic construction:

KARI: But what kind of a role do you think that Norway should take internationally?

JOAR: *Well, we are of course a very small country, almost without meaning,* if you think economically we are completely uninteresting.

KARI: But Norway has lots of oil compared with other countries.

JOAR: yeah, yeah, okay. We are in fact almost at the level of Saudi Arabia. *But it [is] of course an advantage to be meaningless.* It doesn't really matter for us to argue, *they don't bother to get mad at us, because we are so meaningless.* And in that connection, we are a bit you know, peaceful, right. We have been involved in both the Middle East and . . . (here he refers to the Oslo Accords and his second example is not spoken, just given as a gesture of the hand for emphasis).

Note that as the conversation continues he uses Norway's small economy as the example of why it "is meaningless." When I asked him about Norway's oil, he suddenly "remembers" the fact that Norway is, after Saudi Arabia, the second largest oil exporter in the world. Then he explains the strategic advantage of being "meaningless," that other countries don't bother to get upset with Norway. At the end of the passage he adds to the construction of Norway as a nation not worth getting upset with by drawing on the sense of Norway as a "peaceful nation" (referring to the peace prize) and their involvement in the Oslo Accords.

In being small, meaningless and peaceful, he is constructing in our conversation a sense of Norwegian innocence that is very prevalent.

The phrase "Norway is a little land" gives the sense that they are doing "their part" and turns blame back onto those who are "worse," especially the United States, as described earlier. It serves to imply that, "the problem isn't really us. We, in fact, are innocent." . . .

Discussion: Emotions, Emotion Management, and the "Production of Apathy"

Non-response to the possibility of global warming may seem "natural" or "self-evident"—from a social movement or social problem perspective not all potential issues get translated into political action. Yet with a closer view we can understand non-response as a *social process*. Things *could have been* different. Community members could have written letters to the local paper articulating global warming as a political issue, they could have brought the issue up in one of the many public forums, made attempts to plan for the possibility of what the future weather scenarios might bring, put pressure on local and national leaders, decreased their automobile use, asked for national subsidies to cover the economic impacts of the warm winter, or engaged their neighbors, children, and political leaders in discussions about what climate change might mean for their community in the next ten and twenty years. Indeed in other parts of the world things *were* different. The severe flooding in England that fall was linked to global warming by at least some of the impacted residents. People from affected communities in England traveled to the climate talks at the Hague to protest. More recently, three cities in the United States have initiated a lawsuit against the federal government over global warming. Bygdabyingar could have made a similar move—rallied around the lack of snow and its economic and cultural impacts on some level, any level, be it local, national or international. But they did not.

Most of the emotions Bygdabyingar felt in conjunction with information on global warming: fear, guilt and concern over individual and collective identity could have motivated social action. Perhaps in some cases these emotions did generate actions, but they did not generate many. . . .

. . . [E]motions of fear and helplessness contradicted emotion norms of being optimistic and maintaining control. These emotions were particularly managed through the use of selective attention: controlling one's exposure to information, not thinking too far into the future and focusing on something that could be done. Although the range of emotion management techniques appeared to be used across the community, I found these strategies used with more frequency by educators, men and public figures. The emotion of guilt and the fear of being a bad person or desire to view oneself and the collective community in positive light contrasted not only with specific local emotion norms surrounding patriotism, but also the general social psychological need to view oneself in a positive light (i.e., manage identity). Guilt and identity were managed through the use of perspectival selectivity: by emphasizing Norway's small population size and that no matter what they did, Norwegians were not as bad as the "Amerikans."

Conclusion: Emotions, Denial, and Social Movement Nonparticipation

Emotions can be a source of information (Hochschild 1983) and an impetus for social action (Jasper 1997, 1998, Polletta 1998), but my observations in Bygdaby suggest that the desire to avoid unpleasant emotions and the practice of emotion management can also work against social movement participation. Although not normally applied to environmental issues, research on the sociology of emotions is highly relevant to understanding community member's reactions to global warming. While current work in environmental sociology has emphasized the "information deficit model" (Buckeley 2000), my ethnographic and interview data from a rural Norwegian community do not support this interpretation. Instead this research indicates community members had sufficient information about the issue but avoided thinking about global warming at least in part because doing so raised fears of ontological security, emotions of helplessness and guilt, and was a threat to individual and collective senses of identity. Rather than experience these unpleasant emotions, people used a number of strategies including emotion management to hold information about global warming at arm's length.

Emotions played a key role in denial, providing much of the reason why people preferred not to think about global warming. Furthermore, the management of unpleasant and "unacceptable" emotions was a central aspect of the process of denial, which in this community was carried out through the use of a cultural stock of strategies and social narratives that were employed to achieve selective attention and perspectival selectivity. Thus movement non-participation in response to the issue of global warming did not simply happen, but was actively produced as community members kept the issue of global warming at a distance via a cultural tool kit of emotion management techniques.

Note

Norgaard, Kari Marie. 2006. "'People Want to Protect Themselves a Little Bit': Emotions, Denial, and Social Movement Nonparticipation." *Sociological Inquiry* 76 (3): 372–96.

References

Bord, Richard, Ann Fisher, and Robert O'Connor. 1998. "Public Perceptions of Global Warming: United States and International Perspectives." *Climate Research* 11 (1): 75–84.

Bostrom, Ann, M. Granger Morgan, Baruch Fischoff, and Daniel Read. 1994. "What Do People Know About Global Climate Change? I: Mental Models." *Risk Analysis* 14 (6): 959–70.

Brechin, Steven. 2003. "Comparative Public Opinion and Knowledge on Global Climatic Change and the Kyoto Protocol: The U.S. versus the World?" *International Journal of Sociology and Social Policy* 23 (10): 106–34.

Bryn, Steinar. 1994. "The Americanization of Norwegian Culture." Doctoral diss., Department of Philosophy, University of Minnesota.

Buckeley, Harriet. 2000. "Common Knowledge? Public Understanding of Climate Change in Newcastle, Australia." *Public Understanding of Science* 9: 313–33.

Cohen, Stanley. 2001. *States of Denial: Knowing About Atrocities and Suffering.* Cambridge: Polity Press, 2001.

Dunlap, Riley. 1998. "Lay Perceptions of Global Risk: Public Views of Global Warming in Cross National Context." *International Sociology* 13 (4): 473–98.

Eliasoph, Nina. 1998. *Avoiding Politics: How Americans Produce Apathy in Everyday Life.* Cambridge: Cambridge University Press.

Eriksen, Thomas Hylland. 1993. "Being Norwegian in a Shrinking World." In *Continuity and Change: Aspects of Contemporary Norway,* edited by Anne Cohel Kiel, 11–38. Oslo: Scandinavia University Press.

———. 1996. *Norwegians and Nature.* From Official Government website. www.dep. no/odin/english/p30008168/history/032005-990490/dok-bu.html (accessed May 10, 2006).

Gelbspan, Ross. 2004. *How Politicians, Big Oil and Coal, Journalists, and Activists Are Fueling the Climate Crisis—And What We Can Do to Avert Disaster.* New York: Basic Books.

Giddens, Anthony. 1991. *Modernity and Self Identity: Self and Society in the Late Modern Age.* Cambridge: Polity Press.

———. 1984. *The Constitution of Society.* Cambridge: Polity Press.

Halford, Grame, and Peter Sheehan 1991. "Human Responses to Environmental Changes." *International Journal of Psychology* 269 (5): 599–611.

Hochschild, Arlie 1990. "Ideology and Emotion Management: A Perspective and Path for Future Research." In *Research Agendas in the Sociology of Emotions,* edited by T. D. Kemper, 108–203. Albany: State University of New York Press.

———. 1983. *The Managed Heart: The Commercialization of Human Feeling.* Berkley: University of California Press.

———. 1979. "Emotion Work, Feeling Rules and Social Structure." *American Journal of Sociology* 85: 551–75.

Hovden, Eivind, and Gard Lindseth. 2002. "Norwegian Climate Policy 1989–2002." In *Realizing Rio in Norway: Evaluative Studies of Sustainable Development,* edited by William Lafferty, Morton Nordskog, and Hilde Annette Aakre, 143–68. Oslo: University of Oslo.

IPCC (Intergovernmental Panel on Climate Change). 2001. *Climate Change 2001: Synthesis Report.* Cambridge: Cambridge University Press for the IPCC.

Jaeger, Carlo, Ortwin Renn, Eugene Rosa, and Thomas Webler. 2001. *Risk, Uncertainty and Rational Action.* London: Earthscan.

Jasper, James M. 1998. "The Emotions of Protest: Affective and Reactive Emotions in and Around Social Movements." *Sociological Forum* 13 (3): 397–424.

———. 1997. *The Art of Moral Protest.* University of Chicago Press.

Jonassen, C. 1983. *Value Systems and Personality in a Western Civilization: Norwegians in Europe and America.* Columbus: Ohio State University Press.

Kempton, Willet, James S. Bister, and Jennifer A. Hartley. 1995. *Environmental Values in American Culture.* Cambridge, MA: The MIT Press.

Kiel, Anne Cohel. 1993. *Continuity and Change: Aspects of Contemporary Norway.* Oslo: Scandinavia University Press.

McCright, Aaron M., and Riley E. Dunlap. 2003. "Defeating Kyoto: The Conservative Movement's Impact on U.S. Climate Change Policy." *Social Problems* 50 (3): 348–73.

————. 2000. "Challenging Global Warming as a Social Problem: An Analysis of the Conservative Movement's Counter-Claims." *Social Problems* 47 (4): 499–522.

MoPE (Norwegian Ministry of Petroleum and Energy). 2002. *Environment 2002: The Norwegian Petroleum Sector Fact Sheet.* Oslo: Oilje og energidepartmentet.

Polletta, Francesca. 1998. "It Was Like a Fever . . . Narrative and Identity in Collective Action." *Social Problems* 45: 137–59.

Read, Daniel, Ann Bostrom, M. Granger Morgan, Baruch Fischoff, and Tom Smuts. 1994. "What Do People Know About Global Climate Change? II Survey Studies of Educated Lay People." *Risk Analysis* 14: 971–82.

Reed, Peter, and David Rothenberg. 1993. *Wisdom in the Open Air: The Norwegian Roots of Deep Ecology.* Minneapolis: University of Minnesota Press.

Rosa, Eugene. 2001. "Global Climate Change: Background and Sociological Contributions." *Society and Natural Resources* 6 (14): 491–99.

Rosenberg, Morris. 1991. "Self-processes and Emotional Experiences." In *The Self-Society Dynamic: Cognition, Emotion and Action,* edited by Judith Howard and Peter Callero, 123–42. Cambridge: Cambridge University Press.

Swidler, Anne. 1986. "Culture in Action." *American Sociological Review* 51: 273–86.

Thoits, Peggy. 1989. "The Sociology of Emotions." *Annual Review of Sociology* 15: 317–42.

————. 1996. "Managing the Emotions of Others." *Symbolic Interaction* 19: 85–109.

United Nations Development Programme (UNDP). *United Nations Human Development Report 2004: Cultural Liberty in Today's Diverse World.* Data on Norway. hdr.undp.org/reports/global/2004/ (accessed May 10, 2006).

17 Environmental Threats and Political Opportunities
Citizen Activism in the North Bohemian Coal Basin

Thomas E. Shriver, Alison E. Adams, and Stefano B. Longo

Studies of environmental degradation and environmental activism have largely focused on democratic, capitalist societies. But, as this case study of Czechoslovakia shows, state-run, state-planned economic policies can also be severely environmentally damaging. The specific Soviet-influenced communist regimes that existed from the post–World War II era until the 1990s have crumbled, along with much of the state-controlled economic planning they emphasized. This chapter can, however, help us think about environmental activism in the context of other current authoritarian regimes. Shriver and colleagues show that, in this North Bohemian Coal Basin region, when environmental degradation got too extreme to ignore, citizens did take action, even though they lived under a repressive regime. By studying movements such as this one, we move closer to understanding how citizens, when pushed too far, can find the courage to voice opposition to oppressive state policies.

Throughout the period of Communist rule, the North Bohemian Coal Basin of Czechoslovakia became one of the most environmentally devastated areas in the world. After seizing power in 1948, the Communist Party of Czechoslovakia [*Komunistická strana Československa*] imposed strict production schedules throughout the country to meet the heavy industry needs of the Soviet Union. The North Bohemian Coal Basin became instrumental to production efforts because of its vast supplies of low-grade brown coal. The widespread mining operations coupled with the proliferation of coal-fired power plants contributed to profound environmental pollution and alarming human health impacts. These production externalities became so threatening throughout the 1980s that North Bohemian residents fought back against the authoritarian state despite repression and retaliation. This case provides an excellent opportunity to investigate how political and environmental threats can provoke activism, particularly in highly repressive settings. Drawing from literatures in environmental sociology and social movements, our research delineates the relationship between extreme production and citizen protest in repressive contexts.

254

Our analysis focuses on an authoritarian state socialist system, where economic growth was directed solely by state elites. We use the case of the North Bohemian Coal Basin to examine the extreme environmental conditions that resulted from intensifying production in the region, and how residents risked harsh state retaliation to protest these conditions.

The date for this project were collected over a fifteen-year period and come from in-depth interviews with environmentalists and other residents in the region, archival documents, and historical accounts. Our findings highlight how the state's commitment to unlimited production expansion resulted in environmental threats to human life, ultimately provoking North Bohemian residents to protest against the authoritarian regime.

Treadmill of Production, Political Opportunity, and Environmental Threats

Treadmill of production theory has significantly influenced American environmental sociology over the past three decades (Buttel 2004; Dunlap 1997; Foster 2005). Introduced by Allan Schnaiberg, the treadmill framework initially sought to analyze the drivers of environmental degradation in the United States following World War II (Gould, Pellow, and Schnaiberg 2004; Schnaiberg 1980). Schnaiberg (1980) pointed to the allocation of the social surplus toward economic growth, which increased ecosystem disruptions. Since its original formation, treadmill theory has been widely utilized and extended within environmental sociology (Clark and Jorgenson 2012; Gould, Pellow, and Schnaiberg 2004; Gould, Schnaiberg, and Weinberg 1996; Hooks and Smith 2004; Weinberg, Pellow, and Schnaiberg 2000).

Treadmill theorists delineate the ecological contradictions associated with an economic system that is inherently dependent on growth (Schnaiberg 1980; Schnaiberg and Gould 1994). Capitalist systems rely on constant, profit-driven expansion, where capital investment is attracted to industries and firms showing greater profits. This constant effort to enhance profits and remain competitive in a growth-focused economy results in a treadmill of production that drives unlimited expansion. Natural resources are commodified, and their resultant profits are reinvested into more extraction and production (Gould, Pellow, and Schnaiberg 2004). As such, more natural resource withdrawals are required, and increasing additions to the environment in the form of pollution result. Thus, this growth-based dependency creates conditions under which environmental sustainability is unfeasible (Gould, Pellow, and Schnaiberg 2004; Schnaiberg 1980; Schnaiberg and Gould 1994).

In capitalist societies, the collective power of the corporate capitalists, together with the political directives of the state, and to a lesser extent the interests of labor, ensures the constant acceleration of growth (Gould, Pellow, and Schnaiberg 2004).

But, how do we understand drivers of growth in political-economic contexts where production is controlled solely by the state? . . .

Scholars have argued that reactions to the externalities resulting from treadmill production are likely shaped by both local conditions and broader political forces (Gould, Schnaiberg, and Weinberg 1996). Political opportunity theory provides theoretical and analytical direction by highlighting the relationship between political opportunity structures and movement mobilization (Eisinger 1973; Jenkins and Perrow 1977; Tilly 1978; McAdam 1982). In the context of this work, *political opportunity structure* refers to the arrangement of opportunities that signals whether activists should deploy resources for collective challenges. This perspective works to explain how changes in the political structure (such as levels of receptivity to challenges) influence movement mobilization.

A growing number of scholars have examined the role of threats in sparking movement mobilization (Almeida 2003, 2008; Goldstone and Tilly 2001; Khawaja 1993; Osa 2001). Analytically distinct from opportunities, threats are an important aspect of the broader political opportunity structure (Maher 2010). Almeida (2008, 14) argued that both opportunity and threat can lead to mobilization: "Viewing opportunity and threat as ideal types, groups may either be driven by environmental cues and institutional incentives to push forward demands and extended benefits (i.e., political opportunity) or be pressed into action in fear of losing current goods, rights, and safety (i.e., threat)." Threats can take the form of potential state repression or the potential loss of assets such as basic human rights (Almeida 2003; Goldstone and Tilly 2001; Schock 1999). Maher (2010) notes that the severity of the threat is directly linked to the likelihood of challenge: where threats are immediate and severe, citizens are likely to mobilize even in the face of extreme repression. Scholars have identified numerous types of threat and linked them to various forms of activism (Almeida 2003; Goldstone and Tilly 2001; McVeigh 1999; Tilly 1978; Van Dyke and Soule 2002).

Research has recognized *environmental* or *ecological* threats as salient motivators for environmental mobilization (Maher 2010; Van Dyke and Soule 2002). Notably, Johnson and Frickel (2011, 305) theorize the linkages between ecological threats and activism, defining these threats as "costs associated with environmental degradation as it disrupts (or is perceived to disrupt) ecosystems, human health, and societal well-being." Their research provides evidence that environmental threats are correlated with the establishment of issue-specific environmental organizations in the United States. However, our knowledge of the role of environmental threats in inciting mobilization is limited to Western liberal democracies, where ecologically damaging production is driven by a growth coalition (state, corporation, and labor). What is the relationship between environmental threats and mobilization in authoritarian societies where the state controls production?

Nationalizing the Treadmill of Production in Czechoslovakia

Following the end of World War II, the Communist Party of Czechoslovakia [*Komunistická strana Československa* (KSČ)] began maneuvering for political power. Over the next two years, a struggle ensued within the government, and the KSČ ultimately seized power in February 1948. Backed by the Soviet Union, the KSČ set out to restructure the economy around heavy industry, including the production of armaments and machinery for export to other Soviet socialist bloc countries. The party established a centrally planned economy organized around distinct five-year production plans (Brzezinski 1967; Jancar 1970) to ensure continual industrialization and production expansion. These plans were designed by a variety of elite bodies, including the State Planning Commission [*Státní plánovací komise*] (Pavlínek and Pickles 2000), which then directed their successful implementation.

The growth coalition in Soviet-style socialism was entirely state controlled. To accelerate industrialization, the state-socialist growth coalition controlled every aspect of the economy, including planning and the appropriation and the distribution of the social surplus (Resnich and Wolff 2002). . . . The KSČ organized production assets into firms managed by state employees. They assigned the highest priority to industrial investment, including armaments, often disregarding the production of food and other consumer goods. Foreign trade and investment with countries outside the Eastern Bloc was discouraged, and national self-reliance was emphasized (Čornej and Pokorný 2003; Krejčí and Machonin 1996). State-appointed economic planners imposed formidable production quotas to accelerate industrial expansion in Czechoslovakia (Tickle 2000; DeBardeleben 1985; Tickle and Vavroušek 1998). A system of incentives was established to reward over-fulfillment of production goals, and penalties were incurred when schedules were not met (Albrecht 1987; Redclift 1989).

The KSČ's centralized planning coupled with an insatiable compulsion for growth accelerated a state-driven treadmill of production. Elite planning bodies in Czechoslovakia essentially manufactured the conditions for economic growth. In many economic sectors, this led to a ritualistic exercise in which resources and products were shipped back and forth between Eastern European nations merely to satisfy production quotas. A former miner from the northern industrial region described this performance: "During communism, we were mining black coal in Ostrava. And, across the border in Poland, they were mining the same black coal. They were then just selling it to each other to fulfill export and import quotas under the Communist system."

KSČ decision-makers pushed unlimited expansion, which necessitated ever-increasing environmental withdrawals and additions. The political system driving this growth was completely unencumbered by external regulatory pressure; thus, the KSČ did little to mitigate the severe environmental impacts. Systemic inefficiencies further exacerbated environmental degradation in Czechoslovakia throughout the Communist era (Jancar-Webster 1987). State subsidies and distorted pricing systems kept energy and other resource costs artificially low (Pavlínek 1997; Albrecht 1998). Havlíček (1997) found that Czechoslovakian companies required 33 percent more energy than Western companies to manufacture the same products. A longtime North Bohemian resident asserted that the Communist regime "maximized production as much as they could without any concern for the environment. They were just *wasting* resources. They didn't think about future generations."

Production Expansion and Environmental Degradation in North Bohemia

The North Bohemian region of Czechoslovakia . . . was the focus of significant production expansion by the KSČ because of its vast reserves of lignite, a low-grade brown coal. While the region had mined brown coal since the 15th century, these efforts increased dramatically under Communist rule. Lignite is a highly inefficient energy source (Carter 1993; Leff 1997; Pavlínek 1997) that produces extensive ash and sulfur dioxide emissions, which contribute to significant air pollution. Yet, the regime relied extensively on this cheap energy source to fuel ever-increasing production (Albrecht 1987).

Given the region's vast reserves of coal and its central geographic location, the KSČ decided to make North Bohemia a "model of socialism" that would illustrate to the rest of the world how centrally planned economic growth could lead to an advanced, industrially developed society (Glassheim 2006). To realize the region's full productive potential, however, the state needed to bolster the available workforce and intensify workers' commitment to production. Officials enticed workers to the region using a variety of incentives and propaganda campaigns. Recruitment offices were set up around the country. Workers were promised higher wages and greater availability of housing options if they moved to the region. Communist daily newspapers such as *Rudé právo* and *Bratislava Pravda* touted coal miners as "heroes of socialism." The state repeatedly emphasized the importance of coal using statements such as "Prague, don't forget when you fire up the heat, from the sweat of our miners, we're building a new state" (cited in Glassheim 2006, 79).

Five-year planning schedules for coal production had to be met—even exceeded—regardless of demand. Local and regional managers had virtually no influence over their companies' quotas (see Pavlínek and Pickles 2000). The former director of a North Bohemian coal mine explained the strict adherence to coal mining production schedules: "Changing a plan of how much we were supposed to mine was impossible, although we knew that no one needed that coal and no one wanted it" (cited in Pavlínek and Pickles 2000, 106).

Production activities in North Bohemia were largely untouched by environmental regulations, partly because party elites lived far from the locus of pollution problems. Communist officials occasionally made trips to production sites, but were rarely attentive to pollution. The state-socialist growth coalition formally vested the responsibility for the enforcement of environmental policies in the State Planning Commission; however, this agency was *also* responsible for maintaining production. Given these dual functions, the State Planning Commission "encountered severe conflicts of interest when trying to meet environmental protection goals while under pressure to fulfill production quotas. The quotas usually won" (Havlíček 1997, 19). The State Planning Commission was thus expected to overlook environmental violations to ensure economic growth (Havlíček 1997).

As a result of expansive coal mining and the proliferation of coal-fired power plants, North Bohemia became a virtual wasteland, reputed to be one of the most heavily polluted regions in the world (Carter 1993; Pavlínek and Pickles 2000). In the early 1980s, emissions of sulfur dioxide exceeded maximum levels by up to thirteen times at least 80 days out of the year. In 1988, a state environmental status report indicated that more than a quarter of the harmful industrial emissions in the country were concentrated in the North Bohemian region. By 1989, air pollution in Most and Chomutov exceeded safe levels for at least three to four days of every week (Carter 1993). The region's forests were particularly damaged by the excessive air pollution. An elderly resident who had moved away to escape the devastation recalled the effects on North Bohemia's landscape:

> I think the biggest tragedy was around the brown coal plants in the North of Czech. The area around Krušné hory was basically like moon land. There were dead woods and forests as far as you could see.

By the end of the 1980s, more than half of the North Bohemian forests were destroyed or in decline as a result of air pollution (Adamova 1993). Conditions were particularly bad in the winter months, when inversion left pollution hovering over the area. The air pollution was frequently so severe that children were sent to rural summer camps for relief. One respondent had raised a child with asthma in the area, and she recalled the camps:

> Children were sent to "schools in nature," which I think was sponsored by the government and the Ministry of Education. It was a trip for two weeks, but that couldn't compensate for the pollution that we lived in for the entire year.

The effects of the state's shortsighted focus on production were also evidenced in trends in human health. Epidemiologists reported deteriorating health conditions in North Bohemia in the 1960s, and more so in the 1970s and 1980s (Pavlínek and Pickles 2000). Life expectancy in North Bohemia was three to five years lower than in other places in Czechoslovakia (Glassheim 2006) and up to ten years lower than in other developed countries in Europe (Pavlínek and Pickles 2000). Rates of cancer, miscarriages, and birth defects in North Bohemian mining towns were significantly higher than other regions in Czechoslovakia (Carter 1993), and worsened throughout the 1980s. Šrám (2001, 20) noted the health problems in the region:

> The first signs of deteriorating human health in the mining districts of Northern Bohemia were related to an increase in allergies, immune-deficiencies and respiratory illnesses in children. Simultaneously, an increase in birth defects and the rising prevalence of children with low birth weight was observed; especially striking was the shortening of life expectancy for inhabitants of this region as compared to the rest of the country, especially in males with an increase in mortality rates for cancer and cardiovascular diseases. (See also Šrám, Kotěšovec, and Jelínek 1996.)

A prominent environmental activist from North Bohemia described the environmental conditions that he was exposed to from childhood:

> I am from Ústí nad Labem, a city in the North of Czech. This was right next to Bílina, the dirtiest river in the world. There were two coal mines, a power plant,

and a chemical factory that didn't have any filters at the time, so from September until about April it wasn't possible for people to open the windows because their eyes would burn and itch.

Environmental Threats and Political Opportunities: Challenging the Treadmill in North Bohemia

Throughout the Communist period, the tension between the treadmill's environmental externalities and public response was mitigated by the state's use of harsh repression to quash protest (Shriver and Adams 2010; Vaněk 1996). Our analysis focuses on the time period between 1985 and 1989 when citizen activism emerged in North Bohemia. Drawing from political opportunity theory, we delineate the relationship between environmental threats resulting from the treadmill of production and residents' attempts to press the state for action in North Bohemia. Our analysis illustrates how severe environmental threats coupled with developing weaknesses in the regime resulted in a loss of legitimacy for the state and its treadmill politics, and incited residents and workers to rise up in protest.

Escalating Environmental Threats and Initial Challenges in North Bohemia

Environmental conditions deteriorated in North Bohemia throughout the 1980s, and perceptions of environmental threat became widespread. The state remained committed to meeting and even exceeding its successive five-year plans for growth. Yet, the Soviets' inability to continue exporting cheap oil to the region forced Czechoslovakia to rely more heavily on its domestic reserves of North Bohemian lignite (Pavlínek and Pickles 2000).

In the early and mid-1980s, the state demonstrated little tolerance for activities perceived as challenges to its treadmill policies. In 1985, a small group of environmental activists in North Bohemia became targets of the state's heavy-handed approach. Pavel Křivka was a recently graduated university student who worked as a scientist in a local museum in Jiín. Křivka attempted to raise awareness about environmental problems in the region by posting a map of the worst ecological areas in North Bohemia. This subtle form of resistance put Křivka on the watch list of the State Security Service [*Státní bezpečnost* (StB)] (Charter 77 1985a). In February 1985, Křivka wrote a letter to a friend in West Germany detailing his criticism of the Czechoslovakian government for their environmental neglect of North Bohemia. He gave the letter to an acquaintance who was traveling to Yugoslavia and asked him to mail it to his friend in West Germany (Charter 77 1985a, 1985b); however, the StB confiscated the letter at the Hungarian border. The state police launched an investigation of Křivka and his associates. Throughout the spring of 1985, the group was constantly monitored and interrogated by StB agents. The StB conducted searches of their homes and workplaces, and collected materials that were used to manufacture anti-state criminal charges against them (Charter 77 1985b).

The state was determined to repress environmental resistance in North Bohemia and used Křivka and his acquaintances as a cautionary example. In April, Křivka was arrested and brought to the regional court in Hradec Kralove. The prosecution used a variety of charges against Křivka and one of his associates, Pavel Škoda. They were accused of having connections with secret-service agents in West Germany. Prosecutors used Křivka's affiliation with a local choir to accuse him of singing anti-state songs. While the charges were baseless, the state security police manipulated his fellow choir members into testifying against him using threats and harassment (Charter 77 1985c). As a result, the court prosecuted Křivka for undermining the republic and crimes against the state from abroad (Charter 77 1985b). This was a common legal strategy used to impose harsh punishment for any critical civic initiatives (Hodos 1987). The pair was ultimately sent to prison. Škoda received a sentence of 20 months (Vaněk 1996), and Křivka was sentenced to three years in the Plzeň-Bory Prison (Charter 77 1985b, 1985c; Charter 77 1987a), a maximum-security facility notorious for deplorable conditions and prisoner abuse (Schwartz and Schwartz 1989).

To repress burgeoning environmental protest, Czechoslovakian officials obstructed attempts to organize outside state-sanctioned environmental groups. For instance, a group of young people attempted to organize in the heavily polluted Chomutov region of North Bohemia in late 1985. Chomutov was home to extensive mining operations; the resultant air pollution caused severe health problems, including an infant mortality rate that was 12 percent higher than the rest of the nation. The young environmentalists recruited friends and acquaintances throughout the North Bohemian region to help them organize what they hoped would become a viable alternative to the state-sanctioned environmental group, Czech Union for Nature Conservation [*Cesky Svaz Ochráncu Prírody*, Č SOP] (Vaněk 1996). The group sought to establish an independent organization called "Initiative for the Protection of Life and the Environment" to draw attention to the serious problems in North Bohemia.

Similar to the Křivka case earlier that year, the state reacted harshly to this organizing effort. The StB arrested the young activists and subjected them to intense interrogations. The police were particularly harsh on those with suspected ties to other dissident activities (Charter 77 1985d; *Die Presse* 1985). The Charter 77 publication *Informace o Chartě 77* published an article entitled "Termination of Environmental Groups in Chomutov" (1985d), which explained that authorities threatened and harassed the environmentalists over several months, and promised long-term imprisonment if they continued to act outside Č SOP. Moreover, StB agents made veiled threats to harm the children of those involved in the organizing efforts if they did not cease their activities, telling one of the Chomutov group's organizers: "It would be a shame if one of your kids was hit by a car on the way home from school." Even though it was ultimately crushed by the state, this independent environmental initiative represented a significant effort to mobilize in North Bohemia.

Diminishing Legitimacy of the State and Expanding Opportunities for Protests

Although the state remained intolerant of citizen protest, officials were forced to recognize the gravity of environmental externalities in North Bohemia. Elites were faced with a dilemma: How could they appease workers and residents in the region by acknowledging their environmental concerns, while continuing to maintain the legitimacy of the treadmill? Officials initiated a series of conciliations intended to mollify residents while protecting the state's production interests. For example, various "compensating measures" were instituted to offset the effects of pollution on the local population, including sending residents out of the area to temporary vacation homes in the mountains and sending more children to "nature schools" to escape the stifling pollution (Pavlínek and Pickles 2000; Charter 77 1987b; Glassheim 2006). They also developed special schools to improve children's health (Jirat 1984), and provided vitamin-enriched lunches and special foods such as yogurt to strengthen children's immune systems (Pavlínek and Pickles 1998; Pohl 1988). State officials also advised North Bohemian residents to "modify their life-style" based on the severity of air pollution. Pohl (1988, 52) explained: "When the pollution level is high, they advise keeping children, the sick, and old people indoors, airing buildings for no longer than five minutes, and avoiding any physical strain."

These efforts did little to quell residents' environmental fears, and many attempted to relocate to escape the overwhelming pollution. Medical professionals made up a large part of this exodus, particularly pharmacists and physicians (Labudová 1986). In response, the state adopted a constitutional resolution (No. 37) that forbade any organization in Czechoslovakia from employing a physician or pharmacist that had terminated a contract in North Bohemia. However, the resolution backfired, as many medical professionals simply refused to move to the region. The state had long relied on a variety of incentives to entice workers to move to North Bohemia; yet, during the second half of the 1980s, many of these incentives were abrogated due to escalating economic problems in the country (Vaněk 1996). Many workers found themselves earning reduced wages, yet still living with severe pollution.

By 1986, Mikhail Gorbachev's political reforms *perestroika* (restructuring) and *glasnost* (openness) gained traction within governments in the Eastern Bloc. In fact, Gorbachev made a historic visit to Prague in April 1987 to promote the reforms to the conservative leadership in the KSČ. Thousands greeted the Soviet leader, hopeful that their government would adopt these more liberal measures. In addition, in April 1986 the Chernobyl nuclear disaster occurred in Ukraine; however, Czechoslovakian state media uniformly denied the severity of the event. A physician from North Bohemia described how the state's reaction to the event served to further delegitimize the KSČ:

> From my point of view, the last people that still trusted the Communist Party lost trust because of Chernobyl. It wasn't only just Radio Free Europe, or Voice of America, but people could watch West German TV or Austrian TV, and they

discussed the radioactive cloud from Chernobyl. But according to official media in Czechoslovakia, *Rudé právo*, nothing happened! And every day they lied more and people became more and more angry.

In light of these developments, small groups of North Bohemian residents began secretly meeting to discuss environmental concerns and fears in their homes.

These potential political openings coupled with increasing environmental threats galvanized citizens' resolve. For example, air pollution in North Bohemia became particularly acute in early 1987. In February, a group of 162 citizens became so desperate that they risked sending a petition to the Chomutov District offices complaining about air quality and lack of publicly available environmental information. The petition, organized by a resident named Karel Mrázek, detailed residents' fears and concerns about industrial pollution in the region. The petition signers specifically asked that environmental information regarding air quality be released. Archival materials from Charter 77 documented residents' requests to know when it was safe to leave their houses and when they should stay inside (Charter 77 1987c; see also Vaněk 1996). Residents argued that air quality alerts were critical for reducing the "frequency of sickness in the Chomutov region" (Charter 77 1987c).

When the state failed to repress their efforts, residents were further emboldened and their challenges intensified throughout the spring of 1987. The number of residents supporting the petition grew to 300 (Vaněk 1996). Karel Mrázek sent a letter to Prime Minister Ladislav Adamec to stress citizen rights for information. The protestors were astounded when city and government officials responded to their complaints; however, the response was little more than a dismissal of the activists' concerns. The Chairman of the Chomutov District Committee responded by saying that production in North Bohemia would be regulated in the future, and officials were already working to inform the public about environmental issues in the region. He pointed out that pollution control measures had recently been implemented, and argued that the public was fully informed about air pollution levels. A Charter 77 *samizdat* publication explained that he attempted to diminish the activists' grievances by saying that plans were in place to inform pregnant women of environmental conditions in the region (Charter 77 1987d).

In the second half of 1987, party elites publicly validated residents' environmental concerns to further defend the legitimacy of the treadmill of production in North Bohemia. In October, Prime Minister Adamec spoke at a KSČ conference devoted exclusively to environmental problems in the region (Pohl 1988). The environmental conference was covered by the regional newspaper, *Pru̇ boj*, and the national party newspaper, *Rudé právo*. Adamec criticized industrial operations in the region and acknowledged that serious mistakes were made regarding environmental protection. He remarked that factories were not equipped with adequate equipment to reduce emissions, and addressed health problems associated with pollution. Adamec also acknowledged that public confidence in the Communist Party was being undermined due to the state's disregard for human and environmental health. His proposed solution involved increased fines and sanctions against polluting facilities, and for individuals responsible for operating equipment (Pohl 1988; *Rudé právo* 1987).

Residents and activists in North Bohemia strengthened their resolve, putting further pressure on government officials to address the problems associated with excessive production. In April 1987, Charter 77 released an environmental report entitled "So That One Can Breathe" (Charter 77 1987b). Archival research revealed that Charter 77 criticized the state for failing to protect the public from environmental pollution, specifically highlighting the deplorable environmental conditions in North Bohemia. The report also referred to the "nature schools" to highlight the unsustainability of industrial production in the area. In essence, the report pointed to the creation of these special schools as evidence that the state acknowledged the severity of environmental problems.

The state responded to North Bohemian residents' environmental grievances by proposing reformist initiatives and more planning sessions. In April 1987, the Council for the Protection and Improvement of the Environment met in the North Bohemian city of Ústí nad Labem. The conference focused largely on environmental protection and remediation of the North Bohemian brown coalfields and the reforestation of Krušné hory (Prague Domestic Service 1987). In 1988, the Czechoslovakian government held its plenary session and introduced a draft concept for environmental protection and national resources through the year 2000. It called for significant reduction in emissions and air pollution, particularly in areas with "extremely aggravated environmental conditions" such as North Bohemia (*Rudé právo* 1988). However, citizens were increasingly emboldened in their discussions of environmental threats in organized meetings of twenty or thirty people. One of the primary organizers recalled speaking openly despite the close surveillance of the StB:

> Because we knew StB were listening, we decided to speak about everything aloud and say it clearly so that they could hear it . . . As a result, StB was horrified and baffled, as they realized that young people were encouraging each other to speak openly and freely. Interrogations usually followed such discussions, but StB did not know what to do with us.

In April 1989, sensing the escalation of fear regarding environmental threats, Prime Minister Adamec visited Chomutov and proposed significant reduction in coal mining in North Bohemia (*Rudé právo* 1989a). In October 1989, state officials publicly acknowledged the problems associated with the growth imperative that pushed constant production. The Secretary of the Communist Party of Czechoslovakia Central Committee, Ivan Knotek, delivered a speech comparing the country to capitalist states and highlighting the links between industrial production and environmental degradation. Importantly, he cautioned the public not to politicize environmental problems: "However, we cannot agree with having this progress—which is common to all industrially advanced countries—misused today to exert political pressure on the socialist countries, including Czechoslovakia" (*Rudé právo* 1989b, 4).

The Collapse of State Legitimacy: Escalating Protests in North Bohemia

Despite the eleventh-hour efforts to appease residents, the state's legitimacy had completely eroded. Taking their cues from the relaxation of state repression and the increased attention to environmental problems, residents and activists intensified their challenges to the state and protests became a regular occurrence throughout 1989 (Tickle and Vavroušek 1998). In November 1989, cold temperatures and damp weather significantly increased air pollution in North Bohemia. The deplorable conditions prompted a teenager named Zbyšek Jindra to post flyers in Teplice encouraging residents to attend a public protest in Nejedlý Square (Jehlička 2001; Kenney 2002). The flyers called attention to environmental health issues and linked the environmental degradation of the region with social oppression, calling for a revolt against the "inhuman attitude" of leading party functionaries (Vaněk 1996).

The call to protest was successful, and approximately 1,000 people—most of whom had never participated in a public demonstration—arrived in the square wearing gas masks, chanting, "We want clean air!" and "Oxygen!" (Kenney 2002, 228; see also Jehlička 2001; Pavlínek and Pickles 2000). There was no significant police response, and so the protests continued the next day, encouraging even more people to attend. On the second day, state police arrived to suppress the demonstration with billy clubs, water cannons, and dogs (Kenney 2002; Pavlínek and Pickles 2000). Despite the state's efforts, the protests continued into a third day and emboldened residents in five other cities in North Bohemia to engage in their own protests regarding the environmental conditions (Jehlička 2001). On that third day of protest, a provincial party leader promised the demonstrators a public meeting about production and the environmental conditions in North Bohemia (Kenney 2002).

The following week, officials from the municipal Czechoslovak Communist Party Committee held a gathering at the Winter Stadium in Teplice to discuss environmental concerns. Although the stadium held 5,000 people, it overflowed with residents and many were denied entrance. Officials set up makeshift audio equipment to broadcast the forum, but residents continued to arrive at the stadium and were angered that they could not enter. Inside the stadium, officials attempted to mollify the angry crowd, promising residents that remediation measures were being planned to modify production and address environmental problems (Prague Domestic Service 1989).

The protests in North Bohemia coupled with the state's ineffective response signaled further openings in the political opportunity structure (Shriver and Adams 2010; Holy 1996; Pavlínek 1997; Jehlička 2001). Moreover, information about the spread of protests and revolutions throughout Eastern Europe was being disseminated through various underground networks and Western media sources (e.g., Radio Free Europe) (Kenney 2002). Indeed, Solidarity candidates in Poland had successfully won a partially open election earlier in 1989, contributing to the ultimate downfall of the Polish Communist Party. In East Germany, many citizens fled the country throughout 1989. On November 9, remaining residents breached the Berlin Wall, which had stood as a physical reminder of the Iron Curtain and had divided Germany into separate republics. During the same period in Bulgaria, environmental protests ultimately led to more widespread civil protests, which forced the Bulgarian Communist Party to step down (Pavlínek 1997). These momentous events signaled additional openings in the political opportunity structure for Czechoslovakian citizens, and were an important harbinger of collapse for the KSČ. Demonstrations spread throughout Czechoslovakia during the remainder of November 1989. Dissident groups, including the prominent human rights organization Charter 77, played a significant role in fueling large protests such as the gathering of thousands of citizens in Prague's *Václavské náměstí* (Wenceslas Square) on November 21. While some protestors were met with harsh repression, the demonstrations grew in their intensity, and on November 27, 1989, the Communist Party of Czechoslovakia gave in to the will of the people and conceded political power.

Discussion and Conclusions

The case of the North Bohemian Coal Basin in Czechoslovakia provides insight into how environmental threats can incite activism in political settings where state repression directly impedes the counterweight of citizen activism.

Our analysis provides insight into broader patterns of resistance, elite legitimacy, and social control by highlighting the role of environmental threats to explain the protests in North Bohemia. Our findings show that the environmental externalities to state-mandated production in North Bohemia posed an imminent threat that spurred residents to protest despite fears of repression. Environmental threats became clear and visible to the general public. As such, residents in North Bohemia noted fears about their health and their family members' health. People were particularly fearful about respiratory illnesses and cancer among children. Residents also noted the high instances of miscarriages and birth defects. In this way, the pollution became a direct threat to residents' basic rights to health and safety. While previous work has differentiated ecological and political threats, we argue that *environmental threats* can be salient aspects of the political opportunity structure, particularly when pollution is the result of state directives. When the state drives the treadmill of production, resulting in severe pollution that threatens human lives, citizens can be forced to mobilize against the state to demand environmental reform.

In both liberal-democratic capitalist and state-socialist contexts, treadmill elites must maintain their legitimacy to retain control over the public. Our findings show that, as environmental threats became inevasible in the North Bohemian region, state officials made limited concessions to maintain their legitimacy. Yet, the state's actions represented an obvious contradiction. On one hand, the state pressed forward with its program of production expansion in the region. On the other hand, elites recognized the environmental consequences of this system, and initiated several small steps to allay residents' perceptions of environmental threats. Unlike the capitalist growth coalition, where state and corporate elites can deflect blame or refocus attention, the Soviet-style socialist state had nowhere to redirect blame. The workers operating at the behest of the state certainly did not bear the responsibility; thus, the state was ultimately culpable for the environmental problems. While promises of future environmental regulation and remediation were intended to mollify angry and frightened residents, official recognition of environmental problems signaled an opportunity to increase protest efforts. Thus, we argue that state acknowledgment of environmental grievances can act as an opening in the political structure.

While treadmill theory has been applied most often to Western capitalist democracies, we have shown that this framework also holds significant explanatory power in other political-economic systems. Our findings show that, like capitalist societies, the KSČ facilitated accumulation by driving continuous production expansion. . . . However, unlike capitalist societies, the state's authoritarian control over production allowed it to *impose* strict production schedules to meet its five-year plans. A growth hierarchy emerged where state managers and their workers were accountable to higher party elites. Thus, the treadmill logic was advanced by state elites and state power.

Importantly, the political and economic backdrop for the treadmill of production in Soviet state-socialism contrasts directly with Western capitalist societies, where corporate capitalists play a hegemonic role in driving the growth imperative. The configuration of interest parties in growth coalitions in Western capitalist economies promotes the appearance of a system that has checks and balances and minimizes conflicts of interest. The presence of both corporate interests and state regulation in the growth imperative allows for a diffusion of blame for the negative externalities of production. In the case of Communist Czechoslovakia, the state's strict control over production created intrinsic conflict of interests, such as in the case of the State Planning Commission, which was charged with both promoting production expansion and regulating the environmental externalities of production. Moreover, in the absence of private enterprises, the state was unable to sidestep responsibility for the environmental devastation that became apparent in North Bohemia.

Our project points to several pathways for future research on the treadmill of production. First, more research is needed to delineate how environmental threats can function as catalysts for protest in nondemocratic political structures. We have shown that imminent environmental threats to basic human rights can galvanize challenges in repressive states, but how can other types of environmental threats such global climate change impact mobilization efforts in different political and economic contexts? Second, while our findings show how treadmill theory applies to former Soviet-style socialist economies, future research is needed to explore treadmill politics in other authoritarian contexts. We argue that the integration of these theoretical perspectives will facilitate future investigations into these very pertinent research questions.

Note

Shriver, Thomas E., Alison E. Adams, and Stefano B. Longo. 2015. "Environmental Threats and Political Opportunities: Citizen Activism in the North Bohemian Coal Basin." *Social Forces* 94 (2) 699–722.

References

Adamova, Eva. 1993. "Environmental Management in Czecho-Slovakia." In *Environmental Action in Eastern Europe*, edited by Barbara Jancar-Webster, 42–57. Armonk, NY: M. E. Sharpe.

Albrecht, Catherine. 1987. "Environmental Policies and Politics in Contemporary Czechoslovakia." *Studies in Comparative Communism* 2 (3–4): 291–302.

———. 1998. "Environmental Policy in the Czech Republic and Slovakia." In *Ecological Policy and Politics in Developing Countries: Economic Growth, Democracy, and Environment*, edited by U. Desai, 267–93. Albany: State University of New York Press.

Almeida, Paul D. 2003. "Opportunity Organizations and Threat-Induced Contention: Protest Waves in Authoritarian Settings." *American Journal of Sociology* 109 (2): 345–400.

———. 2008. *Waves of Protest: Popular Struggle in El Salvador, 1925–2005*. Minneapolis: University of Minnesota Press.

Brzezinski, Zbigniew K. 1967. *The Soviet Bloc: Unity and Conflict*. Cambridge, MA: Harvard University Press.

Buttel, Frederick H. 2004. "The Treadmill of Production: An Appreciation, Assessment, and Agenda for Research." *Organization and Environment* 17 (3): 323–36.

Carter, Francis W. 1993. "Czechoslovakia." In *Environmental Problems in Eastern Europe*, edited by F. W. Carter, and D. Turnock, 63–88. London: Routledge.

Charter 77. 1985a. "Rozsudek nad Ing. Pavlem Křivkou a ing. Pavlem Škodou" [Judgment of Mr. Pavel Křivkaand Mr. Pavel Škoda]. *Informace o Chartě 77* 8 (12): 14–15.

———. 1985b. "Případ ing. Pavla Křivky z Pardubic" [The Case of Pavel Křivkaof Pardubice]. *Informace o Chartě 77* 8 (10): 10–11.

———. 1985c. "Hlavní líčení v trestní v věci proti P. Křivkovi [The Trial in the Criminal Case against P. Křivka]. *Informace o Chartě 77* 8 (12): 10.

———. 1985d. "Zánik ekologické skupiny na Chomutovsku" [Termination of Environmental Groups in Chomutov]. *Informace o Chartě 77* 8 (8): 26.

———. 1987a. "Podmíněné propuštění P. Křivky" [Conditional Release of Pavel Křivka]. *Informace o Chartě 77* 10 (17): 9.

———. 1987b. "Aby se dalo dýchat" [So We Can Breathe]. *Informace o Chartě 77* 10 (7): 2–10.

———. 1987c. "Dopis Karla Mrázka předsedovi ČSR" [Letter from Karel Mrázek to the Chairman of the Czechoslovak Socialist Republic]. *Informace o Chartě 77* 10 (13): 17–18.

———. 1987d. "Dopis Rady pro Životní prostředí K. Mrázkovi" [Letter from the Council of the Environment to Mr. Mrázek]. *Informace o Chartě 77* 10 (17): 14–15.

Clark, Brett, and Andrew K. Jorgenson. 2012. "The Treadmill of Destruction and the Environmental Impacts of Militaries." *Sociology Compass* 6–7: 557–69.

Čornej, Petr, and Jiří Pokorný. 2003. *A Brief History of the Czech Lands to 2004*. Prague: Práh Press.

DeBardeleben, Joan. 1985. *The Environment and Marxism-Leninism*. Boulder, CO: Westview.

Die Presse. 1985. "Police Crush Private Environmental Group." *Die Presse* in German (September 3). Translation by Foreign Broadcast Information Service. *FBIS Daily Report—Czechoslovakia* (FBIS-EEU-85–172, D5).

Dunlap, Riley E. 1997. "The Evolution of Environmental Sociology: A Brief History and Assessment of the American Experience." In *The International Handbook of Environmental Sociology*, edited by M. Redclift and G. Woodgate, 21–39. London: Edward Elgar.

Eisinger, Peter K. 1973. "The Conditions of Protest Behavior in American Cities." *American Political Science Review* 67 (1): 11–28.

Foster, John Bellamy. 2005. "The Treadmill of Accumulation Schnaiberg's Environment and Marxian Political Economy." *Organization & Environment* 18: 7–18.

Glassheim, Eagle. 2006. "Ethnic Cleansing, Communism, and Environmental Devastation in Czechoslovakia's Borderlands, 1945–1989." *Journal of Modern History* 78: 65–92.

Goldstone, Jack, and Charles Tilly. 2001. "Threat (and Opportunity): Popular Action and State Response in the Dynamic of Contention Politics." In *Silence and Voice in the Study of Contentious Politics*, edited by R. Aminzade, J. Goldstone, D. McAdam, E. Perry, W. Sewell, S. Tarrow, and C. Tilly, 179–94. Cambridge: Cambridge University Press.

Gould, Kenneth A., David N. Pellow, and Allan Schnaiberg. 2004. "Interrogating the Treadmill of Production Everything You Wanted to Know about the Treadmill But Were Afraid to Ask." *Organization & Environment* 17: 296–316.

Gould, Kenneth A., Allan Schnaiberg, and Adam Weinberg. 1996. *Local Environmental Struggles: Citizen Activism in the Treadmill of Production*. Cambridge: Cambridge University Press.

Havlíček, Peter. 1997. "The Czech Republic: First Steps toward a Cleaner Future." *Environment* 39 (3): 16–20.

Hodos, George H. 1987. *Show Trials: Stalinist Purges in Eastern Europe, 1948–1954*. New York: Praeger Publishers.

Holy, Ladislav. 1996. *The Little Czech and the Great Czech Nation: National Identity and the Post-Communist Transformation of Society*. Cambridge, MA: Cambridge University Press.

Hooks, Gregory, and Chad L. Smith. 2004. "The Treadmill of Destruction: National Sacrifice Areas and Native Americans." *American Sociological Review* 69 (4): 558–75.

Jancar, Barbara Wolfe. 1970. *Czechoslovakia and the Absolute Monopoly of Power*. New York: Praeger Publishers.

Jancar-Webster, Barbara. 1987. *Environmental Management in the Soviet Union and Yugoslavia: Structure and Regulation in Federal Communist States*. Durham, NC: Duke University Press.

Jehlička, Petr. 2001. "The New Subversives—Czech Environmentalists after 1989." In *Pink, Purple, and Green: Women's, Religious, Environmental and Gay/Lesbian Movements in Central Europe Today*, edited by H. Flam, 81–94. New York: Columbia University Press.

Jenkins, Joseph Craig, and Charles Perrow. 1977. "Insurgency of the Powerless: Farm Worker Movements (1946–1972)." *American Sociological Review* 42: 249–68.

Jirat, Josef. 1984. "Environmental Problems in North Bohemia Lignite Basin." *Hospodářské noviny* in Czech (February 3). Translation by *Joint Publications Research Service*, April 27 (JPRS-EEI-84–049, 1).

Johnson, Erik W., and Scott Frickel. 2011. "Ecological Threat and the Founding of US National Environmental Movement Organizations, 1962–1998." *Social Problems* 58 (3): 305–29.

———. 2011. "Societies Consuming Nature: A Panel Study of the Ecological Footprints of Nations, 1960–2003." *Social Science Research* 40: 226–44.

Kenney, Padraic. 2002. *A Carnival of Revolution: Central Europe 1989*. Princeton, NJ: Princeton University Press.

Khawaja, Marwan. 1993. "Repression and Popular Collective Action: Evidence from the West Bank." *Sociological Forum* 8: 47–71.

Krejčí, Jaroslav, and Pavel Machonin. 1996. *Czechoslovakia, 1918–92*. New York: St. Martin's Press.

Labudová, Hana. 1986. "Reasons for Such Resistance." *Rudé právo* in Czech (February 7). Translation by Foreign Broadcast Information Service. *FBIS Daily Report—Czechoslovakia*, February 25 (FBIS-EEU-86–037, D10).

Leff, Carol Skalnik. 1997. *The Czech and Slovak Republics: Nation versus State*. Boulder, CO: Westview Press.

Maher, Thomas V. 2010. "Threat, Resistance, and Collective Action: The Cases of Sobibór, Treblinka, and Auschwitz." *American Sociological Review* 75 (2): 252–72.

McAdam, Doug. 1982. Political Process and the Development of Black Insurgency, 1930–1970. Chicago: University of Chicago Press.

Osa, Maryjane. 2001. "Mobilizing Structures and Cycles of Protest: Post-Stalinist Contention in Poland, 1954–1959." *Mobilization* 6 (2): 211–31.

Pavlínek, Petr. 1997. *Economic Restructuring and Local Environmental Management in the Czech Republic*. Lewistown, NY: Edwin Mellen Press.

Pavlínek, Petr, and John Pickles. 2000. *Environmental Transitions: Transformation and Ecological Defense in Central and Eastern Europe*. New York: Routledge.

Pohl, Frank. 1988. "Protest against Environmental Pollution in Northern Bohemia." *Radio Free Europe Research* 13 (3): 49–53.

Prague Domestic Service. 1987. "CEMA Meeting Discusses Cooperation, Environment" Prague Domestic Service in Czech (April 7). Translation by the Foreign Broadcast Information Service. *FBIS Daily Report—Czechoslovakia*, April 9 (FBIS-EEU-87–068, AA1).

———. 1989. "Open Discussion at Teplice on Pollution." Prague Domestic Service in Czech (November 20). Translation by the Foreign Broadcast Information Service. *FBIS Daily Report—Czechoslovakia*, November 21 (FBIS-EEU-89–223, 25).

Redclift, Michael. 1989. "Turning Nightmares into Dreams: The Green Movement in Eastern Europe." *Ecologist* 19 (5): 177–83.

Resnich, Stephen A., and Richard D. Wolff. 2002. *Class Theory and History: Capitalism and Communism in the USSR*. New York: Routledge.

Rudé právo. 1987. "Adamec Condemns Pollution of Environment." *Rudé právo* in Czech (October 9). Translation by Foreign Broadcast Information Service, FBIS Daily Report—Czechoslovakia, October 19 (FBIS-EEU-87–201, 21).

———. 1988. "Government Session on Ecology, Natural Resources." *Rudé právo* in Czech (July 27). Translation by the Foreign Broadcast Information Service. *FBIS Daily Report—Czechoslovakia*, August 4 (FBIS-EEU-88–150, 19).

———. 1989a. "Czech National Council Discusses Environment." *Rudé právo* in Czech (March 28). Translation by the Foreign Broadcast Information Service. *FBIS Daily Report—Czechoslovakia*, April 4 (FBIS-EEU-89–063, 14).

———. 1989b. "Knotek on the Environment." *Rudé právo* in Czech (October 13). Translation by Foreign Broadcast Information Service. *FBIS Daily Report—Czechoslovakia*, October 20 (FBIS-EEU089–202, 4).

Schnaiberg, Allan. 1980. *The Environment: From Surplus to Scarcity*. New York: Oxford University Press.

Schnaiberg, Allan, and Kenneth Gould. 1994. *Environment and Society: The Enduring Conflict*. New York: St. Martin's Press.

Schnaiberg, Allan, David N. Pellow, and Adam Weinberg. 2002. "The Treadmill of Production and the Environmental State." *Research in Social Problems and Public Policy* 10: 15–32.

Schock, Kurt. 1999. "People Power and Political Opportunities: Social Movement Mobilization and Outcomes in the Philippines and Burma." *Social Problems* 46 (3): 355–75.

Schwartz, Herman, and Mary Schwartz. 1989. *Prison Conditions in Czechoslovakia*. New York: Human Rights Watch.

Shriver, Thomas E., and Alison E. Adams. 2010. "Cycles of Repression and Tactical Inno-vation: The Evolution of Environmental Dissidence in Communist Czechoslovakia." *Sociological Quarterly* 51 (2): 329–54.

Šrám, Radim. 2001. "Teplice Program: Studies on the Impact of Air Pollution on Human Health (1991–1999)." In *Teplice Program: Impact of Air Pollution on Human Health*, edited by R. Šrám, 19–30. Prague: Institute of Experimental Medicine, Academy of Sciences of the Czech Republic.

Šrám, Radim, František Kotěšovec, and Richard Jelínek. 1996. "Program Teplice—koncepce mezinárodní spolupráce" [Teplice Program—The Concept of International Cooperation]. *Ochrana ovzduší* 8 (5): 2–6.

Tickle, Andrew. 2000. "Regulating Environmental Space in Socialist and Post-Socialist Systems: Nature and Landscape Conservation in the Czech Republic." *Journal of European Area Studies* 8 (1): 57–78.

Tickle, Andrew, and Josef Vavroušek. 1998. "Environmental Politics in the Former Czechoslovakia." In *Environment and Society in Eastern Europe*, edited by A. Tickle and I. Welsh, 114–45. White Plains, NY: Longman.

Tilly, Charles. 1978. *From Mobilization to Revolution*. Reading: Addison-Wesley.

Van Dyke, Nella, and Sarah A. Soule. 2002. "Structural Social Change and the Mobiliz-ing Effect of Threat: Explaining Levels of Patriot and Militia Mobilizing in the United States." *Social Problems* 49 (4): 497–520.

Vaněk, Miroslav. 1996. *Nedalo se tady dýchat* [It Was Impossible to Breathe Here]. Ústav pro soudobé dějiny, AV Č R, Prague: Maxdorf.

Weinberg, Adam S., David N. Pellow, and Allan Schnaiberg. 2000. *Urban Recycling and the Search for Sustainable Community Development*. Princeton, NJ: Princeton Univer-sity Press.

18 Politics by Other Greens
The Importance of Transnational Environmental Justice Movement Networks

David N. Pellow

The idea that social inequalities are not separate from—but are instead a funda-mental part of—ecological unsustainability is a central theme in environmental sociology. Would wealthy and powerful people really produce hazardous toxins or mountains of garbage if they had to live next door to these? In this reading, David Pellow uses the metaphor of the boomerang to suggest that our actions tend to come back to us. Pellow describes how, with the help of transnational environmental orga-nizations, a small group of activists in Mozambique was able to prevent a Danish development agency from incinerating 900 tons of obsolete toxic fertilizer in south-ern Mozambique.

The race, class, gender, and national inequalities and ecological violence that are at the core of global capitalism underscore a point that many participants in environmental movements often overlook: social inequalities are the primary driving forces behind ecological crises. That is, we should no longer view race, class, and other inequalities as the most important variables in a general model that might explain environmental injustice. Rather they are also the most import-ant factors for theorizing the overall predicament of ecological unsustainability. Social inequalities are, therefore, not just an afterthought of an environmentally precarious society; they are at its root.

There are times when we must be reminded of the inescapable interdepen-dence among human societies and of those interdependencies we experience with broader ecosystems. Thus a close observation of the myriad forms of insti-tutional violence among human communities always reveals the associated vio-lence visited upon ecosystems. Therefore social movements confronting human rights abuses—particularly in the global South—tend to also confront ques-tions of ecological abuse because the domination over people is reinforced and made possible by the domination of ecosystems. But the interdependencies that human and non-human systems share underscore that no one is exempt from the far-reaching impacts of institutional and ecological violence. Thus radical transformative democratization of societies is a critical component in the global effort to achieve environmental sustainability and social justice.

In this chapter I investigate the phenomenon of transnational environmental justice (EJ) movements, specifically considering the work of activists, organizations, and networks that constitute this new formation. Linking environmental justice studies, environmental sociology, ethnic studies, and social movement theory in new ways, and drawing on interviews and archives, I ask how social movements challenge environmental inequalities across international borders. I argue that transnational EJ movement networks do this (1) by disrupting the social relations that produce environmental inequalities, (2) by producing new accountabilities vis-à-vis nation states and polluters, and (3) by articulating new visions of ecologically sustainable and socially just institutions and societies.

Environmental Sociology and Social Inequalities

In this first section of the chapter I consider theories of environmental conflict and link them to theories of social inequality. I begin with Ulrich Beck's "Risk Society" thesis, which contends that late modern society is marked by an exponential increase in the production and use of hazardous chemical substances, producing a fundamental transformation in the relationship among capital, the state, civil society, and the environment. What this means is that the project of nation building and the very idea of the modern nation-state are undergirded by the presence of toxins—chemical poisons—that permeate every social institution, human body, and ecosystem. This toxic modern nation-state also depends upon the subjugation of ecosystems and certain human populations designated as "others"—those who are less than deserving of full citizenship. This process attenuates the most negative impacts of such a system on elites. Toxic production systems produce privileges for a global minority and externalize the costs of that process to those spaces occupied by devalued and marginal "others"—people of color, the poor, indigenous communities, and global South nations. The study of such inequalities, of course, is the foundation of the field of environmental justice and inequality studies (Agyeman, Bullard, and Evans 2003).

Thus, according to Beck, advanced capitalism creates wealth for some and imposes risks on others, at least in the short term. In the long run, however, the problem of widespread global ecological harm ends up returning to impact its creators in a "boomerang effect." That is, the risks of late modernity eventually haunt those who originally produced them (Beck 1999). In that sense, Beck acknowledges environmental inequality in the short term, while also maintaining a global, long-range view of what becomes, to some extent, a democratization of risk. Beck confirms the enduring problem of what other scholars have termed the "metabolic rift"—the disruptions in ecosystems that capitalism produces because of its inherent tendency to expend natural resources at a rate that is greater than the ability of ecosystems to replenish those materials (Foster 2000). These rifts are linked to and reinforce social dislocations and inequalities that siphon wealth upward and restrict the economic and political capacities of the working classes and communities of color. Thus environmental harm is necessarily intertwined with the institutional violence that constitutes race, gender, and class hierarchies.

Building on these ideas from within environmental sociology, I now turn to theoretical developments from within the field of ethnic studies. For well over a century, a number of scholars and public intellectuals have used words like *poison* and *toxic* in speech and writings about racism. This is a powerful way to capture the harm racism does to both its victims and perpetrators or beneficiaries. Many authors have described racism as a *poison* that reveals deep contradictions and tensions in this nation, which have periodically erupted in violence, revolts, and wars over the years. Critical race theorists Lani Guinier and Gerald Torres make use of this terminology in their book *The Miner's Canary*. They write:

> The canary's distress signaled that it was time to get out of the mine because the air was becoming too *poisonous* to breathe. Those who are racially marginalized are like the miner's canary: their distress is the first sign of a danger that *threatens us all*. It is easy enough to think that when we sacrifice this canary, the only harm is to communities of color. Yet others ignore problems that converge around racial minorities at their own peril, for these problems are symptoms warning us that we are all at risk. (Guinier and Torres 2002: 11, emphases added)

Guinier and Torres also introduce a concept they term "political race," which "encompasses the view that race . . . matters because racialized communities provide the early warning signs of *poison* in the social atmosphere" (Guinier and Torres 2002: 12, emphasis added). In other words, political race forces us to think beyond specific instances of culpability and discrimination to produce a broader vision of justice for society as a whole. These concepts push us to rethink and challenge racism because it "threatens us all," not just people of color who may be its primary targets. The concept of political race begins with an emphasis on race and moves to class, gender and other inequalities, so while the principal emphasis is on race, this model is also inclusive of other categories of social difference.

The toxic metaphor for racism—and class and gender domination, for that matter—parallels Beck's "risk society" model in many ways. For example, racism, class domination, and pollution are ubiquitous and deeply embedded in our institutions, our culture, and our bodies. Moreover, while the production of both race, gender, and class hierarchies and toxic chemicals results in widespread harm across human communities and ecosystems, they both can also operate like a boomerang and eventually circle back to impact all members of society through uprisings, social unrest, and other conflicts. Finally, they are also powerful symbols for organizing resistance movements and for bringing people together across social and spatial boundaries. These concepts are helpful for thinking about the power and potential of transnational EJ movements.

Transnational Social Movements for Global Environmental Justice

While the primary focus of this chapter is on anti-toxics struggles, it must be said that the movement for global environmental justice and human rights casts a much broader net. This includes struggles against extractive industries, transboundary pollution and waste flows, free trade agreements, and—more importantly—the ideological and social systems that reinforce such practices, including racism, capitalism, patriarchy, and militarism. For example, hydroelectric dams have catalyzed many communities around the globe where people are fighting water privatization and external control of that most fundamental element on the planet. In response to the massive human rights abuses and environmental impacts associated with large dams, a highly influential and effective international movement emerged to force changes in current dam-building practices. In addition to organizations of dam-affected peoples, this arm of the EJ movement includes numerous allied environmental, human rights, and social activist groups around the world. International meetings in recent years have brought together dam-affected peoples and their allies to network and strategize, and to call for improvements in planning for water and energy-supply projects. Every year, community and activist groups from around the world show their solidarity with those dispossessed by dams on the International Day of Action, a global event organized to raise awareness about the impacts of dams and the value of dam-free and undammed rivers (McCully 2001).

Groups like the International Campaign for Responsible Technology are primarily focused on the social, economic, and ecological impacts of the global electronics industry, from mineral and water extraction for the production of electronics products, to their manufacture, sale, consumption, and disposal. In other words, this particular global EJ network adopts a lifecycle approach to the problem, following the materials and their effects on people and ecosystems (Smith, Sonnenfeld, and Pellow 2006). Many of these EJ movement networks articulate a critique of broader ideological systems of socio-environmental hierarchy that give life and legitimacy to global environmental injustice. Without such critical guiding frameworks, these movements would be limited in their political power and vision.

Numerous transnational social movement organizations (TSMOs) concerned with EJ and human rights issues focus their efforts on a range of state and industrial sectors. Taken together, these global organizations and networks constitute a formidable presence at international treaty negotiations, within corporate shareholder meetings, and in the halls of congresses and parliaments. Even so, they are only a part of the broader global movement for environmental justice. Arguably the most important components of that movement are the domestic local, regional, and national organizations in the various communities, cities, and nations in which scores of environmental justice battles rage every day. Those groups provide the front line participants in the struggles for local legitimacy within TSMOs and their networks. Together, the numerous local grassroots organizations and their collaborating global networks produce and maintain a critical infrastructure of the transnational public sphere.

Social movements must mobilize resources—funds, technology, people, ideas and imagination—to achieve their goals. Transnational social movements are rarely successful if we narrowly define success as a major change in a specific policy within a nation state (Keck and Sikkink 1998). But they are increasingly relevant in international policy debates, as they seek to make not only policy changes in international law and multilateral conventions, but also to change the terms and nature of the discourse within these important debates. These conventions include, for example: the Montreal Protocol (on the production of ozone-damaging chemicals), the Kyoto Protocol (concerning global warming), the Basel Convention (on the international trade in hazardous wastes), and the Stockholm Convention (on the production and management of persistent organic pollutants). In each of these cases, TSMOs are often a critical source of information for governments seeking information about environmental and social justice concerns, and their presence raises the costs of failing to act on certain issues, thus increasing the possibility of government accountability. In a global society where a nation-state's reputation can be tarnished in international political and media venues, transnational social movements can have surprisingly significant impacts. When movements disseminate information to the point that it becomes a part of common wisdom, such "popular beliefs . . . are themselves material forces" (Gramsci 1971: 165). That is, meaning systems can support or challenge systems of structural and material control. This is a critical point because, as cultural studies scholars and urban political ecologists have argued, social movements are struggling over cultural meaning systems as much as they are fighting for improved material conditions and needs (Moore, Kosek, and Pandian 2003).

In other words, the "natural" environment becomes a symbol of meaning for human communities. It can become a symbol of our attachment to—or contempt for—nature, and as a political or cultural tool for mobilizing against people whom hegemonic actors consider inferior and unimportant. The history of the genocide of Native peoples in the United States and the continued practices of environmental racism are just two examples of associating despised human "others" with landscapes and ecosystems that are also targets of extraction, pollution, or selective valuation. On the other hand, for the same reasons, ecosystems play a cultural role in the mobilization of social movements in favor of protecting ecosystems from risks associated with industrialization. As Moore, Kosek, and Pandian (2003) argue, nature is a terrain of power, through which we discursively and materially advance various meanings, agendas, and politics. Thus transnational EJ movement networks challenge environmental inequality by confronting the social forces that produce these outcomes and by arguing for new relationships of accountability vis-à-vis state and corporate actors.

Boomerang Effects

Recall that Beck's "risk society" and Guinier and Torres's "miner's canary" speak to the relational and interdependent character of industrial chemicals and social inequalities through the phenomena of boomerangs. Research on transnational social movement networks reveals that these formations produce their *own* boomerang effects as well. That is, when local governments refuse to heed calls for change, transnational activist networks create pressure that ". . . curves around local state indifference and repression to put foreign pressure on local policy elites. Thus international contacts amplify voices to which domestic governments are deaf, while the local work of target country activists legitimizes efforts of activists abroad" (Keck and Sikkink 1998). It is the interaction between repressive domestic political structures and more flexible structures in other nations that produces this boomerang.

In their influential book *Activists beyond Borders*, Keck and Sikkink (1998) explore the significance of the work of transnational social movement networks. These groups of activists in two or more nations, have, for decades, successfully intervened in and changed the terms of important global and national policy debates, pushed for regulation of activities deemed harmful to social groups, and influenced states to embrace practices that might improve the lives of residents in any given nation. Transnational movement networks often do this by gathering critical information and strategically making it available to publics, governments, media organizations, and other movements in order to force change. These movement networks also achieve such goals through mobilizing support for boycotts, letter writing campaigns, and other forms of protest that shine a spotlight on objectionable institutional practices with the goal of halting or transforming them. Transnational movements frequently take advantage of the multiple geographic scales at which these networks operate and sidestep the barriers that nation-states in one locale may create in order to access the leverage available from within other states—the boomerang. Transnational EJ movements use the boomerang to challenge the power that states and corporations enjoy over vulnerable communities, thus confronting the race, gender, and class inequalities that produce environmental injustices.

What Goes Around Comes Around

Guinier and Torres underscore the importance of "political race" through the metaphor of the "miner's canary," which symbolizes the role of people of color whose oppression is a sign of a poisonous social atmosphere that ultimately threatens all of society, not just those communities that suffer directly from racism. That is, racism creates its own boomerang effects that reveal systems of interdependence and accountability that impact people from all racial and class strata (albeit unevenly). Wars, revolts, uprisings, and social movements spawned, in part, by demands for racial, gender, and class justice against systems of oppression are among the many examples of such a boomerang effect. While mobilizing the boomerangs of transnational social movements, environmental justice activist networks also draw on analyses of race, class, and inequality to unmask the drivers of environmental injustice and to frame a vision of a more sustainable and socially just world.

There are multiple boomerang effects evident in EJ struggles, and I examine two of them here. The first is the way social movements use transnational activist networks to leverage power across international borders to target states and corporations. The second is the boomerang effect of racial and class inequalities and how such hierarchies often harm both beneficiaries and targets/survivors. After presenting a case study in which both boomerangs are in play, I then offer a conceptual framework for thinking through the kinds of social and political accountabilities and interdependencies these stories reveal, and their implications for social movements' dreams of freedom and invigorating new political formations.

Something Toxic from Denmark: Mozambique's Battle with Foreign Pesticides

This story begins in 1998, in Mozambique's capitol city of Maputo, where a Danish international development agency (Danida) funded an effort to incinerate nine hundred tons of obsolete toxic fertilizer and pesticide stocks. This case underscores two major examples of global environmental inequality: the massive export of pesticides, which often leads to surplus obsolete pesticide stocks lying unsecured in warehouses, vacant lots, and fields in global South nations, and the massive export of incineration technology to these nations.

Mozambique has a population of nineteen million people, seventy percent of whom live off the land. Located in Southeastern Africa, it is the world's ninth poorest nation. The country is slowly rebuilding itself after five centuries of brutal colonization by Portugal, followed by seventeen years of civil war, which resulted in the deaths of one million persons. Former independence fighters with the group FRELIMO won the country's first democratic elections in 1994, and UN peacekeeping forces finally departed one year later. Since that time, Mozambique has enjoyed relative peace. Even so, the average Mozambican's life expectancy is just forty years and the citizenry experience grinding poverty on a daily basis. The U.S. ecological footprint is 23.7 acres per capita—and a sustainable footprint in that nation would be 4.6 acres. Mozambique represents the other end of the scale, with an ecological footprint of 1.3 acres per capita. Unfortunately, the reason for this lighter footprint is because there is so much poverty and so little industrialization occurring in Mozambique ("Rich Nations Gobbling Resources at an Unsustainable Rate" 2004). Despite this harsh reality, new civil society organizations are emerging and thriving in this once chaotic place. And the first signs of new civil society growth in Mozambique sprang forth from an international struggle for environmental justice.

A Toxic Discovery: Mozambique as a Risk Society

In 1998, in the capitol city of Maputo, community activist Janice Lemos read a story in *Metical*—an independent local newspaper—about Danida's effort to fund the incineration of obsolete toxic fertilizer and pesticide stocks in a cement factory in the southern city of Matola. Danida sought to donate a hazardous waste incineration facility that would be housed in the cement factory, which the aid agency would also pay to have retrofitted for the operation. Ms. Lemos wrote to the newspaper for more information about the cement kiln incinerator proposal, but none was available. She then contacted Greenpeace International headquarters in the Netherlands, where someone informed her that two U.S.-based toxic waste activists would soon be visiting South Africa, and they might be able to travel to Maputo and Matola if Mozambican community leaders would invite them. With the help of Greenpeace and Oxfam Community Aid Abroad,

Lemos and fellow concerned residents met with the U.S. activists Ann Leonard (then with the group Essential Action) and Dr. Paul Connett (a St. Lawrence University chemistry professor and renowned expert on and opponent of incineration), as well as Bobby Peek, a South African toxics expert and activist (with the Environmental Justice Networking Forum).

The visiting activists were quite concerned because they possessed documentation that cement kiln incinerators produce a range of deadly toxins such as dioxins and furans. In fact, scientists estimate that twenty-three percent of the world's newly created dioxin comes from cement kiln incinerators alone (Puckett 1998). Prior to their arrival, the visiting activists were able to access documents about Danida's plans and had additional information about the proposed project. EJNF's Bobby Peek stated, "Whether or not anybody actually became concerned about the issue . . . we strongly felt that we had the moral obligation to pass on what we knew about the plan, and the real risks of cement kiln incineration. They had the right to know. As we feared, almost nobody had heard about the project at all" (Puckett 1998, 25). This lack of public knowledge was particularly disturbing because Danida has a policy of "actively involving individuals, non-governmental organizations and associations and businesses formally and informally in formulating and implementing environmental policies" (Neilsen 1999). Yet few people in Maputo or Matola had heard anything about the project from Danida. In fact, the foreign visitors were the *only* people at the meeting who had seen a copy of the short environmental impact assessment (EIA) Danida had prepared. Moreover, the report was written in English, although the official language of Mozambique is Portuguese. One local activist remembered, "only a few of us could manage to read the report and . . . do a brief analysis" (Lemos 2004). Connett denounced the entire project. He stated: "In the United States or Canada, those proposing a new toxic waste facility would be obliged to fully discuss all of the alternatives, all of the risks, and would have been required to hold several public hearings before decisions could be made about a particular disposal method. The environmental assessment and public involvement in this project is a sham" (Puckett 1998, 25). For its part, Danida conducted an EIA that concluded no serious environmental impacts would result from the incineration of the pesticides (Mangwiro 1999).

The visiting activists also informed the Mozambican citizens of the questionable record of Waste-Tech Ltd, a South African firm that was to be contracted for the Danida effort. At that time, Waste-Tech Ltd. was seeking to import foreign waste into South Africa—a clear violation of law there—and was the subject of an investigation by the South Africa Human Rights Commission concerning possible abuses in the case of two incinerators it had located within close proximity to an economically depressed community. The firm was also confronted with other legal investigations being conducted by the South African Department of Water Affairs and Forestry.

Mauricio Sulila, one of the local community leaders from Maputo present at the meeting, later told a reporter "When we explained [to others attending the gathering] that the government had decided the factory would burn toxic waste, they became terrified" (Lowe 2003). Local people already suspected the presence of toxic materials at the site because, as Sulila recalled, earlier flooding in the area prompted residents to pump the water into a nearby swamp where, soon afterward, "someone ate a fish caught in this swamp and died" (Lowe 2003). The terror that people experienced at the news of a toxic threat underscores risk society theorists' findings that the dangers of modern industrial pollutants often instill fear and dread among exposed communities (Erikson 1995).

At the meeting with U.S. and South African activists, local residents and community leaders founded an organization to address the problem of environmental hazards in the area. Mr. Mazul, one of the attendees who was also an artist, explained that, since the citizens had been kept in the dark by the Danish and Mozambican authorities at the Environment Ministry, the group should be named *Livaningo*, which translates to "all that sheds light" in Shangaan, one of many languages spoken in that region of the country. Mauricio Sulila was appointed the group's general secretary. Janice Lemos and her sister Anabela joined the group's leadership as well. The development of an activist organization in Maputo also reflected the symbolic or cultural dimension of environmental justice politics. This dimension facilitates people's expression of their sense of concern and care for ecosystems in ways that allow them to convert that sentiment into political action. The use of the word *Livaningo* for the new EJ organization was a perfect example. Embodied in this single name lies an acknowledgement of the local culture, the story of how this community came to be under siege, and an intent to make transparent and improve their situation. "All that sheds light" is also an ecological metaphor for the power of the sun and the power of the community.

Action and Networking at Multiple Scales

Soon after its first meeting, Livaningo grew and enjoyed some influence with the local and national governments. They organized public gatherings and meetings, brought their concerns to local residents and businesses, and made strategic use of the independent press. They also held some of the first public demonstrations in post-revolution Mozambique. Sulila explained, "It is important to say that Livaningo was the first organization in Mozambique to really challenge the government" (Pellow 2007, 174). Livaningo was eventually able to secure the services of a firm that conducted an independent environmental assessment of the project. Anabela Lemos proudly recalled that the firm's "conclusion was completely what we thought from the beginning: under no condition should the cement factory should be turned into an incinerator" (Lemos 2004).

However, the organization faced numerous hurdles in its efforts to oppose the incinerator. For example, the Mozambican government refused to consider Livaningo's independent environmental assessment. Activists then tried to secure an audience with the Danish embassy in Maputo, but they were refused. In the fall of 1998, members of the Danida board of directors visited Maputo, but rejected Livaningo's request for a meeting. In response, the activists elevated the struggle and went to the source. Aurelio Gomes of Livaningo and Bobby Peek of the EJNF in South Africa traveled to Denmark to address the Danish Parliament about Danida's pesticide incineration project. As Gomes stated upon arrival in Denmark, with regard to Danida's earlier refusal to meet with them, "This won't prevent us from voicing our concerns, therefore we've come to Copenhagen today to provide the Danish government with information to justify the immediate halt and rethinking [of the incineration effort]" (Basel Action Network 1998). Although the Parliament granted them an audience, its members made no effort to intervene in the conflict. Despite this rebuff, this was a critical moment in the development of a transnational EJ collaboration, because Mozambican, U.S., South African, and Danish activists were working together in close coordination. Allies such as Greenpeace International and the Joint Oxfam Advocacy Program (JOAP, Mozambique) donated the funding support for these activities. Mauricio Sulila remembered:

> That was great . . . After that, the Mozambique government opened up the door a little. We explained to them that we will not give up, we will not be intimidated. We continued to make pressure, to make noise, to hold international meetings and meetings at the local level. We were working with several organizations, especially Greenpeace Denmark. JOAP's support was fantastic. Say we need to do a demonstration in two days, they were able to provide funds to advertise in the newspaper. When we needed to travel to Denmark, JOAP funded us. It's not a lot of money, but it is at the right time, when we really need it. (Lowe 2003)

When asked later how Livaningo organized so effectively on an international scale, Anabela Lemos stated,

> It is mostly through the Internet. But whenever we campaign, we make some noise here in Mozambique, and at the same time we have the international network. When our government told us to stop complaining, we went to Denmark and we spoke to the people *there*, and we realized that, as a result, they started to listen to us *here*. So we realized then that we couldn't just do a campaign here, but instead we had to work both ways, here in Mozambique and in Denmark. (Lemos 2004)

Activists with Greenpeace Denmark were critical to the campaign's success as well. While the Danish government initially refused accountability for the pesticides, Danish activists took responsibility for their nation's involvement in this conflict. Greenpeace Denmark staff member Jacob Hartmann commented on the inconsistency involved in his government's embrace of the Basel Convention on Transboundary Hazardous Wastes (which prohibits wealthy OECD nations trading with or dumping hazardous wastes in poorer non-OECD nations) while also encouraging the incineration of pesticides in Matola: "Considering that Denmark is one of the countries that have taken the lead on this vital treaty, it makes little sense for Denmark to advocate for an elimination of POPs [persistent organic pollutants] globally while promoting new sources of the worst of them in Mozambique" (Puckett 1998, 26). POPs include the most toxic substances known to science such as dioxins, furans, and polychlorinated biphenyls (PCBs), and are common by-products of incineration.

Denmark's Development Minister, Poul Nielsen, denied that his country was seeking to impose incinerators on Mozambique, but activists found this claim suspect, given that Danida had funded a failed incinerator in India in 1986, and because Denmark was considering financial support for garbage incinerators in Zimbabwe and Tanzania in 1998, the year the conflict in Mozambique ignited ("SA Dumping Plan 'Trashed' by NYC" 1998).

Coalition activists consistently called for the pesticide incineration project to be halted, for the pesticides to be exported to a global North nation, for the wastes to be disposed of using non-toxic non-incineration technology, and for all the costs to be borne by the companies that produced the chemicals in the first place ("Mozambique Activists Win Huge Victory" 2000). And although Livaningo activists were only recently beginning civil society organizing on EJ issues, they were familiar with the problem of environmental injustice, since this was something that has been widespread in the region. Anabela Lemos remarked: "In South Africa, always the dumping sites are near the poor people. And we have a waste dump here in the city and there is a concentration of poor people there, so it's the same thing here. The poor, they always get the waste" (Lemos 2004).

Thus, this transnational coalition of environmental justice activists clearly articulated their opposition to Danida's use of local Mozambican ecosystems as waste repositories. Activists channeled these grievances into a vision of environmental justice that communicated an articulation of the symbolic, cultural, and political dimensions of ecosystems—that is, a viewpoint that challenged the dominant perspective of nature as a site of resource extraction and a place for dumping effluence. Thus, they were deeply engaged in disrupting the social relations that produced environmental injustice in their community and sought new relations of accountability locally, nationally, and transnationally.

The Boomerang in Motion

After two years of campaigning, Livaningo had its first major breakthrough. The Mozambican government agreed to a "return to sender" arrangement and allowed the chemicals to be shipped to a global North nation—the Netherlands—for processing and disposal by hazardous waste treatment firms there ("Mozambique Environmentalists Defeat Incinerator Plan" 2000). While the Mozambican government did have to pay some of the costs, Denmark shared the expenses. And, despite EJ activists' hopes that non-incineration technologies would be used, some of the wastes were indeed incinerated in Europe.

Even so, the EJ coalition achieved its primary goal of "return to sender"—exporting the wastes to a global North nation. Livaningo reached out to a broad group of established TSMO's, including the Environmental Justice Networking Forum (South Africa), Essential Action (U.S.), Greenpeace International (Netherlands, Denmark, and Brazil), the Basel Action Network (U.S.), and Oxfam's JOAP (UK), to amplify its voice and augment any leverage it already had in order to achieve one of the most impressive global-local EJ collaborations in the movement's history. The South African-based EJNF lent a critical African presence to the struggle. No less important was the legitimacy that Livaningo provided for its international partner organizations and activists who might otherwise be viewed as "outside agitators" in Mozambique. And Greenpeace Denmark provided much needed credibility for Mozambican activists confronting the Danish government. Drawing on local, regional, and international activist support, as well as international law and aggressive movement tactics, the coalition succeeded. These external resources were critical to the campaign, but the local activists' level of determination and commitment to the struggle was what ultimately sustained the effort. As Livaningo's Aurelio Gomes remarked, "We have nothing against Denmark, and hope they have nothing against us. We just want them to understand that here in Mozambique, while we may not be wealthy, we will never compromise our health—that is all some of us have" (Puckett 1998, 26). Likewise, Livaningo activist Anabela Lemos commented, "We just decided that we would not fail, although there were many times when it looked as if all hope was lost" ("Mozambique Environmentalists Defeat Incinerator Plan" 2000).

Although Mozambique is a democracy, it is still a young one. The government is still slowly becoming accustomed to the idea of being challenged by civil society groups, whether inside or outside its borders. As Anabela Lemos (2004) commented:

> Mozambique is a country where people are scared to speak out, and still today, but it is getting better. We are going through democracy after so many years. We are the only NGO [non-governmental organization] doing this work. If something is wrong, we speak up, we don't talk just for talking's sake. When we speak, we know we are right and we know we have to say it.

Next Steps for Mozambique: A Broader Vision of Justice and Sustainability

The campaign to halt the incineration project and export the pesticide wastes from Matolo, Mozambique was successful. This was a pleasant surprise for people throughout the international EJ and NGO community as they witnessed activists from one of the world's poorest nations exert uncommon political leverage.

Since this unprecedented success, Livaningo has used the opportunity to broaden its focus beyond toxics to include other environmental justice struggles in the region. As Livaningo expands its work, it is now pursuing projects aimed at introducing ecologically sound waste management systems in health care institutions in Mozambique, in collaboration with international activist groups like Health Care Without Harm. They are also working to oppose harmful "development" projects like the Mpanda Uncua Hydroelectric dam on the Zambezi River in Mozambique, in partnership with TSMOs like International Rivers (based in Berkeley, California). Livaningo is also combating oil extraction efforts by transnational corporations in southern Africa, which would pollute the air, land, and water and return few economic benefits to the people of the region.

Livaningo's victory in the pesticide incinerator case is credited with opening a broader political space for other civil society groups and social movements to work in Mozambique on a host of social concerns. Organizations working on HIV/AIDS, human rights, land rights and global economic justice efforts now enjoy greater support as a result of the political space Livaningo opened. In other words, they forced access into the nation's political process, by challenging and transforming the structure itself. As Anabela Lemos (2004) stated:

> It is true that we opened things up for people in our nation because we are not scared to speak out and to raise our issues. We think we have the right to do so. And I think that civil society has to get involved, we can't just sit on our hands and complain. If something is wrong, we should work for it. I think people should start to realize that to have big changes we have to give a lot up and we have to sacrifice . . . you should not be scared. If you are right, then you have the right to speak, and I think it does make a difference.

Thus, the Danida case allowed activists to build on the success a single environmental justice struggle and expand outward to be inclusive of a greater breadth of environmental and social concerns of civil society. This mobilization also revealed how deeply Mozambique had become a part of the "world risk society" (Beck 1999) through the embrace of ecologically and socially toxic forms of economic development that are rampant throughout southern Africa.

Discussion and Conclusion

The emergence of a transnational movement for environmental justice and the case of Livaningo allows us to think through questions of social hierarchy and the kinds of accountabilities and interdependencies that constrain and enable social and political change from within vulnerable communities. Here I wish to extend Guinier and Torres's "miner's canary" model of a racial metaphor to examine the miner's canary as a *spatial* metaphor. When global South communities are the targets of international environmental injustice (via hazardous waste dumping, illegal waste trading, or resource extraction from the global North), those spaces *and* the people who occupy them constitute the miner's canary. Thus entire communities, nations, and regions are often viewed as disposable or devalued by more privileged actors on the global stage. When social movements mobilize to demand that imported toxics be returned to their points of origin, the receiving nations of the global North serve as a reminder that environmental racism—like racism, class and gender inequality more generally—threatens us all. This analysis allows one to theorize and link the boomerang effects of racism, class inequalities, and social movements across international borders. In this way, the "miner's canary" can signal an impending or potential environmental danger that threatens not only members of vulnerable social groups, but also privileged populations living across vast geographic and social borders. This occurs as a result of the boomerang effects that racial and class inequalities and social movements produce, challenging social hierarchies that create environmental inequality and making hegemonic institutions accountable to vulnerable populations.

The idea of a boomerang effect is productive to theorizing social movements and environmental justice politics because it is a dynamic concept. The boomerang reveals that race and class inequalities and ecological harm associated with global capitalism are not just oppressive of people of color and ecosystems, but may ultimately be unsustainable and hazardous to those who benefit from that system. These race, gender and class inequalities are not just an unfortunate byproduct of an ecologically unsustainable society; they are at its root. Social inequalities are the principal forces driving ecological crises. Thus no one is exempt from racial, class, gender, and ecological violence, and social movements can present important and disruptive challenges to these social forces.

The boomerang is a metaphor. It is also a reminder of the interdependence among human societies and the unavoidable accountabilities we have to each other. The power of the boomerang returns us to the core of the humanenvironment and human-human interactions and the reason why we should be concerned about the various social dimensions of ecosystems: when we harm ecosystems we also perpetrate harms against other human beings, and vice versa (Harvey 1996; Merchant 1980). When we build relationships of respect and justice within human communities, we tend to reflect those practices in our relationships to ecosystems. Transformative, radical restructuring of societies is required to achieve environmental justice, and creative social movements are an indispensable foundation of that process (Speth 2008).

This chapter links environmental justice studies, environmental sociology, ethnic studies, and social movement theory in new ways, by drawing on key concepts and metaphors from these fields to produce new intellectual space for thinking about environmental politics, transnational movements, and social hierarchies. I began with the question: how do social movements challenge environmental inequalities across international borders? I argued that transnational EJ movement networks do this by disrupting the social relations that produce environmental inequalities, by producing new accountabilities among states and polluters, and by promoting a vision of an ecologically sustainable and socially just society. They achieve these ends by mobilizing bodies, information, and the cultural imaginary.

Note

Pellow, David Naguib. 2011. "Politics by Other Greens: The Importance of Transnational Environmental Justice Movement Networks." In *Environmental Inequalities beyond Borders*, edited by JoAnn Carmin and Julian Agyeman, 247–65. Cambridge, MA: The MIT Press.

References

Agyeman, Julian, Robert Bullard, and Bob Evans (eds.). 2003. *Just Sustainabilities: Development in an Unequal World*. Cambridge, MA: The MIT Press.
Basel Action Network. 1998. "Danish Development Project Encouraging Toxic Waste Trade into Mozambique?" Press Release, October 5, Copenhagen.
Beck, Ulrich. 1999. *World Risk Society*. Cambridge: Polity Press.
Erikson, Kai. 1995. *A New Species of Trouble: The Human Experience of Modern Disasters*. New York: W.W. Norton.
Foster, John Bellamy. 2000. *Marx's Ecology: Materialism and Nature*. New York: Monthly Review Press.
Gramsci, Antonio. 1971. *Selections from the Prison Notebooks*. New York: International Publishers.

Guinier, Lani, and Gerald Torres. 2002. *The Miner's Canary:" Enlisting Race, Resisting Power, Transforming Democracy.* Cambridge, MA: Harvard University Press.

Harvey, David. 1996. *Justice, Nature, and the Geography of Difference.* Boston: Blackwell.

Keck, Margaret E., and Sikkink, Kathryn. 1998. *Activists beyond Borders: Advocacy Networks in International Politics.* Ithaca, NY: Cornell University Press.

Lemos, Anabela. 2004. Interview with the author. February 5.

Lowe, Sarah. 2003. "Toxic Waste Victory in Mozambique." *Horizons,* February, Oxfam.

Mangwiro, Charles. 1999. Obsolete pesticides leave Mozambicans with $600,000 problem. *African Eye News Service* (South Africa), July 22.

McCully, Patrick. 2001. *Silenced Rivers: The Ecology and Politics of Large Dams.* London: Zed Books.

Merchant, Carolyn. 1980. *The Death of Nature: Women, Ecology and the Scientific Revolution.* San Francisco: Harper.

Moore, Donald, Jake Kosek, and Anand Pandian (eds.). 2003. *Race, Nature and the Politics of Difference.* Durham, NC: Duke University Press.

"Mozambique Activists Win Huge Victory against Toxic Waste Incineration." 2000. Coalition Press Release, October 5. http://www.ban.org.

"Mozambique Environmentalists Defeat Incinerator Plan." 2000. Environment News Service (ENS), Maputo, Mozambique, October 13.

Neilsen, Poul. 1999. Letter to Livaningo. January.

Pellow, David N. 2007. *Resisting Global Toxics: Transnational Movements for Environmental Justice.* Cambridge, MA: The MIT Press.

Puckett, Jim. 1998. "Something Rotten from Denmark: The Incinerator 'Solution' to Aid Gone Bad in Mozambique." *Multinational Monitor* 19 (12): 24–26.

"Rich Nations Gobbling Resources at an Unsustainable Rage." 2004. Environment News Service (ENS), March 30.

"SA Dumping Plan 'Trashed' by NYC." 1998. Africa News Service, January 8.

Smith, Andrea, 2005. *Conquest: Sexual Violence and American Indian Genocide.* Cambridge, MA: South End Press.

Smith, Ted, David A. Sonnenfeld, and David Naguib Pellow (eds.). 2006. *Challenging the Chip: Labor Rights and Environmental Justice in the Global Electronics Industry.* Philadelphia: Temple University Press.

Speth, James Gustave. 2008. *The Bridge at the Edge of the World: Capitalism, the Environment and Crossing from Crisis to Sustainability.* New Haven, CT: Yale University Press.

PART **VII** CHANGES IN PROGRESS

Ontologies of Sustainability in Ecovillage Culture

19

Integrating Ecology, Economics, Community, and Consciousness

Karen Liftin

Ecovillages are "alternative," intentional communities in which residents commit to socially and ecologically sustainable lifestyles. Such communities exist all around the world, and the author of the following piece, Karen Liftin, spent a year living in and learning about them. While ecovillages represent what some might consider an extreme form of sustainable living, they also offer insights into how everyone might incorporate green principles into their lives. Liftin identifies several lessons that can be exported to outside the ecovillage context, including systemic thinking, low energy living, sharing, good design, and "the power of yes."

19.1 Introduction

After twenty years of teaching international environmental politics, watching the state of the world go from bad to worse, I became convinced that topdown solutions to the unfolding global multi-crisis—as much as we need them—tend to be too little, too late. If the prevailing modalities of everyday life are unraveling our planetary life-support systems, who is pioneering ways of living that could work for the long haul? And, since we know that the nexus of social, economic, and ecological woes that constitute the mounting global mega-crisis necessitates structural transformation, not just lifestyle changes, what might be the political significance of these micro-experiments? This chapter addresses these questions through the study of integrative sustainability practices within the global ecovillage movement.

Ecovillages, which have sprung up in virtually every ecological, socioeconomic, and cultural context imaginable, aim to address the sustainability crisis at the level of everyday life. Only in the 1990s, with the formation of the Global Ecovillage Network (GEN) and the rise of the Internet, did they begin to interact on a regular basis. In the Global North, ecovillages tend to be intentional communities responding to social alienation as much as ecological degradation, while in the Global South they tend to be traditional rural villages concerned with economic sustainability. While ecovillage practices vary according to cultural and ecological context, they are unified by a common commitment to a supportive social environment and a low-impact way of life. To achieve this, ecovillages integrate various aspects of ecological design, permaculture, renewable energy, community-building practices, and alternative economics.

My research led me to spend a year living in ecovillages around the world. I selected on the basis of success, which I gauged in terms of size, longevity, influence, prosperity, and small ecological footprint, because I wanted to see what works. The smallest, Los Angeles Ecovillage had 45 members; the largest, Auroville,[1] had a population of over 2,000. My sampling reflected the diversity of the movement: rural, urban, and suburban; rich, poor, and middle class; religious, secular, and spiritual. What common threads run through this diverse tapestry? Certainly there is a common commitment to sustainable living. Because ecovillages share material resources, their per capita consumption and their income—at least in the Global North—tend to be far lower than their home country averages. Material factors like self-built homes and homegrown food, however, tell only part of the story. A more encompassing explanation for decreased consumption in affluent country ecovillages is the prevalence of sharing—not only of material things like property and vehicles, but of the intangibles that are the essence of community—ideas, skills, dreams, stories, and deep introspection. Human relationships, I found, laid the foundation for ecological sustainability—and, as their members consistently told me, they are also the greatest challenge of ecovillage life. Ecovillages are, as much as anything, laboratories for personal and interpersonal transformation. Indeed, *it is the subjective dimension of ecovillage life that facilitates most of their material successes.*

This suggests that the commonplace understanding of sustainability in terms of the "triple bottom line" or a three-legged stool (comprising ecological, economic, and social considerations) is insufficient because it sidesteps the inner dimension of sustainability, the perennial questions of meaning and belonging that are central to human existence. Consequently, I represent sustainability as a house with four windows: *ecology, economics, community, and consciousness—E2C2.* Each window faces a different direction, thereby presenting a distinctive angle while also disclosing a view of the other three windows. In their holistic approach to sustainability, ecovillages are particularly comprehensible through the four windows of E2C2, but this understanding can illuminate any human endeavor. Like cultures everywhere, ecovillages tend to highlight certain aspects of E2C2 over others, yet each window affords an essential view into any given community. Interestingly, even if an ecovillage starts out with an emphasis on only one or two of these windows, it will generally gravitate toward integrating all four of them.

As this chapter argues, the overlapping material and social practices among ecovillages are themselves symptomatic of a deeper, less tangible common ground. Ecovillagers share the following basic perceptions about the world:

- The gathering global mega-crisis is simultaneously biophysical, social, political, economic, and spiritual.
- The web of life is sacred and humanity is an integral part of that web.
- Being inseparable from nature, we can harmonize our lives with the web of life by tapping into the evolutionary intelligence that brought us to this juncture.
- We are best able to do this in community.
- Saying "yes" is a greater source of power than saying "no."

Across their enormous diversity, ecovillages share a holistic worldview rooted in the assumption that human beings are literally *of the earth* and can therefore access an evolutionary intelligence that links us to the greater whole of life. How they conceive of and access it varies widely, from systems analysis and permaculture to deep democracy and group process to meditation and interspecies communication. At the root of these diverse practices lies a basic ontological commitment to holism and the belief that humans belong to the web of life and, more radically, a commitment to *living as if this were true, and* so, perhaps, making it true. As a consequence, ecovillagers are unusually sensitive to the consequences of their actions, both near and far, and unusually open to sharing. If I had to choose one word to express the essence of ecovillage culture, it would be *sharing—among* themselves but also with the other-than-human world and far-flung others. This radical commitment to sharing is not merely a matter of personal proclivity; rather, it follows inexorably from a holistic worldview. This basic ontological commitment underpins the ecovillage movement, forging a core identity across widely disparate communities.

In applying an integrative approach to sustainability at the level of everyday life, ecovillages bring the ecological, economical, personal, and interpersonal dimensions of life into synergistic relationships with one another. E2C2 takes on a dynamic, self-reinforcing character, with the light from one window reflecting and refracting the light from the others. Sieben Linden's ecological focus, for instance, is primary, but disagreements over what this means in practice prompted them to take up various psychological and spiritual practices. And for self-identified spiritual communities like Findhorn, Damanhur, and Auroville, consciousness is the very soil from which their ecological, social, and economic practices grow.

Ecovillages can be understood as evolutionary laboratories running collective experiments in every realm of everyday life, from agriculture and natural building to interpersonal and even interspecies communication. Yet the "scientists" in these laboratories are not disinterested observers. To the contrary: every ecovillager I interviewed reported extraordinary personal growth through their experiments. Their accounts suggest that when people come together to transform their material and social landscape, they simultaneously enrich their inner landscape; in so doing, they open up new material and social possibilities. Whether their beliefs are secular, religious, or spiritual, the journey entails much the same effort: the work of moving from a fragmented me-centered world to an integrated tapestry of social, ecological, and even cosmological relationships. This inner work is absolutely vital to the outer work which, I believe, is equally true for those of us who may never visit an ecovillage.

In the following sections, I examine the integrative strategies of ecovillages with a twofold aim. In section 19.3, I illuminate the synergistic possibilities that emerge with a strongly integrative approach to E2C2. In section 19.4, I highlight the all-important and oft-neglected dimension of consciousness in igniting and realizing these possibilities. I have selected seven communities with an eye to these aims. Four self-identify as spiritual (Auroville, Damanhur, Findhorn, and Konohana); one (Sarvodaya) is culturally interreligious with a cohesive spiritual worldview; one (Sieben Linden) has an eclectic worldview, with much of the membership shifting over time from a secular to a spiritual worldview; and one is primarily secular (Svanholm). Focusing on the spiritual communities enables me to hone in on the interior dimensions of sustainability; including a transitional and a secular community facilitates comparative analysis. Furthermore, the extraordinary geographic, cultural, and socioeconomic diversity of these seven ecovillages illustrates the movement's global character.

Ecovillages, however, are small and sparsely dispersed, and time is short. The pressing question, then, is: how do we scale up the ecovillage experience? While this question, which I address at length (Litfin 2014), is beyond the scope of this chapter, my conclusions here offer some general principles for scaling up the ecovillage model.

Ecovillages may be understood as a pragmatic response to the material and ideational crises of modernity, one that is grounded in a holistic ontology. Modernity is a historically specific story about: (1) the triumph of human reason over superstition and the vagaries of nature; about history as a progressive march toward the material liberation of humanity; (2) a reductionist approach to knowledge; and (3) the possessive individual, replicated in the sovereign state, as the locus of political authority (Lit:fin 2003, 36). Coupled with industrial technologies and fueled by fossilized "ancient sunlight," this story made sense in 1800, when most Europeans never reached their 30th birthday. With only a billion people on Earth and a vast frontier in the "New World," nature seemed unshakably robust and inexhaustibly abundant. Today, however, the dark side of modernity is inescapable: in the everdeepening disparity between the conspicuous consumption of the North and the grueling poverty of the South; in the myriad forms of pollution that threaten air, land, and· sea; in the mass extinction of species; and in the feverish pursuit of security that seems to generate only greater insecurity.

Combining a supportive social environment with a low-impact lifestyle, ecovillages are pioneering ways of living that transcend the modern dichotomies of urban vs. rural settlements, private vs. public spheres, culture vs. nature, local vs. global, expert vs. layperson, affluence vs. poverty, and mind vs. body. In this sense, they represent a postmodern perspective, but one that constructs a viable alternative rather than merely a deconstruction of modernity. With its simultaneous focus on the global context for their highly localized life practices, along with the creation of GEN as a global social change network, the ecovillage movement is a strong example of "glocalization," or the interpenetration of local and transnational forces. (For an in-depth analysis, see Roudometof 2016.)

While the movement is relatively small, comprising perhaps a few thousand relatively new ecovillages in industrialized countries along with tens of thousands of traditional villages in the developing world that are introducing ecovillage design principles, the ideas and practices informing this movement are spreading rapidly. If the dominant human systems on the planet are not sustainable, as increasingly seems to be the case, then the rise of the global ecovillage movement is of urgent practical consequence. Ultimately, sustainability will not be a matter of choice, for it is the nonnegotiable precondition for inhabiting an ecosystem—including, for a globalized species, a planet—over time.

Ecovillages are as much integrated venues for storytelling as. Ecological laboratories. Most are telling some variation of a simple but profound story that conveys both the current human predicament and its resolution. In a nutshell, the story is that, having come directly out of nature and thus being inseparable from it, we can forge a viable future by tapping into the evolutionary intelligence that brought us to our current juncture. While ecovillagers differ *in* how they describe and access this intelligence, they concur on the basic story line and the fact that we must access a larger intelligence to guide us through these times. Thomas Berry (1999) calls this "the Great Work" of our time; Joanna Macy (2012) calls it "the Great Turning." *Work* and *turning* describe where the story leads but not the story itself, which I prefer to call "the Great Unfoldment." The new story is essentially the narrative of cosmological and biological evolution retold in lyrical terms—and with a sense of urgency befitting the times. The Great Unfoldment unifies a range of apparent dichotomies: humanity and nature, biology and geology, and, for some, nature and spirit. Blending ancient wisdom with contemporary science, this story is cropping up all over the world; ecovillages are enacting it in a highly focused and integrated fashion.

Through their integrative approach to E2C2, ecovillages are instantiating a holistic worldview, a mode of consciousness that is reflected and refracted in the other three windows. We can now catch a few glimpses of ecovillage life with an eye to how their inner work enlivens and magnifies their ecological, economic, and social work toward sustainability. In section 19.4, I offer telling examples from seven of the ecovillages I visited, anecdotes that are intended to be suggestive and evocative rather than conclusive.

19.2 Ecology

Imagination, the capacity to envision an alternative future, is a powerful impetus for social change. As tempting as it might be to focus on cob houses, solar panels, rainwater catchment, and permaculture landscapes, it is more helpful to understand these ecovillage technologies as material expressions of a new story rooted in a holistic worldview. The holism of ecovillage culture helps to reframe: environmental discourse from one of limits and constraint to one of abundance and human potential.

Although Findhorn has been called "the mama ecovillage" because of its size and age, in 1962 its three mystical founders did not have ecology on their minds. Rather, they were "attuning" to divine guidance through prayer and meditation and following this guidance wherever it led. Having never harvesting from "nature spirits." By the early 1970s, their astonishing harvests on Scotland's sandy, windswept soils brought scientists, the media, and thousands of young people to their doorstep. The founders left, but decades later, Findhorn's residents invoke their basic instruction, "Attune to Spirit, attune to Earth" in their daily meditations and work.

As part of a course called "Experience Week," I worked on Findhorn's seven-acre organic farm and experienced "attunement" and another community motto: "Work is love in action." During each morning's "attunement circle," the farm's "focalizer" explained the day's tasks. In silently attuning to our specific task, the point was to discern not what we *wanted* but rather what we were *called* to do. Somehow there was always the right number of bodies for each job. We were encouraged to work in silence and to feel ourselves as "one body." As I harvested beans and shoveled compost, I found myself reveling in the companionship of my coworkers and continually astonished by how much we accomplished.

Half a world away, 15,000 traditional Sri Lankan villages work with Sarvodaya, a highly successful participatory development network whose full name, Sarvodaya Shramadana, means "the awakening of all through the sharing of labor." The basic premise is that by collaborating to meet their needs, villagers simultaneously enhance their material, social, and spiritual well-being. As many of their members told me, "We build the road, and the road builds us." The wellspring of this ambitious work is a simple but powerful peace meditation. For Sarvodaya, social problems—war, poverty, environmental destruction, oppression of women—are rooted not in institutions or even behavior, but consciousness. Meditation is therefore more than a private matter; it is a dynamic force for progress.

One morning at breakfast in Sarvodaya's dining hall in Colombo, I happened to sit with Mr. Mahanama, a senior meditation instructor. He explained Sarvodaya's rationale for including meditation instruction in its sustainable development programs. "If we want to establish peace among ourselves and with nature," he said, "we must first establish peace in our own minds." He offered to teach me Sardodaya's peace meditation that afternoon.

We met at the verdant meditation center in the heat of the day amid the sounds of birdsong and traffic. In an oblique reference to the bloody conflict between Tamil separatists and the Buddhist majority that had already claimed 100,000 lives, Mr. Mahanama first instructed me to honor my own religion or belief system and recognize that every religion is a reflection of the truth. He then guided me through a simple breath meditation. As my mind settled, he pointed out that the air I was breathing had been and will be breathed by millions of sentient beings. After speaking for some time about interdependence, Mr. Mahanama invited me to send compassion to people I like and don't like, to strangers, and finally to all beings past! present, and future. Rather than retreating into an otherworldly bliss, I felt myself embedded in a vibrant web of relationships, both human and nonhuman.

In the ensuing days as I visited Sarvodaya's villages, I had a visceral reference point as villagers described the impact of Sarvodaya's peace meditation on their lives. When I met Harit Priyashanta, for instance, he was the president of Lagoswatte, a model ecovillage built by villagers who were displaced by the 2004 tsunami, but he was once a disaffected alcoholic. He described his transformation through Sarvodaya's peace meditation and the process of building Lagoswatte. I could only imagine the transformative potential of tens of thousands of villages engaging in this practice as they dug wells, built schools, and learned organic farming together.[2]

Across the Bay of Bengal, Auroville's pioneering ecological work is rooted in Sri Aurobindo's world-affirming spiritual injunction, "All life is yoga." Auroville's Earth Institute, for instance, invented a hand-operated machine, the Aurum, to build sustainable homes from compressed earth bricks using soil unearthed for the home's foundation or wastewater treatment system. Auroville is dotted with hundreds of these compressed-earth buildings. Their graceful domes, often painted white on top to reflect the sun's rays, make them an inexpensive and elegant solution to south India's scorching heat. Mud bricks might not sound particularly spiritual, but for Sat Prem Maini, the Institute's director, they are. "I don't see the Earth a formless material without consciousness," he said, "but as Spirit consciously disguised as matter" His comment echoed a core belief among Aurovilians: that biophysical reality is an evolutionary unfoldment of the divine.

The notion that ecological problems are, at root, problems of consciousness is a running theme among ecovillages. In Japan, a country that imports most of its food and grows almost none of it organically, Konohana stands out: it is 100 percent organic and almost fully food self-sufficient. While Konohana's fields at the base of Mt. Fuji constitute the basis of this farming community's economic and social life, they are also the focus of the community's inner life. Their motto, "Before cultivating the field, cultivate the mind," infuses their daily work and nightly "harmony meetings."

One secret to Konohana's success is Konohana-kin (pronounced *keen*), a fermented bacterial brew applied to the soil, fed to livestock, and even ingested by residents. Konohana-kin is based on effective microorganisms (EM), a technique developed by Teruo Higa, a Japanese agricultural scientist, to maximize the production of beneficial bacteria. Konohana developed its own formulation of molasses, brown rice, tofu refuse, bamboo leaves, and pine needles. Konohana-kin, considered by residents as "a gift from the Divine," serves simultaneously as a fertilizer, pesticide, cleaning agent, and preventive medicine. Because it is a dietary staple for the community's goats and chickens, their excrement has no foul odor and antibiotics are unnecessary. Likewise, Konohana members drink Konohana-kin each day. In fourteen years, they informed me, nobody had suffered a major illness.

I asked the community's founder, Isadon, about the relevance of Konohana-kin to the community's larger mission. "Our vision is that human beings will learn to live in harmony with nature," Isadon said. "Here in Japan, people wear masks and put disinfectants in their toilets. It's a violent approach: they are at war with bacteria, but we need bacteria to live. At Konohana, we cooperate with bacteria to make life better."

Some of the most powerful integrative practices I encountered were developed by Joanna Macy, a teacher and activist who popularized socially engaged Buddhism in the West.[3]

Gabi Bott, a Sieben Linden member who was trained by Joanna Macy, helped me to understand this connection. Gabi said the German ecovillage and the "Deep Ecology" workshops integrated her twin passions for spirituality and political activism. Two decades ago, when she worked for the Green Party and as a yoga teacher, she "lived a double life." She could never understand why her two sets of friends, activists and meditators, refused to bring politics and spirituality together. Eventually, Gabi quit both jobs and traveled to California to work with Joanna. "The training changed my life," she said. "Joanna encouraged me to bring this work home to help my country heal from its divided past." When Gabi visited Sieben Linden, she immediately felt at home:

> It was what I'd always wanted: a young community with a huge potential to be holistic. I offer Deep Ecology workshops all over Germany but Sieben Linden is really the perfect place for them.

While Sieben Linden, with its commitment to one-planet living, may be an ideal setting for "the work that reconnects," the same might be said of other ecovillages. Many ecovillagers spoke to me about the synergistic relationship between their ecological practices and their sense of connection to a greater whole. They would no doubt resonate with Wendell Berry's claim that spirituality and practical life should be inseparable:

> Alone, practicality becomes dangerous; spirituality, alone, becomes feeble and pointless. Alone, either becomes dull. Each is the other's discipline, in a sense, and in good work, the two are joined. (1987, 145)

After all, how holistic can a worldview be if it is not given material and social expression?

19.3 Economics

In our economically polarized world, where the average per capita income is roughly $7,000 but extremes of overconsumption and destitution persist, the down-to-earth prosperity of ecovillages demonstrates the possibility of a globally viable happy medium. In the heart of pricey Germany, for instance, Sieben Linden members subsist with pleasure on $12,000/year. Yet frugality is only one component of ecovillage economies. Equally important, they are experts at extricating themselves from the global economy through shared property, collaborative consumption, right livelihood, and a hyperlocal approach to the flow of money—all of which rests upon and reinforces a narrative of belonging.

Only two of the ecovillages I visited were communes, and both enjoyed a supportive cultural context. In Japan, which traditionally reveres family ties, Konohana functions as an enormous family. They live in close quarters, work side by side in the fields, eat their meals at one long table, and discuss the day's activities each night. Like a dose-knit family, Konohana disburses the earnings from its member-owned farm equally. When my translator, Michiyo Furuhashi, came to Konohana, she took an 80 percent pay cut from her work as an environmental consultant for Unilever Corporation, reducing her annual income to $7,000 and her living expenses to $3,000. "Our income is so low that we pay no taxes," Michiyo said. "I've never had so little, but I feel so rich!"

Contrary to the US, where "commune" has negative connotations, the Danish ecovillage, Svanholm, is proud to be one of the last surviving communes. Like Konohana, Svanholm members' assets and earnings go into the common pool. Everybody receives a minimum salary decided at the annual budget meeting. In 2009, it was about $47,000, making Svanholm the most prosperous ecovillage I visited. I wondered about potential free-riders. Birgitte Simonsen, one of the community's founders, assured me that their arduous membership process weeds out the lazy ones. "We probably tum down 80% of those who want to join," she said. "People here need to be able to work and relate well. We need a lot of trust to make Svanholm work, so people need to show they're trustworthy."

Trust; not ideology, is the key to Svanholm's collective economy. This is a constant theme in ecovillages: trusting, earning trust, discerning when to trust. Interestingly, it was at Svanholm, which prides itself in having "both feet on the ground," that I encountered the strongest aversion to spirituality: several members described themselves as "allergic" to it. When I asked Rene Van Dam, one of Svanholm's chief builders, about his spiritual beliefs, he was blunt: "*Phhh!* I don't want bullshit! Yes, we have love and beauty here, but don't call it spiritual. That makes it sound magical, not real." Later in the conversation, however, he waxed eloquent about humanity's place on Earth: "We're like a collection of micro-organisms on this super-organism! But that's biological, not spiritual."

While Svanholm avoids spiritual language, I suspect that some of its experiments in communal living are more effective in softening the boundaries of ego and moving from a story of separation to one of belonging than lofty meditation practices. The community's ability to stay the course over nearly four decades is largely due to the social trust that comes from sharing income and property.

To my surprise, most jobs in ecovillages are quite ordinary: cooks, housecleaners, carpenters, plumbers, web designers, beauticians, farmers, accountants, teachers, and so on. Unlike the anonymous relationships that pervade mainstream jobs, however, ecovillage jobs are about real relationships with people and resources. Consequently; the same money can circulate for quite some time. The yoghurt maker buys milk from the dairy farmer, who buys vegetables from the crop farmer, who gets her hair cut by the hairdresser; who pays a community accountant to keep her books, and so on. Some ecovillages go a step further: they mint their own currency. Damanhur's *credito,* for instance, has been a primary instrument in revitalizing the surrounding economically depressed valley. Spiritually, members refer to the *credito* as "clean money" because it is not based on violence and greed. In a more practical vein, it enables the community to develop its internal economy.

In *Wealth and Nature,* John Greer enumerates three economies. The primary economy comprises Earth's biophysical systems; the secondary economy conjoins these with human labor to generate goods and services; the tertiary economy constitutes the monetary flows that facilitate the exchange of goods and services. In truth, what most of us consider the economy—the secondary and tertiary economies—rests upon a multilayered gift economy of symbiotic relationships. From the biotic food web of soil to parenting to Wikipedia, the modus operandi of a gift economy is, in colloquial terms, "pay it forward" (Liftin 2010). While gift economies are marginalized in today's world, anthropologists consider them to be the bedrock of, culture.

I found some intriguing ecovillage experiments in gift economics. Sarvodaya's foundational premise, for instance, is *shramadana,* the gift of labor. While Auroville, which aspires to a nonmonetary economy, is very far from this goal, it has spawned some promising innovations. At Indus Valley restaurant, customers pay whatever they wish for their wholesome vegetarian meals, which means that some pay nothing. After nine years, the restaurant is flourishing. At Pour Tous ("For All" in French) Distribution Service, Auroville residents pay a small monthly fee to take whatever food and household goods they need. Everybody I interviewed praised the new system.

For the vast majority of us who are tied to the market economy, a gift economy may sound foreign. Yet each one of us is the unwitting beneficiary of a mind-bogglingly complex gift economy. The obvious question, then, is: how do we offer our own gifts to sustain this marvelous flow of gifts? The answer spans every dimension of E2C2.

19.4 Community

As evolutionary laboratories, ecovillages are running a range of relational experiments. I continually heard that human relationships were both the most challenging and most rewarding aspect of community life. "Being here is like being in a fire," one ecovillager told me. "Your lack of trust, your anger, your family neuroses—everything that separates you from the world comes out here! Getting over our individualistic culture means remaking ourselves." By standing in the fire of community, ecovillagers are rewriting the story of separation with their own lives. Even if their original intention was to live sustainably, the choice to do so in community throws them into a transformational cauldron.

Dieter Halbach, a former leader of the German peace movement and a cofounder of Sieben Linden, described this transition. The divide between spiritual and political communities, he said, ran deep during the 1970s and 1980s. Intent upon transforming society only after attaining enlightenment, spiritual communities were generally hierarchical and lacking in economic transparency. Political communities, on the other hand, were more egalitarian but frequently dissipated their energies on lengthy meetings:

> Because of my bitter experiences in politics and communities, I saw from the outset that we needed someone to help us cultivate our inner culture. So I brought

in a· friend, a Buddhist psychotherapist and an organic fanner. Now we're able to accept some hierarchy. We've learned that when we find the right person for the job and trust them, things flow better. This frees up time and energy to give back to the larger society. Sieben Linden started out as political, but now we're bridging the divide. It's very exciting. We're now in a position to help spiritual communities, and they're calling on us.

As I travelled through ecovillages, I found myself increasingly fascinated by this nebulous yet vital quality called *trust*. What is trust? How is it created—and destroyed? Sharing may well be essential to sustainability, but so long as we have a choice, sharing only makes sense in the context of trust. If we assemble a list of best ecovillage sustainability practices, every one of them is greatly enhanced by trust: car-sharing, co-ownership of property, collaborative consumption, community food production. My journey to ecovillages around the world was in many ways a study of the myriad ways of building trust. Across the board, open communication was the key.

When we come together authentically, transcending and including our individuality, something is born that is far greater than the sum of its parts— something we seem hardwired to want: a culture of belonging. By virtue of their highly integrative approach to E2C2, ecovillages are pioneering new stories of belonging—which, I would argue, is the psychosocial counterpart of sustainability. When we relate as integral parts of a greater whole, we automatically experience a greater sense of belonging, but we are powerless to create that culture alone. For that, we need one another.

19.5 In Sum

In these necessarily brief glimpses into ecovillage life, we see some of the synergistic possibilities of an integrative approach to E2C2. Ecological practices like natural building, organic farming; and frugality simultaneously express and reinforce stories of cosmological belonging. Likewise, the subjective and intersubjective dimensions of ecovillage life are crucial to their material and social successes. And the very act of coming together with others forges a collective field of consciousness such that every ecovillage has its own cultural norms and shared stories. Equally important, the inevitable personal and interpersonal challenges create transformational cauldrons. As ecovillagers become increasingly able to witness and embrace the relativity of diverse perspectives (including their own), conflict becomes a means to expanded awareness. Consequently, they experience progressively widening circles of identity and ever more integrated states of consciousness, which in turn ramify into the community's myriad experiments.

19.6 Scaling It Up

In many ways, my global ecovillage journey was a paradoxical one: I was an international relations scholar acutely aware of the global nature of our problems, yet I was touring micro-communities in search of seeds for a viable future.

I am convinced that these seedlings have something to teach us at every scale of human existence. Time is far too short to construct ecovillages for 7 billion people but not too short to apply their lessons everywhere, from our individual homes to our imperiled planetary household.

Social purpose, I believe, is one key. For humans, it is like water is for fish: the largely unexamined: intersubjective medium in which we swim. Ecovillagers have chosen to swim in a different medium by reorienting their lives around the core purpose of sustainability. With sustainability as a core purpose—one that is fully consistent with other objectives like happiness and democracy—we can then tease out the basic principles of ecovillage living and begin to scale them up. In order to export the ecovillage experience, we need a few simple principle that reflect the integrative character of that experience while attending to larger social and economic realities. Taking sustainability as a core purpose, I suggest the following five principles:

1. *Systemic thinking* is vital to ecovillage success stories. Material infrastructure, for instance, can promote and amplify social trust arid ecological sustainability. Examples are already proliferating: edible landscapes in parks; suburban "landless peasants" growing food in their neighbors' backyards; bike-friendly cities. This integrative approach to E2C2 differs sharply from prevailing piecemeal approaches to city planning, national policy, and international institutions—but change is afoot.

2. *Subsidiarity,* the idea that decisions should be made at the lowest level practicable, has its roots in democratic theory, Catholicism, and international law. An ecological reading of subsidiarity would meet human needs with minimal resource consumption and waste—which would entail substantial relocalization. Yet if we wish to sustain a global civilization, then our reading of subsidiarity might be something like, "Export your photons but leave your molecules at home."[4] On this measure, ecovillagers are pioneering a low-energy cosmopolitan identity,[5] which could underpin a nested hierarchy of democratic governance in a sustainable global civilization.

3. *Sharing* is the essence of both ecology and community. Ecovillages offer models for sharing everything from property and cars to self-governance and skills. As we scale up, the sharing expands to include schools, parks, roads, waterways, the atmosphere, the biosphere, the Internet—anything that sustains us in common. Full-cost accounting and other policies that protect "the space between" simultaneously foster ecological, social, and economic sustainability.

4. *Design* with an eye to the future is a vital element of ecovillages and their scalability. Even with planning, communities—especially those least responsible for the problems—will face enormous stress from climate change and energy descent. As Ozzie Zeher demonstrates, even with an all-out effort, renewables cannot fuel today's energy consumption, much less a world of 10 billion people hoping to live like the Global North. Yet, as Zeher admits, renewables will eventually supply most of humanity's energy needs. The catch is that we need to *create* the kind of society that *can* be powered by

renewables (Zeher 2012, 342), which means translating ecovillage design strategies into every scale of human organization.

5. *The power of yes* is greater than the power of no. Focusing on the most practical issues of life, ecovillages embody a kind of hands-on, do-ityourself (DIY) politics. They are creating parallel structures for self-governance within the prevailing social order while demonstrating how to live well with less. Whether it is one person stringing a backyard clothesline or the hundreds of cities banding together at the global level to promote sustainability (as in the International Consortium of Local Environmental Initiatives), every positive example is compelling—and likely to become more so in the future.

We need not live in an ecovillage to learn from them. With these five principles and our compass pointed toward sustainability, we are equipped to scale up the ecovillage model to every human scale, from the neighborhood to global governance. Ecovillages can be understood as adaptive responses to an unfolding evolutionary crisis. Just as the story of separation is reaching the end of its tether, these emerging experiments are enacting a new story of belonging by interweaving ecology, economics, community, and consciousness. Being the ecological oddity that we are, the one that can separate itself from the whole (at least in our own mind), we humans must now *consciously* integrate ourselves into the tapestry of life. In this endeavor, we can look to ecovillages as forerunners.

Notes

Liftin, Karen. 2016. "Ontologies of Sustainability in Ecovillage Culture: Integrating Ecology, Economics, Community, and Consciousness." In *The Greening of Everyday Life: Challenging Practices, Imagining Possibilities,* edited by John M. Meyer and Jens M. Kerstin, 249–64. Oxford: Oxford University Press.

[1] Auroville calls itself a universal township, not an ecovillage. Because of its renowned sustainability practices and its prominence within the GEN, I chose to include it in my study.

[2] For a recording of Sarvodaya's peace meditation, see the book's website: <http://ecovillage book.org/sarvodaya-peace-meditation/>.

[3] Joanna Macy's first encounter with socially engaged Buddhism came in the 1960s through her development work in India and Sri Lanka. Her first book, *Dharma and Development,* introduced Sarvodaya's peace meditation and its groundbreaking village programs to the West.

[4] This slogan has been attributed to Denis Hayes, founder of Earth Day.

[5] On the concept of low-energy cosmopolitanism, see Quilley 2011. Whether or not a sustainable global civilization is an oxymoron is a vital question, one beyond the scope of this chapter.

References

Berry, Thomas. 1999. *The Great Work: Our Way into the Future.* New York: Bell Tower.

Berry, Wendell. 1987. *Home Economics: Fourteen Essays.* San Francisco, CA: North Point Press.

Dawson, Jonathan. 2006. *Ecovillages: New Frontiers for Sustainability.* Totnes: Green Books.

Evemden, Neil. 1993. *The Natural Alien: Humankind and the Environment.* Toronto, ON: University of Toronto Press.

Gilman, Robert. 1991. "The Ecovillage Challenge." *Context Institute.* http://www.con text.org/iclib/ic29/gi1man1/ (accessed April 9, 2016).

Litfin, Karen. 2003. "Towards an Integral Perspective on World Politics: Secularism, Sovereignty, and the Challenge of Global Ecology." *Millennium: Journal of International Studies* 32 (1): 29–56.

———. 2010. "The Sacred and the Profane: The Politics of Sacrifice in an Ecologically Full World." In *The Ecological Politics of Sacrifice,* edited by J. Meyer and M. Maniates, 117–43. Cambridge, MA: The MIT Press.

———. 2014. *Ecovillages: Lessons for Sustainability.* Cambridge: Polity Press.

Macy, Joanna. 2012. "The Work that Reconnects." *Joanna Macy Reconnects.* http://www.workthatr-econnects.org/ (accessed April 9, 2016).

Quilley, Stephen. 2011. "Entropy, the Anthroposphere and the Ecology of Civilization: An Essay on the Problem of 'Liberalism in One Village' in the Long View." *Sociological Review* 59: 65–90.

Roudometof, Victor. 2016. *Glocalization: A Critical Introduction.* London: Routledge.

Zeher, Ozzie. 2012. *Green Illusions: The Dirty Secrets of Clean Energy and the Future of Environmentalism.* Omaha, NE: University of Nebraska Press.

Plans for Pavement or for People? 20
The Politics of Bike Lanes on the "Paseo Boricua" in Chicago, Illinois

Amy Lubitow, Bryan Zinschlag,
and Nathan Rochester

Alternative transit, including bicycling, is increasingly promoted as a solution to reduce global climate change emissions and air pollution levels. Though bicycling rates in the United States remain low, many cities are seeing an increase in bicycle commuters. Bicycle commuting also provides an important form of alternative mobility for lower-income citizens who do not own cars. However, the creation of bike lanes in cities can be politically complicated in myriad ways. Through a case study of the largely Puerto Rican Humboldt Park neighborhood in Chicago, this chapter explores what can happen when cycling infrastructure, such as bike lanes, is seen by residents as a tool for gentrification. The authors show how community engagement is critical for promoting bike infrastructure that is reflective of the needs of marginalized communities and how a community-led bike training center and repair shop facilitated such engagement.

Introduction

In the context of stagnant wages, growing traffic congestion and shrinking government budgets, bicycling has become an increasingly popular and affordable form of urban transportation in the USA (Beehner 2013). In addition to the economic factors that continue to steer urbanites towards bicycling, personal values associated with sustainability have often come to be reflected in transit choices in the USA (Hoffman, 2015; Pucher et al., 2010). Increasing bike ridership, coupled with the state sponsorship of transit-oriented sustainable development projects, have encouraged many American cities to develop plans for bike infrastructures. These plans are typically touted as urgent and beneficial for all residents (Parkin et al., 2012). An "if you build it, they will come" mentality, which presumes that more bike infrastructure will invariably produce more cyclists, has often led to the "fast tracking" of bike projects (Dill and Carr, 2003;

Lugo, 2013). In certain communities this fast tracking may be met with suspicion and resistance, particularly if rapid infrastructure development is not preceded by a robust community engagement process (Checker, 2013; Lubitow and Miller, 2013).

Such was the case in 2003 when the Chicago Department of Transportation (CDOT) proposed development of bike lanes along a stretch of Division Street in the Humboldt Park neighborhood. Known as Paseo Boricua ("Puerto Rican Promenade"), the area is the business district and cultural center of the USA's second largest Puerto Rican community (Wilson and Grammenos, 2005).[1]

Adorned with a pair of 60-foot high Puerto Rican flags and large murals depicting Puerto Rican heritage, Paseo Boricua is embraced as a symbol for community self-determination and a home base for resistance against gentrification. The proposal to install the bike lane was met with community resistance and a veto from then-Alderman Billy Ocasio, the preeminent political figure of Chicago's 26th Ward (which contains Humboldt Park).[2]

Given this complex neighborhood context, this paper considers how community engagement with bicycling as a form of economic development can mediate perceptions and experiences of gentrification in the Humboldt Park neighborhood. Drawing from interviews with community bicycle advocates, city officials and transportation planners in Chicago, we explore the tensions surrounding gentrification, neighborhood identity, and cycling facilities.

Urban Context: Bicycling Trends in the USA and Chicago

Although ridership in the USA appears to be distributed relatively evenly across income groups, it is predominantly a white practice. In 2010, non-Hispanic whites (66% of the US population) made 77% of all bike trips. However, other racial and ethnic groups' share of bike trips increased almost 50% (from 16% to 23%) between 2001 and 2009 (US Department of Transportation (US DOT)), 2010). Slowly but surely bicycling in the USA appears to be increasing in popularity among communities of color, many of which are disproportionately likely to be stretched thin by the rising cost of owning and maintaining automobiles. The city of Chicago, the site of the current study, provides a glimpse into these trends and the politics and practices surrounding them.

Chicago, Illinois is the USA's third-largest city and one of the nation's most diverse (US Census Bureau, 2015).[3] In recent years Chicago has made significant investments in bike infrastructure and has seen corresponding increases in usage. Between 2000 and 2010 its bicycle commuter rate more than doubled, increasing from 0.5% to 1.3%. The 2010 rate was significantly higher than its big-city counterparts New York, New York (0.8%) and Los Angeles, California (0.9%), but significantly lower than cities such as San Francisco, California (3.5%) and Portland, Oregon (6.0%). In 2012 Bicycling Magazine ranked it as the 5th most bicycle-friendly city in the USA (Chicago DOT, 2012: 10). In keeping with these trends, in 2012 the city announced a plan to increase its network of on-street bikeways to 645 miles by 2020, nearly tripling its current network size.

Mayor Rahm Emanuel introduced the plan with the claim that his vision "is to make Chicago the most bike-friendly city in the United States" (Chicago DOT, 2012: 7).[4] Emanuel has articulated that the cycling plan is motivated, in part, by the desire to attract technology companies and "entrepreneurs and start-up businesses" to Chicago (Davies, 2012).[5]

This framing of cycling infrastructure as an economic development tool to attract young, affluent residents to the city may shed some light on why some residents of the Humboldt Park neighborhood drew a connection between bike facility development and neighborhood displacement. The perception of bike lanes as "white lanes of gentrification" speaks to broader concerns about how changes to the built environment may be a catalyst for undesirable neighborhood changes and residential displacement.[6] In what follows we explore these dynamics and community efforts to respond to, and engage with, cycling infrastructure expansion.

Counterpublics: Alternatives to Dominant Urban Planning Dynamics?

Urban infrastructure projects in the USA routinely take on an air of neutrality; roads, bridges, apartment complexes and bikes lanes promote an agenda related to economic growth that obscures the very real power differentials inherent in these decisions. This "post-political" city is one where those in power minimize or simplify complex political demands emerging from cultural and social discourse (Swyngedouw, 2009; Zizek, 2002). Carr (2012) has voiced concern that public input processes intended to democratize city planning have instead been manipulated in ways that marginalize community voices while privileging elite interests. This body of scholarship suggests that urban infrastructure development is largely the domain of elites and public engagement is at best illusory and, at worst, a problematic distraction. However, other scholarship emphasizes mechanisms by which the public can potentially overcome opaque planning processes.

Neighborhood Context: Humboldt Park, Chicago

The Humboldt Park neighborhood in Chicago . . . has long been home to one of the largest Puerto Rican populations in the USA despite a long history of displacement, reclamation and, most recently, gentrification. Displacement has pervaded the Puerto Rican experience since US colonization in 1898, not only from the island of Puerto Rico but within the USA as well (Flores-Gonzalez, 2001; Rinaldo, 2002). Low-wage work brought waves of Puerto Ricans to Chicago in the 1940s, where they settled downtown. The construction of a number of universities, hospitals (and the related influx of white workers) in the inner city forced the community consistently westward (Flores-Gonzalez, 2001). By the 1970s, West Town (which borders Humboldt Park to the east) became home to a large percentage of Chicago's Puerto Rican population, and organized resistance against further displacement coalesced (Betancur, 2002; Rinaldo, 2002).

In the late 1960s and 1970s, public and private sponsors of urban redevelopment threatened the close-knit Puerto Rican community and generated organized resistance. As more Puerto Ricans in Chicago joined the struggle for self-determination, community organizations led a variety of actions to express discontent. As Betancur details, "These included confrontations with city hall and police over services and police brutality. . . ." and "development of a large network of local organizations and service agencies controlled by Latinos . . .," among others (2002: 797). Despite these and other efforts, West Town gentrified rapidly; between 1970 and 1980, the percentage of Chicago's Puerto Rican population that resided in West Town fell from 42% to 25% (Betancur, 2002). Puerto Ricans, largely displaced from West Town, began to concentrate and establish roots in Humboldt Park.

Humboldt Park and Paseo Boricua: Establishing Cultural Space

Although a considerable amount of the Puerto Rican population (25%) remained in West Town beyond the 1970s, the Humboldt Park neighborhood quickly became the primary home of Puerto Ricans in Chicago. In particular, a stretch of Division Street in Humboldt Park known as La Division has grown to embody the community's organized efforts to establish deep roots in inner-city Chicago. Since the 1980s, La Division has emerged as the home of Puerto Rican businesses and offices for advocacy organizations and politicians. Efforts by this community to permanently stake a claim to this space are a continuation of those made in West Town, but are now inspired by a greater understanding of the forces that pushed Puerto Ricans westward. While family, community and cultural offerings attract Puerto Ricans to live near La Division, organizations such as the Puerto Rican Cultural Center (PRCC) and the Division Street Business Development Association (DSBDA) work to make it affordable and viable to stay for the long term (Wilson and Grammenos, 2005).

The progress of La Division only accelerated in the 1990s as it became the increasingly clear center of the Puerto Rican community. In fact, it has become more commonly known as Paseo Boricua distinguishing it in name from adjacent stretches of Division Street that expand into other Chicago neighborhoods. This renaming was considered another step in the mission to make this area of Humboldt Park "a recognizable economic, political, and cultural space for Puerto Ricans" (Flores-Gonzalez, 2001: 9). The same can be said for the two mammoth Puerto Rican flags over Paseo Boricua, which serve simultaneously as the gateways and boundaries for the neighborhood. Puerto Rican community-building discourse in Humboldt Park remains intimately related to resistance against gentrification that has advanced westward from the city center for decades, disproportionately displacing Puerto Ricans and other minority populations in its wake (Betancur, 2002).

Current Study

Given this historical context, the current study explores how community members and city officials understand and respond to calls for enhanced bike infrastructure in Humboldt Park. To clarify these dynamics, interviews were conducted with community members, city officials, bicycling advocates and city staff at the CDOT.

Methodology

Seventeen interviews were conducted over the course of two weeks during the summer of 2013. . . . Interviews were further supported by attendance at a two-hour meeting of the Mayor's Bicycle Advisory Council Meeting (overseen by CDOT staff and attended by more than a dozen Chicago-area organizational representatives appointed to the council). Additional observations were collected through four days of active participation at Humboldt Park's community bike shop and education center, West Town Bikes/Ciclo Urbano (WTB/CU). Field notes were taken immediately following these events to help inform the analysis of interview data.

Interview participants reflect a diverse range of ages, racial and ethnic groups, and a relative balance between genders. Participants were recruited either because they played a prominently public role in the neighborhood (e.g. local Alderman), were a visible public figure related to bicycling in Chicago (e.g. professional, paid, bicycling advocate or a city planner), or were active participants at WTB/CU. All interviewees reflect perspectives of persons who are knowledgeable or engaged with bicycling or decision-making in the city of Chicago.

The "White Lanes of Gentrification"? Race, Class, and Bike Lanes in Chicago

In 2012, a series of blog posts on Grid Chicago featured content about whether or not new bike lanes proposed for Division Street in Humboldt Park could be considered the "white lanes of gentrification" (Greenfield, 2012a). The blog posts started an online debate in which community members and officials weighed in on the meaning and purpose of bike infrastructure in the neighborhood. For some, because bicycling in Chicago has been dominated by white, middle-class residents and lower-income white "hipsters," there was a perception that expanding bike lanes into minority communities symbolically paved the way for gentrification.

One interviewee described the distinctions that many interviewees made regarding the complex relationships between white cyclists and neighborhood change:

> People who are involved in bicycle advocacy [often] identify as a "cyclist," and quite often that comes from a sport and recreational background, and . . . not always, but too often, very competitive and very elitist. When you have

[residents] viewing cyclists that way, there's little interest in providing a facility in your neighborhood for those people . . . It's always been white guys who've come and like, invaded them and said, "We're here to save you, we're here to help you." And yes, Puerto Ricans in Humboldt Park have made a stance [against gentrification]. And certainly bicycle facilities, bikes for a very long time, have been seen as a recreational activity of privileged white people . . .

A Puerto Rican politician representing the neighborhood reflected upon these dynamics at length; extended portions of this interview are included to demonstrate his perceptions of the type of people moving into the neighborhood, coupled with his concerns about bike lanes utilizing public space in unsettling ways:

A lot of these bikers that have come in to this neighborhood, look you know, they're not affluent . . . They're hipsters by and large . . . [and] the other group that follows them . . . are the more affluent yuppies that you find.

And so I have resisted a lot of what they would wish to have. I created a transportation committee . . . I did a study to calm down the traffic on Humboldt Boulevard . . . and when we were doing this the city department [said] "Why don't you consider also a bike lane?" I said I don't want that because . . . the bikers that will take the bike lane between Division and North Avenue, they will continue. So I said "No I'm not gonna do that, for safety [reasons]." I always use the safety reason. Because then they will take over. They want to take over. You know, now it's this lane, then it's this other lane, they remove the cars from the streets, [then it's] "let's give up the whole street to walking," you know. I know that's the way I see it. And I think there is a lot of resent [*sic*] for it. I hear a lot from residents complaining about these people, they think that they own the streets, you know, they don't respect [us].

The excerpts above are illustrative of an ongoing perception of bike lanes as a significant mechanism of gentrification and a means by which public space is appropriated in service of neighborhood "outsiders." In the latter quote there is also evidence of city-level requests for bike lane development and, implicitly, this respondent notes that "safety" is leveraged as a rhetorical strategy to resist city-imposed infrastructure; the complex politics of gentrification are not brought into the official city-level conversations or contestations of the bike lanes. Rather, safety concerns are judged to be a more effective mechanism to challenge bike facility development.

However, this perception of bike lanes as a driver of gentrification is quite nuanced. According to Jose Lopez of the PRCC, it is not so much the bike lanes themselves, but the process by which the lanes are implemented and built that is deemed problematic:

It was not that we and Alderman Ocasio opposed the lanes, but at the time it was viewed as a process that mostly involved white people . . . Our attitude was that we would support the lanes as long as there was community engagement in terms of how the lanes would connect with the community's own projects and ideas about cycling. We never really got a response from the city and that was the problem. (Greenfield, 2012b)

Interviewees who were residents of the Humboldt Park area underscored the perception that meetings regarding infrastructure decisions often failed to actively engage local Puerto Rican residents:

> On big projects community input [can be] mandated as part of the plan. And how is that executed? The Department of Transportation sends out a notice to residents of that ward and set up their own meeting, and who shows up to it? . . . outsiders who are bicycle advocates who want to ride through those neighborhoods. They aren't necessarily residents of those neighborhoods, and I don't think are the best voice of those neighborhoods . . . I find that often planners and engineers make plans for pavement and not necessarily plans for people.

In part, the abbreviated decision making processes preceding the development of the Division Street bike lanes can be connected to the nature of bicycling initiatives in Chicago. Mayor Rahm Emanuel's attempts to implement an ambitious expansion of bicycle lanes in the city have been spurred on not only by the availability of federal funding for such projects but also by the seemingly widespread acceptance of bike projects. As such, the city has undertaken some of the most rapid infrastructure development in the USA.[7] One white bicycling advocate summarized the generalized acceptance of bicycling:

> So . . . you typically get a response when you say like, "Well I'm not quite sure bike lanes are the way to go." . . . It's like, "What? What's wrong? These are the best things we've had in years, and, why are you against us?"

This advocate recognized the taken-for-granted nature of bicycling infrastructure in Chicago and acknowledged the sense of apprehension among certain groups when community members voiced resistance to planned infrastructure changes.

As a result of the formalized promotion of bicycling infrastructure and an informal perception of such developments as universally accepted, bike facilities in Chicago have often been installed quite quickly. While efforts may be made to conduct community outreach, this may not adequately capture community desires. One minority community organizer summarized this dynamic:

> In Chicago . . . there's no process for communities for public meetings . . . there's no mechanism by which community members . . . especially in underserved communities, where people can be asking for what they need in a proactive way. What often happens is either the Department of Transportation or the city department, they may have an opportunity to apply for . . . the federal money that's available . . . and they're pushing for a network of better streets for everybody, but . . . on the community level there may not be a mechanism that's proactively telling them where each of these resources should go and how they should be used.

This quote highlights the constraints that city bureaucracies face in attempting to utilize hard-won public funds for infrastructure development while also conducting appropriate and meaningful outreach to communities. Such engagement takes time and energy and can curtail the trajectory of public spending on these projects, particularly if the proposed project is not widely accepted (Lubitow and Miller, 2013).

A failure to conduct adequate outreach when attempting to fast track a bicycling infrastructure project can have negative implications. Six interviewees cited the same story centered on the rapid implementation of a bike lane in an area close to Humboldt Park. In this situation, the city painted bike lanes that directly interfered with a space utilized for church parking on Sundays. Interviewees described how community members were angry about the intrusion into this space and suggested that a lack of input from clergy and local residents led to conflict over the newly striped roadway. One Puerto Rican participant summed up this dynamic:

> [CDOT] didn't do enough research about where they were putting these things . . . they took a space that was a community space . . . and without talking to them, they plopped something down. And that was really bad for everybody.

This story highlighted how the city's failure to adequately engage the community ultimately wasted resources that might have been used in other areas where the lanes would have been more appropriate. As the participant above noted, this created a situation in which bicyclists lost out; three participants reported that churchgoers now park illegally in the bike lanes on Sundays. It also created a dynamic in which some community members became resentful of bike lane development, both because of the lack of outreach and because of the rapid, hierarchical implementation of infrastructure.

Rapid infrastructure implementation and a lack of formalized participatory structures is further complicated by Chicago's unique political environment. The city's 50 legislative districts ("wards") are each represented by an elected Alderman with control over a range of localized decisions (on issues such as safety, public health and even taxes). In 2003, Alderman Billy Ocasio originally vetoed the Division Street bike lane in Humboldt Park. On the one hand, Chicago's system allows community residents some capacity to influence neighborhood-level changes by contacting their Alderman directly. On the other hand, the nature of Aldermanic privilege allows infrastructure decisions to be made with little or no input from residents.[8] A quote from one community organizer highlights the challenges of this decision-making structure:

> We don't have a formal process for community input. Decisions about projects are kind of decentralized, they're made by the local elected officials essentially. So one Alderman in one community might say, "Hey, this project is happening," or "We're interested in this project, let's have a public meeting." Another one might say, "You know what, go ahead, don't worry about it, I don't want a public meeting." So that obviously is a problem, and so what may happen is then you have community groups or residents who don't know about a project and didn't provide any input.

This quote illustrates the variable nature of the power of Alderman to determine what happens in neighborhoods and demonstrates how community residents can be both central to certain decisions and entirely excluded from others.

In Humboldt Park, the establishment of bike facilities reveals the complex nature of public space; community concerns regarding economic development and ongoing gentrification are mediated by a broader lack of transparent mechanisms to determine how to use and augment the street. Although Puerto Ricans in the area have a dynamic history of bike ridership, broader economic trends and a complex political environment have contributed to a perception that bicycle facilities are, symbolically and perhaps literally, a means of elite outsiders colonizing the public space that Puerto Rican residents have consistently fought to maintain.[9]

Mediating Gentrification? Community Engagement Models, Economic Development

The previous section explored the complexities of the symbolic meaning of bike lane expansion, along with the relative lack of public input regarding infrastructure decisions. This section explores how grassroots community organization and economic development around bicycles can encourage a diverse array of residents to engage in bicycling initiatives.

WTB/CU, a community-led bicycle training center and repair shop, evolved out of the non-profit Bickerdike Redevelopment Corporation's "BickerBikes" program that began in 2004. Although the program has served Humboldt Park and adjacent neighborhoods since its inception (and its parent organization has served the area since 1967), it didn't open a storefront on the Paseo Boricua until 2009. With an understanding of the community resistance that culminated in Alderman Ocasio's 2003 veto of bike lane installation, WTB/CU's organizers secured the blessing of community leaders in Humboldt Park before opening the shop (Greenfield, 2012a).

As a community organization, WTB/CU prioritizes education and outreach to community members. Classes and training not only help to teach youth how to repair bicycles and work with others in a business setting, but WTB/CU also teaches adults how to feel safe riding. Many interviewees suggested that the Puerto Rican community sometimes felt that biking was for white outsiders; thus, WTB/CU engaged in significant outreach to both minority adults and youth.

Two community bicycling advocates who worked at WTB/CU noted that travel by bike was a cheap and liberating form of transportation for youth:

> [Compared with public transit], it is almost always faster to take a bike . . . And so teaching people that a bicycle doesn't have to be a last resort but a very legitimate form of transportation around the city of Chicago. That idea of giving access to something that otherwise seems inaccessible is a huge part of what we talk about to the kids [who attend our programs].
>
> A lot of the kids who come through our programs have never been to downtown Chicago, have never been to the lake. . . . We take them on rides to the lake, to the zoo, to places, to universities and community colleges, so they see that not having $2.25 in their pockets does not preclude them from getting to where they need to go and is not an excuse.[10]

A third community organizer with ties to WTB/CU noted the significance of teaching community members how to engage with bikes:

> The city puts too much emphasis on infrastructure and not enough on education, "if you build it they will come" will only go so far . . . I mean . . . in addition to having infrastructure, it's equally important to have education programs and outreach.

While education and outreach were a clear mechanism for enhancing community engagement and interest in bicycling and city infrastructure, for most research participants, overcoming a top-down, exclusionary planning process could be achieved by appealing to the economic development capacities of bicycles.

When asked how to start to transform the centralized planning structures in Chicago, interviewees consistently called for a grassroots, community-led rather than city-imposed approach. They maintained that leveraging the connections, expertise and trust of established individuals and organizations in a neighborhood can generate more appropriate and open planning processes. . . .

In addition to the general belief that grassroots advocates are more in touch with how to encourage the mass appeal of bikes to a diverse group of residents, interviewees routinely highlighted the economic opportunities provided by bicycles. A white male organizer/advocate commented:

> . . . my intuition is that the notion of sustainability for communities of color, where it's going to be most appealing is where we can connect it to economics and jobs. Because that is something that is much more relevant and much more current, really.

Considering the socioeconomic constraints that many of Chicago's predominantly non-white communities face, development that provides economic opportunity will be well-received by at least some community members. In their experience, multiple organizer/advocates found success by framing language related to bikes in this manner. One interviewee, a white male from Humboldt Park, described the economic frame as a means to connect bikes to much larger issues in his community:

> I realized that bikes could be a great tool . . . bikes fit into community building far better than just pursuing bicycle advocacy and activism. That for me anyhow, and the work that I was doing with youth programs, that there are much bigger issues that the community was dealing with, such as housing and education, healthcare, jobs, that bikes could be a part of, and that there was much larger, much greater interest from the community in addressing those issues . . .

However, a white male organizer/advocate spoke to the idea that bikes are understandably not always the main priority of residents of these neighborhoods:

> . . . there are larger concerns in these communities that don't have bicycle facilities, and so how can they even be concerned about like whether or not they have facilities . . . I am really concerned about the health and wellness of my community in a much different way. Not whether or not it's safe to jog with my dog, but whether or not it's safe for my children to walk to school without being ran down by out of control traffic or being shot at.

Interviewees suggested that so long as its value is not overstated and the magnitude of more basic local safety concerns is acknowledged, bike infrastructure may be embraced as a viable community improvement. In particular, advocates in Humboldt Park focused their discussions on economic opportunity and safer streets, while still emphasizing the importance of respecting the neighborhood's cultural identity and connecting bikes to this identity. . . .

As a storefront on Humboldt Park's Paseo Boricua and a member of the DSBDA, WTB/CU has been careful to respect the cultural identity of the neighborhood and consider it in all of its work. One of its employees, a white male, describes how the organization's very name is a practice in celebrating the cultural heritage of the neighborhood:

> It was a huge victory when not only were we allowed to move onto Division Street, but we were invited and we work very, very closely with the Puerto Rican Cultural Center and also the [DSBDA]. DSBDA is a special service . . . we help promote business along Paseo Boricua . . . that supports the mission and vision that Paseo Boricua is supposed to be. And so West Town Bikes moving onto Division Street needed to be in line with that. And this is the reason that we named our sales shop Ciclo Urbano, because we really wanted to reflect the neighborhood that we're in, where we're at and who we serve.

Resulting in part from the close relationship that WTB/CU has maintained with trusted organizations in Humboldt Park, many of the neighborhood's residents have embraced bikes and bike infrastructure. The early resistance to bike lane construction through the Paseo, which began in 2003, was due to it being perceived as yet another imposition by a largely white-led city government. Furthermore, this imposition was considered strongly supported by public and private developers seeking to expand the city center westward without regard for the history of Puerto Ricans in Chicago. In a drastic shift from this original stance, many interviewees described the burgeoning bike culture in Humboldt Park with a sense of ownership. One Puerto Rican community member, also an employee of WTB/CU, suggested that her organization has played a significant role in this development:

> So I feel like West Town Bikes is a big part of why everybody now rides bikes. Like I never used to see, like [my co-worker] said that it used to mainly be predominantly be white people riding bikes, and now it's mixed and there's all types of bikes. . . .

One WTB/CU organizer who had grown up in Humboldt Park reflected on the way that the bike shop and its programs had generated alternatives:

> Ciclo Urbano/West Town Bikes . . . being on Division Street and kind of creating this business that's gonna help un-gentrify . . . [and not] a boutique or some type of high end clothing, or just a random juice bar . . . or something like a oxygen shop (laughing). So this brought another world into this community . . . Bringing bikes . . . and explaining and teaching how to use a bicycle . . . and how much a bike can change your life, how much a bike can mold your future, and how you can use it as a tool to make your life better.

One Puerto Rican worker at WTB/CU who had completed the youth programs and was now a leader in the organization described how the economic development model had transformative potential:

> I like to educate my younger generation on how much it's a choice to change their lives, and how much it's a individual responsibility. And it does involve

a social responsibility, but at the same to change the bigger you gotta change yourself. There's a word, derived from gentrification, and I like to use it, it's (gente)rification . . . (gente)rifying comes from gente, which means people, and (gente)rification is when people from their own community open up businesses, change their community for their own people.

These examples are suggestive of a public space being reformulated in service of economic needs and community interests. WTB/CU represents an innovative attempt to build community power, politically and economically, through education and employment opportunities. We suggest, below, that this space is also illustrative of a vibrant counterpublic that has the capacity to alter the urban landscape.

Discussion

Much of the popular discourse surrounding the rapid integration of bicycles into Chicago's pre-existing infrastructure has characterized it as a universal public good. Bike lanes in Chicago, as well as the recent Divvy bike-share program, have been overseen by a series of official planners with a universalized vision of what public space means. The dominance of technical knowledge necessitates that political demands for cultural recognition and community voice must be minimized to allow for the unencumbered implementation of a limited set of choices for street designs (Swyngedouw, 2009; Zizek, 2002).

These top-down mechanisms do little to offset anxieties and tensions about both the literal and symbolic means by which gentrification occurs. This study reveals that, whether or not bikes are catalysts for gentrification, for many people bikes do symbolize a type of gentrification in a meaningful and important way.

High-end economic and real estate development in and around Humboldt Park has generated situations in which minority residents reported feeling marginalized. The emergence of WTB/CU in this space represents an important mechanism through which a critical counterpublic has taken shape. The bike center's appeals to local cultural identity, when merged with economic development strategies designed to build and directly serve the neighborhood, have generated a new public space in which youth and residents are politically and socially engaged with the politics on their street. These dynamics have been present on Division Street for decades through the continual struggle to stave off encroaching development, yet WTB/CU presents a unique example of a counterpublic because it creates a space to directly engage with hegemonic planning processes and to offset some of the negative economic impacts of development.

Through job training and education, WTB/CU has created a space that is largely directed by and for minority youth from the Humboldt Park neighborhood. In this space, youth are empowered to learn new skills, to teach others, and to become reliable co-workers and leaders. The space established by WTB/CU allows for minority youth to generate their own environmental, political and social ethos. In turn, this regroupment has created the capacity for many

WTB/CU participants to engage with broader efforts to challenge hegemonic decision-making processes in Chicago. WTB/CU employees and advocates have been actively engaged in larger events and discussions about transit, the environment, and health and safety, and have also been tasked with serving on advisory panels and city-wide programs. For example, the Executive Director of WTB/CU serves on the Mayor's Bicycle Advisory Committee that makes decisions about bike infrastructure and planning. Additionally, youth from WTB/CU have begun to participate in broader city-wide training and educational programs.[11]

A space created and maintained by minority youth harnesses the economic benefits of encroaching gentrification for direct community benefit, while contributing to the establishment of community-level power and voice as Humboldt Park residents' perspectives and opinions are increasingly solicited and integrated into city plans for cycling. Although WTB/CU may not be dramatically transforming the landscape of urban planning in Chicago, the organization represents a vital space for community identity and political voice to grow.

We conclude here by suggesting that the top-down approach to decision-making present in Chicago is unlikely to adequately reflect the interests of all residents. If bicycling is to be the environmentally friendly, healthy and sustainable transit solution that is has the potential to be, decision-making processes at the city level must also consider how to enhance community engagement. Although WTB/CU represents a powerful effort to build engagement and voice, urban planners must take seriously community concerns by generating opportunities for a more participatory public sphere.

We suggest that planners remain mindful that changes to street design, particularly the implementation of bike facilities, are often intimately tied to community concerns about ongoing neighborhood changes. Bicycles have great potential to revolutionize how people use and interact with public space, but a truly just and socially sustainable bike infrastructure must incorporate community concerns and avoid strictly technological, universalized assumptions about the use and value of bicycles.

Notes

Lubitow, Amy, Bryan Zinschlag, and Nathan Rochester. 2016. "Plans for Pavement or for People? The Politics of Bike Lanes on the 'Paseo Boricua' in Chicago, Illinois." *Urban Studies* 53 (12): 2637–53.

[1] We utilize "Division Street" and "Paseo Boricua" interchangeably; although Division Street is a larger street, both terms are used to describe the same segment of street between Western and California Avenues.

[2] Chicago's legislative branch consists of 50 "Aldermen," each representing one of the city's "wards." Using the unwritten but historically entrenched rule of "Aldermanic privilege," Chicago Aldermen can veto any development project within their jurisdiction (Keefe, 2013).

[3] As of 2010, the city was 32.9% African American, 31.7% non-Hispanic white and 28.9% Hispanic or Latino. 2009–2013 data estimate a median household income of US $7,270 and a 22.6% poverty rate (US Census Bureau, 2015).

[4] The plan was compiled through community input gathered at eight public meetings and via the establishment of nine Community Advisory Groups. These meetings were described as "a tremendous success" with about 200 people participating at least once (Chicago DOT, 2012: 20).

⁵ Chicago's ambitious bike-share program, Divvy, is another example of this expansion. Implemented in June 2013, by March 2015 the program included 300 bike stations available to 33% of the city's population. A 176 station expansion will soon increase population coverage to 56%, but still does not extend south of 75th Street (Vivanco, 2015).

⁶ The association of bicycle infrastructure with white encroachment into minority neighborhoods was given voice in Portland, OR at a meeting to update the community plan for a rapidly gentrifying neighborhood, when an African American community leader referred to a proposed bicycle lane as the "white lanes of gentrification" (Portland DOT, 2001: 21). The referential quote has since been used to describe similar perceptions of the bike lane on the Paseo Boricua (Greenfield, 2012a).

⁷ The Federal Aid Highway program has slowly built up the funds available for pedestrian and bike facilities in the USA, devoting nearly US$4.5 billion since 2009 (US DOT, 2014).

⁸ In fact, the bike lane development that interfered with church parking was approved by that neighborhood's Alderman; in this instance, this politician made a unilateral decision.

⁹ The Schwinn Bicycle factory was, for decades, a presence on Chicago's West Side. Today, the Chicago Cruisers club in Humboldt Park celebrates the local practice of fixing up low-rider bicycles (Greenfield, 2011).

¹⁰ When this interview was conducted, US$2.25 was the cost of a one-way bus or train ticket in Chicago.

References

Beehner, L. 2013. "Cycles of Protest: How Urban Cyclists Act Like Insurgents." *Theory in Action* 6 (2): 52–86.

Betancur, J. J. 2002. "The Politics of Gentrification: The Case of West Town in Chicago." *Urban Affairs Review* 37: 780–814.

Bogdan, R., and S. K. Biklen. 2007. *Qualitative Research for Education: An Introduction to Theory and Methods*. Boston, MA: Pearson/Allyn and Bacon.

Carr, J. 2012. "Public Input/Elite Privilege: The Use of Participatory Planning to Reinforce Urban Geographies of Power in Seattle." *Urban Geography* 33 (3): 420–41.

Checker, M. 2013. "Wiped Out by the 'Greenwave': Environmental Gentrification and the Paradoxical Politics of Urban Sustainability." *City & Society* 23 (2): 210–29.

Chicago Department of Transportation. 2012. Chicago Streets for Cycling Plan 2020.

Cox, K. R., and A. Mair. 1989. "Urban Growth Machines and the Politics of Local Economic Development." *International Journal of Urban and Regional Research* 13 (1): 137–46.

Davies, A. 2012. "Rahm Emanuel Thinks Bike Lanes Will Attract Tech Companies to Chicago." *Business Insider*, December 5.

Dill, J., and T. Carr. 2003. "Bicycle Commuting and Facilities in Major U.S. Cities: If You Build Them, Commuters Will Use Them." *Transportation Research Board* 1828 (1): 116–23.

Flores-Gonzalez, N. 2001. "Paseo Boricua: Claiming a Puerto Rican Space in Chicago." *Centro Journal* 8 (2): 7–23.

Fraser, N. 1990. "Rethinking the Public Sphere: A Contribution to the Critique of Actually Existing Democracy." *Social Text* 25/26: 56–80.

Glaser, B. G., A. L. Strauss, and J. A. Strauss. 1967. *The Discovery of Grounded Theory; Strategies for Qualitative Research*. Chicago: Aldine Publishing Company.

Greenfield, J. 2011. "The Chicago Cruisers, a Puerto Rican Bike Club, Celebrates the Schwinn." *Chicago Grid*, August 2.

———. 2012a. "Bike Facilities Don't Have to Be the 'White Lanes of Gentrification'." *Chicago Grid*, May 10.

————. 2012b. "Jose Lopez Offers the PRCC's Perspective on the Paseo Bike Lanes." *Chicago Grid*, May 11.

Gregory, S. 1994. "Race, Identity and Political Activism: The Shifting Contours of the African American Public Sphere." *Public Culture* 7: 147–64.

Harvey, D. 2008. "The Right to the City." *New Left Review* 53: 23–40.

Hoffman, M. L. 2015. "Recruiting People Like You: Socioeconomic Sustainability in Minneapolis's Bicycle Infrastructure." In *Incomplete Streets: Processes, Practices and Possibilities*, edited by S. Zavestoski and J. Agyeman, 139–53. London: Routledge Taylor & Francis Group.

Keefe, A. 2013. "Pregnancy Tests? Pigeon Poo? What Chicago Aldermen Really Do." In *Curious City*. http://www.wbez.org/series/curious-city/pregnancy-tests-pigeon-poo-what-chicago-aldermen-really-do-107648 (accessed March 22, 2015).

Lubitow, A., and T. R. Miller. 2013. "Contesting Sustainability: Bikes, Race and Politics in Portlandia." *Environmental Justice* 6 (4): 121–26.

Lugo, A. E. 2013. "CicLAvia and Human Infrastructure in Los Angeles: Ethnographic Experiments in Equitable Bike Planning." *Journal of Transport Geography* 30 (1): 202–7.

Parkin, J., S. Ison, and J. Shaw (eds.). 2012. *Cycling and Sustainability*. Bingley: Emerald Group Publishing Limited.

Pucher, J., J. Dill, and S. Handy. 2010. "Infrastructure, Programs, and Policies to Increase Bicycling: An International Review." *Preventative Medicine* 50 (1): S106–S125.

Rinaldo, R. 2002. "Space of Resistance: The Puerto Rican Cultural Center and Humboldt Park." *Cultural Critique* 50: 135–74.

U.S. Census Bureau. 2015. *State and County QuickFacts: Chicago (City), Illinois*. http://quickfacts.census.gov/qfd/states/17/1714000.html (accessed March 31, 2015).

U.S. Department of Transportation. 2010. *National Household Travel Survey 2009*. http://nhts.ornl.gov (accessed April 12, 2015).

Vivanco, L. 2015. "Divvy's Expansion Begins, but Doesn't Hit All Corners of Chicago Map, Activists Say." In *Redeye Chicago*. http://www.redeyechicago.com/news/redeye-divvy-expansion-begins-april-20150401-story.html (accessed April 1, 2015).

Wilson, D., and D. Grammenos. 2005. "Gentrification, Discourse, and the Body: Chicago's Humboldt Park." *Environment and Planning D: Society and Space* 23: 295–312.

21

Campus Alternative Food Projects and Food Service Realities
Alternative Strategies

Peggy F. Barlett

Food is currently an exciting arena for research and activism. Peggy Barlett has written extensively on campus sustainability, and in this chapter, she examines how colleges and universities are leading a charge to insist that the food they purchase and serve to students is sustainably and equitably produced. Using case studies and interview data, Barlett identifies two ideal typical strategies educational institutions use to frame and guide their sustainable food purchasing. Some schools opt for a "relational" approach, whereby connections with local people—farmers, for example—are prioritized. Others adopt a "metrics" strategy, which emphasizes certifications and numerical targets. Barlett encourages readers to think not so much about which approach is better but more about how to find the best fit for the college or university's institutional culture.

Introduction

"Colleges and universities are leading the sustainable food movement and have been for a while," observed Roberta Anderson (personal communication, June 8, 2010) of the Food Alliance; they create "spaces of possibility" (Goodman, DuPuis, and Goodman 2011: 4). Local food initiatives seek to rebuild forms of economic relations that are no longer purely profit driven but include as well dimensions of environmental care and social justice characterized by direct market links between producers and consumers, rather than links between wholesalers, brokers, or processors (DeLind 1999; Lyson 2004). From the consumer's perspective, the lack of knowledge created by "food from nowhere" is replaced by "food from somewhere" that includes place-based dimensions of meaning and identity (McMichael 2009; Trubek 2008). The potential impact is large; the food service industry's annual revenue is over \$40 billion (IbisWorld 2016).

. . .

. . . [T]wo different (campus) strategies to enact sustainable food projects are identified: *relational approaches* that build personal ties with local farmers or cooperatives and *metrics approaches* that emphasize purchases that meet sustainability criteria verified by third-party certifications. Though aspects of these two strategies can be combined and many schools do pursue a hybrid approach, their implications are distinct and can be usefully explored as alternatives.

This discussion will begin by describing the relational and metrics approaches in greater depth, using a case study for each. Then, it will link the dimensions of variability in campus food service structures and the concerns of food service personnel to the ways they constrain alternative food strategies. This institutional context will . . . lead to some recommendations for campus action and further research.

Method

This analysis is based on interviews with individuals from twenty-six colleges and universities, at diverse stages of building campus sustainable food projects: ten small liberal arts colleges, eight large or medium public universities, and eight private research I universities. On-campus visits at eighteen schools were expanded with semi-structured, open-ended interviews with scholars, sustainability directors, or food service personnel—and in three cases, student food movement leaders. . . .

. . . This report also draws on the author's ten years as chair of the Emory University Sustainable Food Committee. . . . .

Two Strategies and Two Examples

The Relational Approach

Relational approaches have emerged among schools that seek to have a local and more systemic impact in support of an alternative food system. A relational approach seeks values-based supply chains based on trust relationships between dining service personnel and nearby or regional farmers (Stevenson and Pirog 2008). Schools can adopt a relational approach for the fresher, tastier product that comes direct from a local farm (Vilma et al. 2015), but in most cases, a primary goal is to build an alternative food network that supports more environmentally sound production methods, small family-owned and -operated farms, and regional economic development through diversified agriculture. Social justice concerns about farm workers may also be involved. Distribution can be carried out by direct deliveries to the campus loading dock—where farmers are paid directly—or by cooperative deliveries by groups of farmers or through conventional distributors. Some institutions have addressed difficulties with sustainable food deliveries by working with food hubs that bulk and pre-process items for institutional convenience. Financial gains from cutting out corporate distributors can help provide a higher price to farmers without harming the dining hall budget.

A relational approach emphasizes seasonality where possible. From September to Thanksgiving at one Midwestern college, students are provided lots of fresh broccoli: "It's astounding the amount they will eat," said the sustainable food director. Then, "students slide right on to winter squash, heavier sauces." Seasonal foods often have lower prices, which offer savings for the university, but selling all of a harvest can be helpful for local farmers as well. . . .

The relational approach can incubate fragile new projects and shepherd growers through the early stages of adaptation to new crops. At one liberal arts college, a food service director recounted how his previous farming experience allowed him to distinguish and encourage environmentally-friendly practices in his farm visits. His expertise gained him farmers' trust, as he sought to listen to their challenges and devise solutions. He helped several farmers develop stronger CSA networks, to support their summer crop sales, and farmers then directed available produce to the college in the other seasons: "Now they're going gangbusters." Farmers who resist the expense, hassle, and standardization of formal certifications but who have meritorious practices can be brought into the purchasing relationship without the need for third party oversight (Hassanein 1999). Relational approaches can also ignore certain quality issues—such as "slugs on the produce"—which may not be acceptable to a food service contractor.

A relational approach seeks particularly to expand economic opportunities in the region, often by promoting value-added products. One school contracted with both a local farmer for tomatoes and then a local processing firm to turn them into salsa for the college's use. Another school subsidized a small dairy to buy yogurt-making equipment in order to add local yogurt to menus. Rebuilding the food processing infrastructure is also necessary in most areas. Meat and grain processing plants, storage facilities, and small value-added processing operations have often disappeared with food system concentration and have to be rebuilt.

The concern that local food projects may not attend sufficiently to farm workers or other social justice concerns is clearly valid for some of the schools using a relational strategy. Because of the nature of the college-farm relationship, some leaders report it was difficult to make queries, especially about worker pay. In the Northeast, one food activist said raising concerns about worker pay was "a taboo subject" with farmers and was not included in her school's project, to preserve the trust in those relationships.[1] A Midwestern college prioritized the economic benefit of a local food network and did not screen for environmental or social practices: "We didn't want to come down on conventional farmers . . . it would cast a black cloud over what we want to do."

A relational strategy can scale up to have regional impact (Ruhf and Clancy 2010). Some institutions seek to build food innovation districts that link universities in public-private partnerships that include job training, food hubs, and other development tools (Center for Regional Food Systems 2015; Galarneau, Millward, and Laird 2013). Relational approaches to campus food can go beyond campus-farmer relationships to include faith-based groups and other non-profits, businesses, and governmental entities. Wider governmental planning conversations link food access and quality with transportation and housing decisions (Hamm 2015). Several statewide and multi-state regions have also seen the emergence of coordinated planning, such as Michigan's Good Food Charter (2016), the Minnesota Food Network (2016), and the New England Farm and Sea to Campus Network (FINE 2016). Though it is early days to assess the impact of these regional efforts, they are proof that conversations begun within higher education can have effects at larger scales.

Relational approaches are not always local. In the case of one New England college, the school's alumni connections with a fishing family in Alaska led it to prioritize salmon purchases from that source, out of respect for its environmentally responsible practices (and despite the firm's lack of formal certification of those practices). Some campuses develop Direct Trade relationships with a particular coffee farm and thereby support more favorable market relationships with a known supplier in a developing country.[2] Because such long distance trust relations are less common than building local/regional ties, this analysis will focus on the latter.

The Metrics-Based Approach

A metrics-based approach involves the commitment of a college or university to spend a specified portion of its food service budget on products having certain characteristics or carrying particular certifications. It allows the institution to redirect purchase dollars and take an ethical stand for food system change, re-embedding environmental or social values previously excluded from conventional economic market relations (Kloppenburg, Hendrickson, and Stevenson 1996; Paxson 2013). For example, the director of Cal Dining at University of California Berkeley said, "We wanted to incorporate organic products into our program because it's the right thing to do for our community, and our customers were asking for it" (Greensfelder 2006). In the process of demanding alternative products, the metrics approach begins to challenge the lack of transparency about where food comes from and how it is produced that characterizes the conventional system, thus contributing to some expanded level of food system accountability (Hatanaka 2014: 5).

Perhaps the most widely-used system of metrics on campuses today is the Real Food Challenge (2016), which encourages student-led auditing of all campus food purchases for two months (or longer) and a calculation of the percentage of "real food," as measured by several goals: local/community based, fair, ecologically sound, and humane. Roughly two dozen certifications and qualities are used in the assessment of real food, and over thirty-five campuses have committed to a goal that 20 to 40 percent of their total purchases will fulfill at least one of the real food criteria by a particular date (often 2020). Criteria such as "within 150 miles," "all employees receive living wage," Rainforest Alliance certified, and Food Alliance certified are used to determine the percentage of real food in a three-tier system of "green light" (best representation of real food), "yellow light" (not as strict but counts as real food), and "red light" (does not count as real food) (Real Food Challenge 2016). Nearly 200 institutions have participated in the Real Food Challenge.

Metrics-based approaches require new tracking systems, and institutional distributors have to adjust to these new demands. In the ten-school University of California system, not only dining halls but fast food franchises are asked to track certifications, which provides the opportunity for new market nodes and greater transparency. If metrics are adopted in a participatory and thoughtful way, such as when campus forums or classes discuss the issues, educational impacts are even wider and can extend to academic partners as well.

A metrics-based strategy can address a range of social, economic, and environmental concerns, and campus purchases add to market momentum in other sectors. For example, fair trade products have seen substantial market growth and widening circles of participation over the last two decades; there are now certified producer organizations in sixty-three countries, and certified products sold in sixty countries (Hatanaka 2014: 6; Lyon, Ailshire and Sehon 2014). For over twenty years, sales of certified organic foods in the United States have also grown by double-digits (Organic Trade Association 2015). The Fair Food Campaign of the Coalition of Immokalee Workers in Florida and efforts of Compassion in World Farming and other animal welfare groups have led both fast food chains and retail multi-nationals to shift to different production systems for tomatoes and eggs. With immediate benefits to farm workers and animal welfare, such large-scale interventions may also provide the benefit of these products to nearby neighborhoods served by these corporations. In general, however, metrics are not able to take account of social justice issues in farm production since few certifications address labor justice issues (Anderson 2008). And, in a different kind of scrutiny, in one student fieldtrip to a seafood distributor, "local" seafood was observed actually to be imported.

Increased local and sustainable purchasing using a metrics approach can sometimes increase dining purchase costs, but it does not always result in higher dining fees for students. Yale University began with a grant to support its Sustainable Food Initiative but also managed the higher costs of locally grown, sustainable, and some organic food by reducing the wide variety of options offered at each meal, lowering waste, adding more vegetarian options, and emphasizing lower-cost seasonal items. Some schools choose to increase student meal plan fees as a result of their sustainable food efforts, but many have not (Barlett 2011). One school allocated a 10 percent increase in annual food expenditures to the project in the first year to offset the added expenses. In later years, the subsidy was phased out through judicious use of seasonal purchases and in-house processing. Students at several schools have spearheaded trayless dining, with the understanding that savings through reduced food waste would be allocated to local and sustainable procurement.

Market impact can be significant from one school's actions, but metrics can be even more powerful when adopted by multiple institutions (Hignite 2009; University of California 2015). An example of market impact of sustainable seafood standards can be seen in the inclusion of Louisiana wild caught shrimp on the Monterey Bay Aquarium's Seafood Watch "avoid" list. State law prohibited Louisiana agents from inspecting for fishers' compliance with use of federal turtle exclusion devices, and the Oceana environmental group indicated that commitments to Seafood Watch guidelines among 13,000 stores and restaurants—and presumably some campus dining services, as well—led to a boycott of Louisiana shrimp (Alexander-Bloch 2015). Citing this market disadvantage, in July 2015, the Louisiana legislature repealed the prohibition, and Seafood Watch shifted its rating to "good alternative," thereby expanding the market opportunities for Louisiana shrimp.

Two Examples

The University of Montana can serve as a useful example of an institution that has embraced the relational approach. Students working with professor Neva Hassanein and dining director Mark LoParco began in 2003 to explore ways to strengthen sustainable agriculture and economic development in the Missoula area (Hassanein and LoParco 2013). From 2003 to 2016, the University of Montana's local purchases grew from $80,599 to $936,800 (Trevor Lowell, personal communication, March 7, 2017). Over 100 farms now participate, and the proportion of the dining budget directed to the Farm to College state-wide program grew from 3 percent to 26 percent (Ian Finch, personal communication, June 17, 2015; Trevor Lowell, personal communication, March 7, 2017).

The relational strategy supports negotiations that take into account the mutual needs and constraints of both growers and the institution. For example, to expand the university's supply of sustainably produced, grass-fed, and grass-finished beef, dining personnel met with two local ranchers to look at the possibility of expanded production with attention to available land and breeding stock, coordination with the packer, and delivery schedules. Within three years, the university was purchasing 300 percent more meat from them, and the ranchers benefited from a fixed price and a secure market. Less meat was coming from industrial feedlots, with their heavy environmental footprint, and the university benefited from the taste and more healthful nutritional profile of the grass-fed meat (American Public Health Association 2003; Clancy 2006). Overall, the University of Montana's relational approach to supporting an alternative food system in their area strengthened farmer cooperatives, diversified available farm products, and expanded the number of viable small- and medium-sized farms in the region. Dining service personnel report high student satisfaction with food quality, which has been achieved with stable or declining overall food expenditures, in contrast to other similar institutions that have not used Farm to School strategies (Mark LoParco, personal communication, June 17, 2015; Trevor Lowell, personal communication, March 7, 2017). The financial performance of the program contradicts the widespread assumption that for farmers to benefit from an alternative food system, consumers, and especially students on a meal plan, have to be charged more. A study of eighty-five California colleges and universities found that shifts to local purchasing did increase prices paid to farmers in most cases (from 0% to 35%), but overall food service budgets were able to adjust (Feenstra et al. 2011).

. . .

Emory University can provide an illustration of a metrics-based strategy—and also its evolution. Guided by innovations at Yale University and antedating the Real Food Challenge, Emory's Sustainable Food Initiative was based on an approved list of certifications and sources, with a goal that 75 percent of total purchases, for both campus dining and the university's hospitals, be either sustainably-grown or locally-grown. Early on, a two-day forum engaged almost 100 students in choosing which metrics to prioritize, and a Sustainable Food

Committee comprised of dining service staff, faculty, graduate and undergraduate students, and administrators debated and recommended specific purchasing guidelines to meet the institution's goals. Instead of using a mileage radius as many schools do, the committee recommended defining "local" in two tiers, with Georgia-grown as first priority and the eight-state Southern region second (Emory University Sustainable Food Committee 2013). This regional definition recognized that major crops come seasonally to Atlanta from north Florida and then from seven other nearby states, a significant improvement in food miles when replacing products from California or Mexico. In fact, as Emory's primary distributor, Sysco, began to provide improved tracking for Emory's program, it found that sometimes onions or other products from both Florida and California were available simultaneously in the warehouse, and they were able to fill Emory's order with the more local product.

Criteria were also developed to encourage purchases of fair trade coffee, humanely raised meats, sustainable seafood, and other certifications, including USDA organic. This metrics-based approach gave Emory Dining and its corporate partner, Sodexo, maximum flexibility to take advantage of market opportunities as they arose. As more sustainably-produced meats and dairy became available, the goals allowed room for new suppliers, and a new organic distributor became an important partner in the food chain. Over time, the dining service utilized a range of suppliers, with considerable change from year to year. Over the first nine years of the Emory program, purchases that conformed to at least one of the desired criteria grew from 1 to 2 percent in 2006 (as estimated by the Sodexo sustainability coordinator at the time) to a high of 26 percent in 2012–2013 (based on tracking of invoices and product codes for the year).

In 2014, as Emory's food service contract came up for rebid, the metrics approach shifted to be less flexible. Several particular certified products were specified in the request for proposals for residential dining, non-branded retail, and catering. Most notable is the demand that all ground beef on campus be grass-fed, a requirement made possible by a rapid increase in supply in the state. Humane certification was a requirement for eggs, and antibiotic administration only for disease treatment was required for chicken, reflecting not only concerns about human and animal welfare but also water pollution. Monterey Bay Aquarium's good and best seafood or Marine Stewardship Council certification was required for all fish and seafood purchases.

This revised metrics approach is now being implemented by Emory's new contractor, Bon Appetit, and the strategy has expanded in two years to include twenty relationships with local farmers. . . . Emory's purchases of either sustainably-grown or locally-grown food reached 38 percent in 2015–2016, for a total of over $2 million in purchases. . . .

Matching Strategies with Food Service Structures

Appropriate strategies for contributing to a more sustainable food system are constrained by the particulars of each campus's food service structure (Kimmons et al. 2012), including commitment to sustainability goals, food service mission, contractual constraints, staff capacity, financial situations, and geographical location. Public institutions and especially land grant schools often have a long tradition of support for local agriculture, and extending that support from research and extension to direct purchase for dining halls can be seen as part of the university's mission. Other schools have embraced local/sustainable food service goals, which can support dining innovations. But where sustainability commitments are weak, shifts in dining procurement have less traction (Conner et al. 2011). Likewise, the mission of the dining service itself is important. For some schools, dining is expected to be a significant revenue generator, similar to parking, and thus any proposal that might raise food costs is unwelcome. At one university, campus sustainability leaders were told, "If it's not cost-neutral, we can't think about it." Costs in such cases can refer to staff time as well as to dollars.

Other schools expect dining to contribute to a vibrant campus community life. Adding tastier food and/or an educational component around sustainable food can be an enhancement to the mission. For those schools that embrace ethical commitments and social justice, alternative food projects build on those commitments. . . .

Another factor of institutional variability is whether a school's food service is contracted or self-operated. Almost half of all institutional food services are provided by the "big three" corporations: Aramark, Sodexo, and Compass Group (which includes Bon Appetit) (Fitch and Santo 2016). In addition, purchasing flexibility can be limited by availability from local distributors. The two largest United States distributors—Sysco and US Foods—are estimated to control 75 percent of the national market for broadline distribution services (Fitch and Santo 2016). Several student activists reported the roadblock of being told, "Sysco doesn't carry it. . . ." Corporate food service providers mostly make their profits from negotiating favorable long-term price points for key food items and by aggregating purchases from multiple campuses in their negotiations. Such business models require advance menu planning, leaving less opportunity to respond to seasonal local surpluses or favorable prices.[3] . . .

Dining service staff capacity also varies widely and can have a major impact on the ability of an institution to innovate in purchasing, especially using a relational strategy. Some food service personnel have personal interest in sustainability, agriculture, or environmental issues and are pleased to incorporate them into their daily work life. Others see the critiques of the alternative food movement as an attack on the excellence of United States farming or on their food service and are hostile to change. Some find the headaches of locating new suppliers or coping with insurance requirements to be overwhelming. Or, as was the case in a Midwestern college, challenges with local supply and new insurance arrangements are overcome with the cooperation of several administrative units.[4]

As we have seen, relational strategies are assisted by staff with agricultural knowledge, who can enter easily into conversation with farmers and understand farm constraints. Others, especially chefs with only urban experience, may have neither the expertise nor the interest, making a metrics strategy more feasible. The traditional reward structures and training for food service administrators may not include the skills for either strategy, and demands for alternative food projects may threaten a leader's competence or performance evaluation. Of course, some food services respond to the challenge by hiring new people. National food service conferences have also greatly expanded training opportunities over the last decade, and all of the big food service corporations have developed templates for increasing local food purchases. In general, for campus staff with less interest, knowledge, skill, or institutional support, the metrics approach is more easily integrated into dining service habitus, involving mainly substitutions of suppliers and easy menu adjustment. In terms of institutional accountability, both approaches can be strong if they track purchases well.

Institutional geography, scale, and financial circumstances also vary widely and affect the feasibility of purchasing shifts. Location presents some limits to local sourcing, where desert environments, extremes of cold or heat, and availability of appropriate soils and slopes present agronomic constraints. Low population density or thin transport infrastructure can also restrict regional food system expansion. For a small liberal arts college located in an agrarian county, where faculty and staff are neighbors and friends with farmers who would like to do business with the college, building and sustaining relationships to support an alternative supply chain are more feasible and supported by existing ties. Though many adjustments of insurance, timing, and payment have to be negotiated, the college benefits not only from fresh produce, dairy, and meats but also from having the reputation of benefiting the local economy. A small or large school in an urban area, however, particularly one surrounded by sprawling suburbs, can face a greater challenge in finding farmers with whom to partner and overcoming barriers of transport, congested delivery routes, and unfamiliarity on both sides. One food service director in such an urban area described how his school's commitment to a mileage radius severely limited local purchases because there were so few farms in the area.

Scale can be a constraint simply from the amount of product needed as well. Small farms with organic vegetables or pastured poultry may not be able to provide adequate volume for a big campus. Larger cafeterias may also be unwilling to deal with non-standard portion sizes from multiple vendors, favoring the ability of conventional distributors to meet large orders. Food hubs that bulk and pre-process produce from smaller farms seek to meet this challenge, but distribution remains a significant impediment to purchases of sustainably grown and locally-produced food in many areas (Heiss et al. 2015). Schools that are struggling to survive financially are also less likely to innovate, but some have turned to improved quality in fresh, local food as a way to attract students. Each of these institutional contexts affects the feasibility of a relational or metrics approach.

. . .

Conclusions and Opportunities for Action

. . .

The implications of the analysis presented here are that campus leaders who wish to strengthen their food service connections to the alternative food movement might well begin with broad dialogues to assess whether their goals are mainly reduction of environmental harms from agriculture, farm and food worker justice, social resilience, economic vitality for the region, or enhanced community and individual health. With that clarity, strategies using metrics or a relational approach can be matched to institutional context and locale, and plans for policies and accountability designed accordingly.

The review here of relational and metrics approaches reveals that a relational approach offers strong advantages, but that a "pragmatic, incremental" metrics approach also can be effective (Goodman, DuPuis, and Goodman 2011: 145). Particularly where schools are small scale or favorably located geographically, or mandated to serve the economic development of the state, there can be opportunities for relational approaches that rebuild social integration with local farmers. Some leaders may wish to push for a combined approach that prioritizes certain metrics for out-of-state purchases. Institutions with high staff turnover, unfavorable location, or unenthusiastic administrations can redirect purchases using a metrics approach that embeds ethical concerns, especially if they can be written into contracts and policies. As many regions move towards resilience planning and are concerned about food access, the goals of an alternative food system can be carried forward into wider communities. Schools that seek to support a more reflexive food purchasing process can explore alliances with academic researchers, student internships, and governance structures that allow for democratic discussion and mutual education.

When it comes to the role of higher education to promote a wider conversation about desired futures, a relational approach has an advantage. It can more easily support bioregional planning and coordinated assessment of public and private investments to include a healthy food system along with other societal goals. Encouraging such new partnerships and information flow is best served by affiliative relationships, and the breadth of potential partners is one of the strengths of the relational approach (Heiss et al. 2015; Stevenson and Pirog 2008). Within diverse social and agricultural contexts across the country, regional cooperation can serve to discern appropriate goals, scales, and needed investments to support more resilient regional food systems.

In all these opportunities, the organizational capacity and mission of the school's food service has to be taken into account, recognizing that multiple goal orientations are common, within each individual as well as within any one institution. The diversity of alternative food strategies, projects, and institutional structures in higher education means that exemplars are available from different stages and with different templates for action; there is no need for any school to reinvent the wheel. . . .

The range of experiments now underway is also a call for improved research, especially on the full effects of alternative food policies and strategies; on student consumption habits; farmer viability; patterns of inequity by gender, race, or class; and the long-term viability of institutional commitment. Shared information on impacts, roadblocks, strategies, and mistakes will allow an assessment of the extent to which core values are being enacted. Institutional change can be hard to document, especially when confidential records are involved, but there is an urgent need to document and analyze emerging process. As we pour billions of dollars into agri-food systems that degrade soils, poison rivers, underpay workers, and transfer power to corporate conglomerates, this analysis of higher education's sustainable food projects shows that inclusive approaches and multiple strategies are available to build momentum toward new practices, paradigms, and politics.

Notes

Barlett, Peggy F. 2017. "Campus Alternative Food Projects and Food Service Realities: Alternative Strategies." *Human Organization* 76 (3): 189–203.

[1] There is also a justice dimension embedded in the standardization of food characteristic of large-scale food services. A desire to offer each patron a consistent portion of food for a consistent price is one component of food system regimentation and the desire for predictability. A form of fairness to the consumer is also linked to greater efficiencies and profit for the corporation, with resulting rigidities for the producer.

[2] Because there is no third party certification, however, direct trade cannot guarantee whether favorable prices are translated into farmer profits or farm worker earnings.

[3] Not all corporate business models are based on standardization, however; Bon Appetit allows each campus locale and chef to adjust menus and purchases to respond to such unexpected opportunities.

[4] One school circumvents requirements to choose lowest cost by adding a "freshness" component to food requirements.

References

Alexander-Bloch, Benjamin. 2015. Seafood Watch Removes Louisiana Shrimp from "Avoid" List. http://www.nola.com/environment/index.ssf/2015/07/seafood_watch_removes_louisian.html (accessed July 7, 2015).

American Public Health Association. 2003. Precautionary Moratorium on New Concentrated Animal Feeding Operations. http://www.ajph.org/legislative (accessed November 11, 2007).

Anderson, Molly D. 2008. "Rights-based Food Systems and the Goals of Food Systems Reform." *Agriculture and Human Values* 25 (4): 593–608.

Barlett, Peggy F. 2011. "Campus Sustainable Food Projects: Critique and Engagement." *American Anthropologist* 113 (1): 101–15.

———. 2012. Bring Heritage Breeds to Holiday Table. http://www.cnn.com/2012/11/21/opinion/barlett-thanksgiving-heritagefood/ (accessed November 22, 2012).

Center for Regional Food Systems. 2015. Michigan State University. http://foodsystems.msu.edu/activity/info/food_innovation_districts (accessed July 10, 2015).

Clancy, Kate. 2006. *Greener Pastures.* Cambridge, MA: Union of Concerned Scientists Publications.

Clancy, Kate, and Kathryn Ruhf. 2010. Is Local Enough? Some Arguments for Regional Food Systems. http://farmdoc.illinois.edu/policy/choices/20101/2010108/2010108.pdf (accessed March 18, 2017).

Conner, David S., Andrew Nowak, JoAnne Berkenkamp, Gail W. Feenstra, Julia Van Soelen Kim, Toni Liquori, and Michael W. Hamm. 2011. "Value Chains for Sustainable Procurement in Large School Districts: Fostering Partnerships." *Journal of Agriculture, Food Systems, and Community Development* 1 (4): 55–68.

DeLind, Laura B. 1999. "Close Encounters with a CSA: The Reflections of a Bruised and Somewhat Wiser Anthropologist." *Agriculture and Human Values* 16 (1): 3–9.

DuPuis, E. Melanie. 2000. "Not in My Body: rBGH and the Rise of Organic Milk." *Agriculture and Human Values* 17 (3): 285–95.

DuPuis, E. Melanie, and David Goodman. 2005. "Should We Go 'Home' to Eat? Toward a Reflexive Politics of Localism." *Journal of Rural Studies* 21 (3): 359–71.

DuPuis, E. Melanie, David Goodman, and Jill Harrison. 2016. "Just Values or Just Value? Remaking the Local in Agro-Food Studies." In *Between the Local and the Global: Confronting Complexity in the Contemporary Agri-Food Sector*, edited by Terry Marsden and Jonathan Murdoch, 241–68. Bingley, UK: Emerald Group Publishing Limited.

Emory University Sustainable Food Committee. 2013. Sustainability Guidelines for Food Service Purchasing. http://sustainability.emory.edu/page/1008/SustainableFood (accessed July 10, 2015).

Feenstra, Gail, Patricia Allen, Shermain Hardesty, Jeri Ohmart, and Jan Perez. 2011. "Using a Supply Chain Analysis to Assess the Sustainability of Farm-to-Institution Programs." *Journal of Agriculture, Food Systems, and Community Development* 1 (4): 69–84.

Fitch, Claire, and Raychel Santo. 2016. *Instituting Change: An Overview of Institutional Food Procurement and Recommendations for Improvement*. Baltimore, MD: Johns Hopkins Center for a Livable Future.

Galarneau, Tim, Suzanne Millward, and Megan Laird. 2013. Farm to School Efforts: Innovations and Insights. http://escholarship.org/uc/item/83p8p2b3 (accessed July 9, 2015).

Goodman, David. 2004. "Rural Europe Redux? Reflections on Alternative Agro-Food Networks and Paradigm Change." *Sociologia Ruralis* 44 (1): 3–16.

Goodman, David, Melanie DuPuis, and Michael Goodman. 2011. *Alternative Food Networks: Knowledge, Practice, and Politics*. New York: Routledge.

Greensfelder, Liese. 2006. New Organic Dining Option a First for U.S. Campuses. University of California Berkeley News, April 3. http://www.berkeley.edu/news/media/releases/2006/04/03_organic.shtml (accessed May 24, 2017).

Hamm, Michael W. 2015. Regional Food Systems for Improved Resilience. http://foodsystems.msu.edu/resources/regional_food_ systems_for_resilience (accessed July 9, 2015).

Hassanein, Neva. 1999. *Changing the Way America Farms: Knowledge and Community in the Sustainable Agriculture Movement*. Lincoln: University of Nebraska Press.

———. 2003. "Practicing Food Democracy: A Pragmatic Politics of Transformation." *Journal of Rural Studies* 19 (1): 77–86.

———. 2008. "Locating Food Democracy: Theoretical and Practical Ingredients." *Journal of Hunger and Environmental Nutrition* 3 (2): 286–308.

Hassanein, Neva, and Mark LoParco. 2013. Cultivating Food Democracy. http://tedxtalks.ted.com/video/Cultivating-food-democracy-Neva;search%3Aneva%20 hassanein (accessed November 4, 2013).

Hatanaka, Maki. 2014. "McSustainability and McJustice: Certification, Alternative Food and Agriculture, and Social Change." *Sustainability* 6 (11): 8092–112. http://www.mdpi.com/20711050/6/11/8092 (accessed March 17, 2015).

Heiss, Sarah N., Noelle K. Sevolan, David S. Conner, and Linda Berlin. 2015. "Farm to Institution Programs: Organizing Practices that Enable and Constrain Vermont's Alternative Food Supply Chains." *Agriculture and Human Values* 32 (1): 87–97.

Hignite, Karla. 2009. "Green Cuisine." *Business Officer Magazine*, June–July. http://www.nacubo.org/Business_Officer_Magazine/Magazine_Archives/June-July_2009/Green_Cuisine.html (accessed July 9, 2015).

IbisWorld. 2016. Food Service Contractors in the US: Market Research Report. https://www.ibisworld.com/industry/default.aspx?indid=1681 (accessed March 7, 2017).

Kimmons, Joel, Sonya Jones, Holly H. McPeak, and Brian Bowden. 2012. "Developing and Implementing Health and Sustainability Guidelines for Institutional Food Services." *Advances in Nutrition* 3 (3): 337–42.

Kloppenburg, Jack, Jr., John Hendrickson, and G. W. Stephenson. 1996. "Coming in to the Foodshed." In *Rooted in the Land*, edited by William Vitek and Wes Jackson, 113–23. New Haven, CT: Yale University Press.

Lyon, Sarah, Sarah Ailshire, and Alexadra Sehon. 2014. "Fair Trade Consumption and Limits to Solidarity." *Human Organization* 73 (2): 141–52.

Lyson, Thomas A. 2004. *Civic Agriculture: Reconnecting Farm, Food, and Community.* Cambridge, MA: Tufts University Press.

McMichael, Philip. 2009. "A Food Regime Analysis of the 'World Food Crisis.'" *Agriculture and Human Values* 26 (4): 281–95.

Minnesota Food Network. 2016. Charter. http://mnfoodcharter.com/ (accessed June 6, 2016).

Organic Trade Association. 2015. State of the Industry. https://www.ota.com/resources/market-analysis (accessed July 9, 2015).

Paxson, Heather. 2013. *The Life of Cheese: Crafting Food and Value in America.* Berkeley: University of California Press.

Real Food Challenge. 2016. Real Food Guide. http://www.realfoodchallenge.org/sites/g/files/g809971/f/201403/Real%20Food%20Guide%20Version%201.0%20March%202014_0.pdf (accessed March 17, 2016).

Ruhf, Kathryn, and Kate Clancy. 2010. It Takes a Region. . . . Exploring a Regional Food System Approach: A Working Paper. Northeast Sustainable Agriculture Working Group. http://nesawg.org/sites/default/files/NESAWGRegionalFoodSystem FINALSept2010.pdf (accessed June 6, 2016).

Stevenson, G. W., and Rich Pirog. 2007. "Values Based Supply Chains: Strategies for Agrifood Enterprises of the Middle." In *Food and the Mid-level Farm: Renewing an Agriculture of the Middle*, edited by Thomas A. Lyson, G. W. Stevenson, and Rick Welsh, 119–43. Cambridge, MA: The MIT Press.

Stevenson, G. W., Kathryn Ruhf, Sharon Lezberg, and Kate Clancy. 2007. "Warrior, Builder, and Weaver Work: Strategies for Changing the Food System." In *Remaking the North American Food System: Strategies for Sustainability*, edited by C. Clare Hinrichs and Thomas A. Lyson, 33–62. Lincoln: University of Nebraska Press.

Trubek, Amy B. 2008. *The Taste of Place: A Cultural Journey into Terroir.* Berkeley: University of California Press.

University of California. 2015. University of California Sustainable Practices Policy. http://ucop.edu/sustainability/policies-reports/index.html (accessed July 9, 2015).

Vilma, Helene, Ivette A. López, Lurleen Walters, Sandra Suther, C. Perry Brown, Matthew Dutton, and Janet Barber. 2015. "Perspectives of Stakeholders on Implementing a Farm-toUniversity Program at an HBCU." *American Journal of Health Behavior* 39 (4): 529–39.

From the New Ecological Paradigm to Total Liberation
The Emergence of a Social Movement Frame

22

David N. Pellow and Hollie Nyseth Brehm

Contemporary mainstream U.S. environmental activism is rooted in incremental change, and though its activities did lead to notable policy successes in the 1960s and 1970s, including the establishment of the Environmental Protection Agency, mainstream organizations, such as the Sierra Club, have been criticized for being reformist and exclusive. Since the mid-twentieth century, many new environmentally oriented social movements have developed to embrace ecological goals while promoting more inclusive and radical agendas. In this final chapter of the reader, the authors Pellow and Brehm describe how a new social movement conceptualization, "total liberation" is currently emerging out of several mid-twentieth-century developments. This new approach transforms the limited goal of reducing harm to the environment and people to a more holistic goal of reordering human-nonhuman relationships toward equality for all beings. Activists subscribing to the total liberation frame espouse an ethic of justice, are anti-authoritarian, oppose capitalism, and believe in the use of direct action tactics. The total liberation frame, a work in progress, emphasizes that social injustices and environmental crises are interwoven and that one set of problems can't be solved without solving the other.

Introduction

In 2001, Earth Liberation Front (ELF) activists nailed metal spikes into hundreds of trees in Washington State's Gifford Pinchot National Forest to protest the U.S. Forest Service's decision to sell them to a timber company. Activists sent a communiqué to several media outlets shortly after, which read, in part:

> This timber sale contains 99 acres of old growth and is home to spotted owls, grizzly bear, lynx, wolf, goshawk, just to name a few of its many inhabitants. This is truly a beautiful area, unfortunately one of the last of its kind because of the system we all live under. We want to be clear that all oppression is linked, just as we are all linked, and we believe in a diversity of tactics to stop earth rape and end all domination. Together we can destroy this patriarchal nightmare, which is currently in the form of techno-industrial global capitalism. (ELF 2001a)

The emergence of the ELF and Animal Liberation Front (ALF) in the 1980s and 1990s marked a new focus within U.S. ecological politics that involved forms of radical analysis and action that we had rarely seen in environmental or animal rights movements until that point.[1] By the late 1990s, segments of these movements were converging around new ideas and tactics, producing a broader discourse that linked ecology, social justice, and animal rights.

Groups like ELF and ALF primarily seek to combat environmental degradation and animal exploitation. These activists believe that the exploitation of ecosystems and nonhuman species calls for immediate, direct action. They reject structured, bureaucratic approaches and instead target what they see as the roots of the problem. Radical earth and animal liberation movements have gained global visibility and notoriety in recent decades, causing significant property and economic damage to laboratories, slaughterhouses, power lines, fur farms, and industrial agricultural facilities through arson, animal rescue/liberation, and vandalism (Best and Nocella 2004, 2006). Through these actions and the discourse that supports them, activists question what they view as the violence of inequality, capitalism, state power, and speciesism. Regardless of their success, the vision of these radical social movements can shed light on present social arrangements, highlighting how they may be seen as unjust in ways that most members of society have never considered. And while these movements often reflect different emphases, we propose that they are developing a new collective action frame: total liberation.

The total liberation frame combines important elements from other key movement frames, including the New Ecological Paradigm (NEP), deep ecology, ecofeminism, and the Environmental Justice Paradigm (EJP), which will be explained in more detail below, to chart a new course for social movements concerned with a broader vision of ecological politics.

Social Movements and Frame Transformations

Social movements are carriers of beliefs, ideologies, and ideas, and frames are the "signifying work and meaning construction" that movement activists routinely practice (Benford and Snow 2000: 614). These frames provide an intellectual grounding and moral compass for movement communities, guide their actions, and offer a way of constructing ideas, values, and social significance for activists (Benford and Snow 2000). What is sociologically significant about radical movements, then, is the power of their vision of change—the frames they produce—and what they suggest about a society's potential for facilitating both oppression and freedom.

Interest in the framing process in relation to social movements has been a key aspect of sociological scholarship. Social movements construct collective action frames as their participants negotiate a shared understanding of the problems they seek to address, how the problems should be handled, and the motivation behind their actions (often called diagnostic, prognostic, and motivation frames, respectively) (Snow and Benford 1988). Such framing generally involves strategic efforts to link the interests of social movements to those of their constituents.

These efforts, termed "frame alignment processes," include *frame bridging* (linking two or more frames), *frame amplification* (embellishment or clarification of existing values), *frame extension* (extending a frame beyond its primary interest), and *frame transformation* (changing old understandings and/or generating new ones) (Snow et al. 1986). While each frame alignment process is key to framing, this article focuses on frame transformation.

Benford and Snow (2000) note that few movement studies have addressed the concept of frame transformation. Frame transformation occurs when "new ideas and values . . . replace old ones" and "old meanings, symbols, and so on are discarded [and] erroneous beliefs and misframings are corrected" (Taylor 2000: 512). More broadly, frame transformation involves a "general reframing of the issues" within a movement (Taylor 2000: 512). . . .

Because of the lack of empirical inquiry regarding frame transformation, other aspects of its application and elaboration remain highly undeveloped as well (Benford and Snow 2000: 625). For example, existing literature (Snow and Benford 1988; Benford and Snow 2000; Ulsperger 2002; Futrell 2003) assumes that a movement undergoes a frame transformation by transforming only a single frame. That is, a movement reimagines itself by transforming one frame into a new frame. Yet, this is a limiting view of how movements actually function. It is well known that movements often learn from and are influenced by other (historical and contemporary) social movements (Wang and Soule 2012). Accordingly, we contend that frame transformation can also be the result of a movement drawing on and transforming several frames, perhaps even from multiple movements, thus multiplying the various sources and resources it might use to reassess its own frames. In fact, many social movements began precisely because of a need to fuse two or more existing frames. The environmental justice movement, for example, emerged when activists linked civil rights, public health, and ecological protection frames and challenged the dominant NEP that undergirds the mainstream environmental movement (Bullard 2000). Thus, we seek to further develop the concept of frame transformation, and to do so, we rely upon the radical environmental and animal rights movement.

Methodology

We began our study in 2008 with the goal of understanding how radical environmental and animal rights activists seek to effect change in the United States. We rely upon (1) 88 semistructured interviews; (2) fieldwork at conferences, activist gatherings, and other events; and (3) content analyses of thousands of pages of newsletters, magazines, journals, Web sites, and related publications produced by activists.

Frame Transformation: From What?

The total liberation frame resulted from a combined transformation of four of the most important frames from the 20th- and early 21st-century ecological movements in the United States. These traditions emerged in the context of intensified industrialization and urbanization that occurred after World War II and include the NEP, deep ecology, ecofeminism, and the EJP.[2] . . .

The NEP (New Environmental Paradigm)

The NEP emerged during the 1960s as a response to the loss of ecosystems and nonhuman species that were sacrificed through the growth of industrialization and urbanization (Dunlap and Van Liere 1978; Catton and Dunlap 1980; Milbrath 1984). . . . William Catton and Riley Dunlap (1980) contend that the NEP emerged as a response and in opposition to what they term the Dominant Western Worldview, the driving force behind modern environmental crises that views humans as fundamentally distinct from other species and views the earth as a source of infinite resources. The NEP challenges these views but stops short of extending those critiques to capitalism or the elite classes (Humphrey, Lewis, and Buttel 2002) and is therefore a reformist approach that seeks incremental change.

Deep Ecology

In 1972, the Norwegian activist philosopher Arne Naess introduced deep ecology, which sees humans as merely a single species on a planet with millions of other species that have intrinsic value. Naess contrasted deep ecology with shallow ecology, which seeks to protect and/or improve the health and affluence of humans in industrialized countries (Naess 1973). Deep ecology sought a shift in western values, pushing the western concept of the self from anthropocentricism toward biocentrism. Moving beyond the NEP's focus on a balancing of needs between humans and nonhumans, a deep ecological approach decentered human beings entirely (see Devall and Sessions 1985). For Naess, deep ecology could only be realized in a society marked by the absence of domination between humans and nonhuman species.

Ecofeminism

Much like deep ecology, ecofeminism—which emerged in the 1970s—proposes a politics that recognizes human interdependency with all other beings. However, many ecofeminists charge that deep ecology naively encourages a "oneness" or boundary-free relationship among living beings in a way that ignores actual social differences and histories of exploitation (Warren 1990).

Ecofeminism, therefore, was largely a response to the interrelated problems of ecological unsustainability and patriarchy (MacGregor 2006). This umbrella term encapsulates a range of perspectives whose "basic premise is that the ideology which authorizes oppressions such as those based on race, class, gender, sexuality, physical abilities, and species is the same ideology which sanctions the oppression of nature" (Gaard 1993: 1; see also King 1989). Ecofeminism "calls for an end to all oppressions, arguing that no attempt to liberate women (or any other oppressed group) will be successful without an equal attempt to liberate nature" (Gaard 1993: 1). What makes ecofeminism a distinct body of ideas is its position that nonhuman nature and naturism or dominionism (domination of nonhuman nature) are feminist concerns (Warren 1997: 4). . . .

The EJP (Environmental Justice Paradigm)

Beginning in the 1990s, many scholars argued that the emerging EJP went even further than the NEP, deep ecology, and ecofeminism by emphasizing the unequal environmental burden that historically marginalized human populations confront. Thus, they argued that social justice and human rights for people of color, working class people, Indigenous peoples, and women must be at the center of environmental discourse (see Capek 1993; Taylor 2000; and Walker 2009). While ecofeminism drew attention to the conceptual and philosophical links between the domination of humans and nonhuman nature, the EJP further concretized these ideas in material terms. It foregrounds the relationship between social inequality and the environment and encourages dominant institutions to address these issues.

The EJP nonetheless preserves a hierarchy of people over nonhuman nature as it "urge[s] people to be responsible *stewards* of the earth" (Taylor 2000: 541, emphasis added). Furthermore, while environmental justice activists have demonstrated a willingness to challenge state and corporate policymaking, the movement is ultimately rooted in a reformist model of social change that accepts the fundamental legitimacy of those institutions, including the legal system (Benford 2005).

The Total Liberation Frame

The total liberation frame broadens and challenges the boundaries and assumptions of the NEP, deep ecology, ecofeminism, and the EJP by encompassing a wider intersection of concerns linking social justice, dominant institutions, and ecological politics. It is the result of a frame transformation that began during the 1990s among radical environmental and animal rights activists and groups who were influenced by the politics of social justice that permeated many other social movements, social change organizations, and academic disciplines on university campuses across the United States. Ideas and concepts like intersectionality (Crenshaw [1991] 1994), multiple and linked oppressions, and social privilege took hold in many of these spaces and had a noticeable effect on the language and practices of social movements. For example, at numerous activist gatherings, we found literature presenting perspectives on countering state repression and racial, gender, sexual, class, and species oppression. These perspectives were echoed in activist workshops, in our interviews, and in activist publications. The idea that we can no longer understand, analyze, or resist a single form of oppression in isolation from other forms materialized in feminist and antiracist activist and intellectual circles across the nation, and we saw these ideas appear in the writings, speeches, and actions of radical animal rights and environmental activists (Jones 2004; Sheen 2012).

Based on our data and analysis, we propose that the total liberation frame is comprised of four elements: (1) an ethic of justice and anti-oppression for people, nonhuman animals, and ecosystems; (2) anarchism; (3) anticapitalism; and (4) an embrace of direct action. . . .

Justice and Anti-Oppression for All

Beginning in the late 1990s, there was increasing evidence of a convergence between radical earth and animal liberation activists around a call for justice and anti-oppression politics focused on people, nonhuman animals, and ecosystems (Molland 2006: 56).[3] This perspective developed through a set of theoretical propositions that sought to understand the harm that humans perpetrate against ecosystems and nonhuman animals as reflective of and linked to systems of oppression *within* human society. An ELF communiqué from May 1997 began, "Welcome to the struggle of all species to be free. We are the burning rage of a dying planet" (ELF 1997).

For many activists, an ethic of justice was linked to a language of "rights" and liberation, contending that ecosystems, animals, and people should be free from oppression and harm. Since radical movements claim to confront the roots of a problem, members generally defined justice as the elimination of the conditions that produced the injustice in the first place. In addition, justice for ecosystems includes a focus on justice for nonhuman animals. Our interviewees also felt a need to focus on animals because they are unable to directly speak for themselves.

Numerous interviewees told us that making the links between the exploitation of ecosystems or nonhuman animals and humans required activists to recognize the role of privilege—especially their own—as humans and as members of largely white, middle class social movements. As one activist succinctly explained, "We don't live in a vacuum. Racism occurs in the same world that animal exploitation occurs." Similarly, a popular slogan among many activists is "animal liberation, human liberation, one struggle, one fight!" For these activists, the existence and acceptance of inequalities within human society and between humans and other living beings (speciesism) are major drivers of the abuse across various species and ecosystems.

One veteran Earth First! activist remembered a movement gathering years ago when one aspect of this problem arose:

> A lot of people were basically indulging in their white, upper-middle-class, primarily male privilege who really didn't want to confront these issues. . . . And I as an Earth First!er [felt] that we need to be dealing with wilderness . . . issues while at the same time, confronting oppression within our circles simultaneously, or we will fail at both.

Accordingly, most movement events we attended entailed workshops on challenging social oppression; recognizing privilege; working in solidarity with people of color, immigrants, and Indigenous peoples; and answering questions like

"How does the movement privilege white, male, middle-class, and able-bodied youth?"

Many activists drew direct connections between the harm visited upon ecosystems and other beings. One activist wrote the following in the *Earth First! Journal*: "If we don't have trees, then there would be no animals. If there were no animals, then there would be no us" (Earth First! Journal 2002a). This focus on the interconnectedness of justice for ecosystems, animals, and people stems from seeing exploitation and injustice as having similar root causes. According to one interviewee, ". . . it's really the same infrastructure and it's really the same societal norms that contribute to the exploitation of animals, the planet, people." Thus, while some activists focused on a single cause because of time and resource constraints, they remained cognizant that ecosystems, animals, and people are connected.

Nicoal Sheen, who works for the North American Animal Liberation Press Office, summed up the concept of total liberation as follows:

> Total liberation provides agency for all. Every single being on this planet deserves their liberation and freedom from social constructs that limit who we are, how we live and how we interact with each other—human and other-than-human. Total liberation as a concept and in praxis recognizes that our oppression is inextricably linked and must be fought on all fronts. (Sheen 2012)

Insisting on justice for ecosystems, nonhuman animals, and people, this perspective draws directly from and issues a challenge to other movement frames that tend to place a greater emphasis on just one or two of these elements. While earlier movement frames focused mainly on what activists saw as the need for justice for one element of society (mainly ecosystems, animals, or people), the total liberation frame represents a transformation because it integrates concerns for human society with nonhuman species and ecosystems. For example, both the NEP and the EJP are ultimately anthropocentric, while deep ecology is decidedly biocentric. Like deep ecology, ecofeminism prioritizes concerns about nonhuman species and ecosystems, but deep ecology lacks serious attention to forms of oppression like racism and heteropatriarchy. The total liberation frame transforms these and takes the position that any effort to favor either anthropocentrism or biocentrism is a false choice because the rights and needs of people, ecosystems, and nonhuman animals are linked and must be integrated. This perspective also underscores the ways in which activists view human society as inseparable from nonhuman natures (Jerolmack 2012; Wachsmuth 2012), with humans being responsible for the exploitation of nonhumans and human activists feeling obligated to stop such abuse.

Anarchism

Social movements that reject virtually any form of hierarchy organically lend themselves to anarchism, and radical environmental and animal rights movements are frequently shaped by such views. By anarchism, we mean a theory of governance that is anti-authoritarian and premised on mutual assistance and cooperation. Anarchists are not only critical of, but also generally opposed to, the development of states. . . . The vast majority of interviewees stated that they harbored, in one activist's words, "an inherent distrust of government" and its institutions. One environmentalist who has been active in Earth First! shared:

> Philosophically, politically, the closest label that would fit on me would be anarchist. I believe that which governs best, governs least. . . . And, you know, historically, that used to be called "being an American." But now it apparently makes me an anarchist, and if so, all right, I'll raise the black flag. . . . Reading Jefferson and Franklin and the rest, it seems that they, too, would have been anarchists and dragged away by the Patriot Act if they were around today.

One interviewee referenced a well-known ELF activist in order to make his point that using the legal system to pursue individuals and institutions responsible for harming ecosystems is counterproductive for anarchists: ". . . like my friend Craig Rosebraugh says, 'you're using evil to counteract evil' because the criminal justice system is certainly evil. . . . We don't want stronger criminal justice laws . . . because they aren't about justice; they're about keeping certain subsets down."

Much of this distrust stems from the connection activists make between the state and hierarchy—a force they see as deeply oppressive—and the state's direct involvement in the destruction of wilderness and wildlife. One interviewee explained:

> . . . the reality in the world is, if you look at the loss of human life, Republican and Democratic parties are responsible for more mass death than anything else on the planet. And the planet itself is threatened by their interests. So it's not surprising that, to protect their own interests, they create an infrastructure that supports what they do and opposes anybody who disagrees.

The rejection of hierarchy at the foundation of anarchism is also important for the structure and functioning of these radical movements. One activist told us about his nonhierarchical organization's approach to everyday work and campaigns: "It was a beautiful anarchist campaign, it was so decentralized and it was about information sharing, and there was a spectrum and diversity of tactics—nobody was going to tell you [that] you could or could not do this." This type of horizontal organizational structure, which focuses on democratic decision making and shared power, was common in the movements and groups we studied (for an exploration of this phenomenon in the Occupy Movement, see Maharawal 2013). It also represents a rejection of the way that many large environmental and animal rights organizations operate, with bureaucratic command structures and hierarchies that mirror the way corporate and state institutions function.

It is noteworthy that the anarchist sentiment among many members of these movements has been strengthened in recent years because of state repression against more radical factions of environmental and animal rights networks. In 2005, a prominent federal government official declared radical earth and animal liberation movements the number one domestic "terrorist" threat in the United States (Schuster 2005). Since then, state and federal governments have passed legislation, sponsored law enforcement efforts aimed at neutralizing these movements, and engaged in surveillance, infiltration, intimidation, and imprisonment—a range of practices that has become known as the Green Scare (Potter 2011).[4]

Many activists we interviewed interpreted the Green Scare as an abuse of government power against social movements exercising constitutionally protected activities. One environmentalist argued that there is a double standard at work when corporations that damage ecosystems are celebrated for creating jobs, while environmentalists trying to protect those ecosystems are labeled "terrorists":

> It's ironic that the people [industry] blowing up entire mountains point at the people [activists] and say, "That's a bad idea," and saying, "Oh, you're eco-terrorists for daring to protect our watersheds." We're the new Communists. . . . It's dehumanizing, and that dehumanization makes it easier to justify repression and attacking First Amendment rights.

Overall, anarchism is a core component of the total liberation frame, linking concerns with hierarchy among people, nonhumans, and ecosystems to a distrust of states and the power they embody. Anarchism's antipathy for state institutions is complemented by its attempt to offer alternatives that involve a different way of thinking, living, governing, and decision making—a cultural project in which coercion is minimized and cooperation is emphasized (Maharawal 2013).

Anticapitalism

Anarchist movements are often deeply anticapitalist. Likewise, many social movements committed to anti-oppression and justice for vulnerable populations tend to be critical of, if not opposed to, capitalism. If capitalism is a system of production and social relations that is inherently hierarchical and predicated on the exploitation of working-class labor and ecosystems (Foster 2000), then the earth and animal liberation movements have many reasons to oppose it.

As many of our interviewees pointed out, capitalism requires continuous feed stocks of ecological wealth and nonhuman animals, workers to ensure the flow of those resources, and consumers to purchase and consume the products at the end of the cycle (Schnaiberg and Gould 2000).[5] This "treadmill of production" also demands social compliance from workers, citizens, and consumers and thrives on the intensification of social hierarchies and militarism (Hooks and Smith 2004). As one interviewee articulated, "Our government, our country was founded upon exploitation. And so the capitalist system . . . treats not only workers, but the environment and animals as, essentially, commodities to be exploited. Our entire culture is based on exploitation of human and non-human animals." Similarly, another interviewee declared that "the global economy treats people and the environment basically the same: [as] exploitable resources."

As these statements demonstrate, an anticapitalist sentiment builds on the anarchist rejection of hierarchy and extends to the exploitation and objectification of ecosystems, animals, and people. For example, at the TWAC, we read activist literature that stated: "In the dominant-western-capitalistic worldview, nature is objectified. . . . Women are also fragmented into parts to be used for recreation and profit (in ways that men are not)."

One environmental activist linked his anarchist embrace of cooperation and mutual assistance to what he viewed as its perversion under capitalism:

> We should be more cooperative, but we're in a system that profits from us *not* cooperating with each other. . . . So when you look at a corporation . . . basically, a corporation is a model of forced cooperation for the benefit of a few, not for the collective people that are trying to work together.

This anticapitalist perspective is often focused on a single corporation or corporate leader. Activists regularly protest large firms that are household names (like ExxonMobil, Wal-Mart, and McDonald's) for allegedly harming people, nonhumans, and/or ecosystems. For example, Home Depot has been the target of an environmentalist campaign in which Earth First! has been involved. As an activist involved in this effort reported:

> . . . Home Depot is the largest retailer of old-growth forest products in the world. On the shelves of over 700 stores you can find products ripped out of the heart of every major threatened forest on the planet. . . . The continued sale of products derived from these forests must stop in order to turn the tide of mass destruction. (Earth First! Journal 1999: 25)

Many activists also share the belief that states and corporations are frequently too closely aligned, a hallmark of anticapitalist politics. The North American Animal Liberation Press Office explained the intersection of the above concerns:

> The crisis in the natural world reflects a crisis in the social world, whereby corporate elites and their servants in government have centralized power, monopolized wealth, destroyed democratic institutions, and unleashed a brutal and violent war against dissent. Corporate destruction of nature is enabled by asymmetrical and hierarchical social relations, whereby capitalist powers commandeer the political, legal, and military system to perpetuate and defend their exploitation of the social and natural worlds. . . . (NAALPO 2008: 10)

Radical environmental and animal rights movements demonstrate strong opposition to the routine violence, suffering, and exploitation that capitalist institutions are said to mete out to ecosystems, nonhumans, and human communities. They also voice considerable dislike for the inequalities that capitalism produces and appears to thrive on. Whether it is corporations or states aligned with them, these social movements reject any system of commerce and governance that creates hierarchies and requires the appropriation of life and labor in order to sustain itself.

The total liberation frame's anticapitalism is a direct challenge to and transformation of other ecological movement frames that tend to engage in criticism of capitalist institutions but ultimately accept their legitimacy, such as the NEP and the EJP's promotion of "green capitalism" and encouragement of collaboration with corporations whose leaders claim "green," "socially responsible," and/or "humane" practices (Seager 1994; Torres 2007). Deep ecology has traditionally concentrated on inward spiritual and cultural aspects of ecological politics, offering little in the way of a direct challenge to capitalism. And while ecofeminism is not explicitly anticapitalist, its opposition to all forms of oppression indicate that the logical implication of that frame suggests a likely support of such a viewpoint.

The anticapitalism within the total liberation frame also explicitly links the fate of human and nonhuman populations and ecosystems, detailing the ways that all species have a shared experience of abuse under this system and therefore a common interest in ushering in its abolition. Activists articulating the total liberation frame do not view capitalism simply as an economic system; it is a system that depends on humans and nonhumans to serve as raw materials, workers, and consumers, thus producing common oppressions and the inextricable meshing of our collective experiences. Thus, the total liberation frame is a transformation of other ecological movement frames in that it readily declares an anticapitalist perspective that pays equal attention to all manifestations of oppression among humans and across the human/nonhuman species divide.

Direct Action

Every activist we interviewed for this study supported taking action to free and defend animals and ecosystems from captivity, harm, and destruction at the hands of governments and corporations, and each activist expressed the view that it was her or his duty to intervene against injustice. For these activists, direct action can take many forms, including mobilizing people to prevent or advocate a particular policy or practice, property damage, and personal confrontation.

Movement activists regularly engage each other in writing and in oral discussions at gatherings about what direct action entails. Writing in the *Earth First! Journal*, one group of advocates stated:

> Direct action—action that either symbolically or directly shifts power relations— is an essential transformative tool. . . . Direct action, if only for a moment, seizes leadership and thus injects into the public sphere a competing discourse—a strand of a new reality that has the ability to ripple outward. (Earth First! Journal 2002b: 4)

In other words, direct action is not simply about confronting authorities and forcing them to follow a particular path. It is also about transforming power relations in society so as to ensure that future practices will arise from a different worldview than the current dominant one.

For many activists, direct action is the most effective method to attack the root of the problem. As the ELF Web site states:

> The quality of our air, water, and soil continues to decrease as more and more life forms on the planet suffer and die. . . . How much longer are we supposed to wait to actually stop the destruction of life?. . . . There is also a certain intelligence and logic to the idea that with one night's work, a few individuals can

accomplish what years of legal battles and millions of dollars most likely did not. (Pickering 2007: 56)

For many others, direct action is viewed as one of several tools in the movement repertoire. On this point, one activist who served time for arson told us that there are many tactics that movements must draw on to be effective: "I think that if we're going to create real change . . . we really need to recognize that it's going to take a combined effort of lobbying, lawsuits, civil disobedience, and illegal direct action to create the social pressure we need to create change. . . ." Thus even some radical activists agree that, at times, working within the system can produce results.

In particular, property destruction is a major topic of conversation among radical environmental and animal rights activists. A high-profile activist served nine years for arson at a car dealership—an action that he and a colleague took to protest climate change and the ecological consequences of U.S. "car culture." As we spoke, he indicated his support of property destruction in defense of environmental sustainability:

> In terms of inspiring people around the world, [our action was] incredibly successful. . . . I mean, it was a symbolic thing that we wanted to do to bring attention to an issue that we felt was important. And . . . not only did we accomplish that, here we are years later and around the world, people are now doing this on a semi-regular basis. They're recognizing that car culture is connected to the oil industry. . . .

Similarly, an animal liberation activist told us that direct actions involving property destruction have resulted in significant movement gains: "Justin Samuel and Peter Young raided a fur farm and that fur farm . . . went out of business . . . that's one instance where a corporation capitulated completely to animal liberation."

Activists practicing direct action also make conceptual links between harms visited upon ecosystems and animals and injustices facing human beings. The following communiqué reports a direct action that reflects that vision:

> . . . members of the Earth Liberation Front descended upon the Old Navy Outlet Center . . . [and] smashed . . . plate glass windows and one neon sign. This action served as a protest to Old Navy's owners' involvement in the clear cutting of oldgrowth forest in the Pacific Northwest. . . . Old Navy, Gap, Banana Republic care not for the species that call these forests home, care not for the animals that comprise their leather products, and care not for their garment workers underpaid, exploited and enslaved in overseas sweatshops. (ELF 2001b)

Thus, direct action is a core part of radical environmental and animal rights movements' tactical repertoire and a critical component of the total liberation frame. Direct action also represents a major challenge to existing ecological frames that tend to rely on less confrontational approaches and tactics that are generally within the bounds of the law (with the exception of civil disobedience), amounting to a key element of the frame transformation that total liberation constitutes. The NEP, deep ecology, and the EJP call for activists to practice reformist and moderate tactical methods for achieving social change, none of which present fundamental challenges to dominant social institutions. In addition, ecofeminism retains a lack of clarity regarding direct action, but it is clear that many ecofeminists do support such efforts (Jones 2004).

Discussion

Thus far, we have documented that a frame transformation is occurring and outlined the key elements of the frame.

A key reason for the frame transformation is the eruption of socio-ecological crises and the perceived need to respond to them, as almost every activist with whom we spoke cited the following trends. According to leading scientists, damage to ecosystems over the last 50 years was more severe than during any other time in history, as the health of coral reefs, fisheries, oceans, forests, and river systems declined precipitously, while climate change indicators, species extinction, and air and water pollution rose dramatically. Paralleling these trends is the large-scale increase in factory farming and industrial animal production, consumption, and experimentation that results in the slaughter of billions of nonhumans each year. The threats to planetary sustainability and continued massive exploitation of nonhuman species through industrial agriculture, chemical testing, and entertainment have been widely reported and have rippled through activist communities. The effects of such activity have had enormously negative impacts on ecosystems and nonhuman species (United Nations 2005), and the total liberation frame emerged, in part, as a response.

The second reason for the frame transformation is the palpable frustration by radical activists with mainstream ecological movements' orientation, values, and tactics. . . .

. . . [A]nti-oppression politics are rooted partly in the experiences of many radical ecology activists learning from tensions between mainly white and middle-class mainstream environmental and animal rights movements and Indigenous communities and people of color who perceive racially offensive and culturally insensitive campaigns, tactics, language, and behavior by mainstream activists. Because of these tensions and accompanied frustrations, the radical movements studied here have decided that one of the most important approaches to movement building should involve developing anti-oppression principles and practices within their ranks.

In fact, the founding of Earth First! was in direct response to radical environmentalists' perception that mainstream groups compromised too readily with state and corporate leaders on policy goals. Speaking about mainstream approaches to environmental change through compromising with the U.S. Forest Service, Earth First! cofounder Howie Wolkie once stated, "We played the game, we played the rules. We were moderate, reasonable, and professional. We had data, statistics, and maps. And we got fucked. That's when I started thinking, 'Something's missing here. Something isn't working'" (Best and Nocella 2004: 38). Today, the rejection of the mainstream as elitist and out of touch continues to exert a major impact on the contours and direction of radical ecological movements.

A third reason for the frame transformation is the influence of other social justice movements. These other movements' ideas, beliefs, actions, and successes have shaped radical ecological movements and given form to much of the total liberation frame. Many activists in radical ecological movements draw inspiration from other social movements, including the Civil Rights, Black Power, American Indian, Irish Republican, Abolitionist, Luddites, and Industrial Workers of the World movements of the 19th and 20th centuries (Pickering 2007: 57–8). For example, many radical earth and animal liberation activists cite MOVE, which was a revolutionary group in Philadelphia led by African-American men and women who lived in a communal residence, practiced a vegetarian diet, and spread an anticapitalist, anti-authoritarian, antiracist, antimilitarist message. They publicly protested police brutality, animal abuse, and racism, and were eventually bombed by the Philadelphia police department, killing multiple members of the community. Those who survived were placed in federal prison, many of whom are still there at the time of this writing. The MOVE prisoners are widely viewed as political prisoners, and radical animal liberation and earth liberation groups publicly support the group by publishing interviews with MOVE members, listing those in prison on "political prisoner support" pages in publications, and featuring MOVE members as speakers at conferences (N.A. 1999).

Many activists specifically cited other movements during interviews. . . .

. . . (A) North American Animal Liberation Press Office (NAALPO) Press Officer Jerry Vlasak spoke to us about his inspiration from historic social justice movements:

> We see this as a struggle comparable to other liberation struggles: slavery, Algerian resistance, anti-Apartheid, and the resistance to the wars in Iraq and Afghanistan—all are struggles of oppressed people and nonhumans for freedom. We don't see a difference. Speciesism is something we don't endorse, just like we don't endorse racism or sexism or ageism or homophobia. . . . Every successful liberation struggle has always used illegal means. Nelson Mandela was a lawyer who tried to use legal means to fight Apartheid in South Africa, and when it didn't work he broke the law and went to jail. . . . This is a perfectly legitimate form of resistance. When everything else has been tried, then people need to do something else.

The above data reveal that the frame transformation associated with total liberation began and continues to unfold as a result of the perceived expansion of socio-ecological crises and associated threats to humans, nonhumans, and ecosystems; frustration with mainstream ecological movements; and influences from other social justice movements.

Building on the NEP, deep ecology, ecofeminism, and the EJP, the ecological movements were transformed from reducing harm to the environment, public health, and nonhumans to an effort to achieve a complete reordering of human/nonhuman relations toward equality for all beings.

Notes

Pellow, David N., and Hollie Nyseth Brehm. 2015. "From the New Ecological Paradigm to Total Liberation: The Emergence of a Social Movement Frame." *The Sociological Quarterly* 56: 185–212.

[1] We include radical environmental and animal rights movements under the banner of *ecological politics* because these activists are focused on the relationship of human beings to the broader nonhuman world (see White 2003).

[2] While we propose these were key influences, arguably many other ideas influenced the frame and the activists utilizing it. We cannot review all influences in one article, but we would point out a key example—social ecology. According to its founder, Murray Bookchin, hierarchy within human society predates and is at the root of the human domination and control of nonhuman nature (Bookchin 2005). Thus, social ecology calls for the eradication of hierarchy in order to produce ecologically sustainable societies (Torres 2007).

[3] While we are not suggesting that the animal and earth liberation movements are merging, there is evidence of collaboration and information sharing between these movements. In many editions of radical animal liberation publications like *No Compromise* there are listings of "ecodefense" political prisoners alongside those of animal liberation prisoners and interviews with and articles by earth liberation activists who regularly draw links between the issues driving each movement. Conversely, in every *Earth First! Journal* published over the last several years, there are updates on various animal liberation campaigns, and listings of animal liberation prisoners.

[4] The strength of these movements is, even according to many prominent activists, waning in the wake of government crackdowns, with much of the movements' energies being channeled into supporting imprisoned activists.

[5] While we do not present a specific anticapitalist movement frame (since our interviewees are not specifically influenced by anticapitalist scholarship per se), there is clearly a great deal of overlap on this topic between the total liberation frame and the works of Marxist environmental sociologists we reference in this article (see Foster 1999, 2000; O'Connor 1994; Schnaiberg and Gould 2000; Torres 2007).

References

Benford, Robert. 2005. "The Half-Life of the Environmental Justice Frame: Innovation, Diffusion, and Stagnation." In *Power, Justice and the Environment: A Critical Appraisal of the Environmental Justice Movement*, edited by David N. Pellow and Robert J. Brulle, 33–53. Cambridge, MA: The MIT Press.

Benford, Robert D., and David A. Snow. 2000. "Framing Processes and Social Movements: An Overview and Assessment." *Annual Review of Sociology* 26: 611–39.

Best, Steven, and Anthony Nocella II, eds. 2004. *Terrorists or Freedom Fighters? Reflections on the Liberation of Animals*. New York: Lantern Books.

———. 2006. *Igniting a Revolution: Voices in Defense of the Earth*. Oakland, CA: AK Press.

Bookchin, Murray. 2005. *The Ecology of Freedom: The Emergence and Dissolution of Hierarchy*. Oakland, CA: AK Press.

Bullard, Robert. 2000. *Dumping in Dixie: Race, Class, and Environmental Quality*. Boulder, CO: Westview.

Capek, Stella. 1993. "The 'Environmental Justice' Frame: A Conceptual Discussion and an Application." *Social Problems* 40: 5–24.

Catton, William R., and Riley E. Dunlap. 1980. "A New Ecological Paradigm for Post-Exuberant Sociology." *American Behavioral Scientist* 24: 15–48.

Crenshaw, Kimberlé. 1994. [1991] "Mapping the Margins: Intersectionality, Identity Politics, and Violence against Women of Color." In *The Public Nature of Private Violence*, edited by Martha Albertson Fineman and Roxanne Mykitiuk, 93–118. New York: Routledge.

Devall, Bill, and George Sessions. 1985. *Deep Ecology: Living as If Nature Mattered*. Salt Lake City, UT: Gibbs M. Smith.

Dunlap, Riley, and Kent D. Van Liere. 1978. "The 'New Environmental Paradigm.'" *Journal of Environmental Education* 9: 10–19.

Earth First! Editorial Collective. 2007. "EF! Anti-Oppression Policy." *Earth First! Journal*, September–October.

Earth First! Journal. 1999. Mabon (August–September), p. 25.

———. 2002a. August–September, p. 13.

———. 2002b. Samhain (October–November), p. 4.

Earth Liberation Front (ELF). 1997. Communiqué. May.

———. 2001a. Communiqué, July 27.

———. 2001b. Communiqué. March 5.

Foster, John Bellamy. 1999. "Marx's Theory of Metabolic Rift: Classical Foundations for Environmental Sociology." *American Journal of Sociology* 105 (2): 366–405.

———. 2000. *Marx's Ecology: Materialism and Nature*. New York: Monthly Review Press.

Futrell, William. 2003. "Framing Processes, Cognitive Liberation, and NIMBY Protest in the U.S. Chemical-Weapons Disposal Conflict." *Sociological Inquiry* 73 (3): 359–86.

Gaard, Greta. 1993. *Ecofeminism: Women, Animals, and Nature*. Philadelphia, PA: Temple University Press.

Haines, Herbert. 1984. "Black Radicalization and Funding Civil Rights, 1957–1970." *Social Problems* 32 (1): 31–43.

Hooks, Gregory, and Chad L. Smith. 2004. "The Treadmill of Destruction: National Sacrifice Areas and Native Americans." *American Sociological Review* 69: 558–76.

Humphrey, Craig R., Tammy L. Lewis, and Frederick H. Buttel. 2002. *Environment, Energy, and Society: A New Synthesis*. Belmont, CA: Wadsworth.

Jones, Pattrice. 2004. "Mothers with Monkey Wrenches: Feminist Imperatives and the ALF." In *Terrorists or Freedom Fighters? Reflections on the Liberation of Animals*, edited by Steven Best and Anthony J. Nocella II, 137–56. New York: Lantern Books.

King, Ynestra. 1989. "The Ecology of Feminism and the Feminism of Ecology." In *Healing the Wounds: The Promise of Ecofeminism*, edited by Judith Plant, 18–28. Philadelphia, PA: New Society Publishers.

MacGregor, Sherilyn. 2006. *Beyond Mothering Earth: Ecological Citizenship and the Politics of Care*. Vancouver: University of British Columbia Press.

Maharawal, Manissa. 2013. "Occupy Wall Street and a Radical Politics of Inclusion." *The Sociological Quarterly* 54 (2): 177–81.

Mies, Maria, and Veronika Bennholdt-Thomsen. 1999. *The Subsistence Perspective: Beyond the Globalized Economy*. London, UK: Zed Books.

Milbrath, Lester W. 1984. *Environmentalists: Vanguard for a New Society*. Albany: State University of New York Press.

Molland, Noel. 2006. "A Spark that Ignited a Flame: The Evolution of the Earth Liberation Front." In *Igniting a Revolution: Voices in Defense of the Earth*, edited by Steven Best and Anthony J. Nocella II, 47–58. Oakland, CA: AK Press.

Naess, Arne. 1973. "The Shallow and the Deep, Long-Range Ecology Movement." *Inquiry* 16: 95–100.

North American Animal Liberation Press Office (NAALPO). 2008. North American Animal Liberation Press Office Newsletter 4 (1) January. https://animalliberationpressoffice.org/NAALPO/ (accessed August 2009).

O'Connor, James. 1994. "Is Sustainable Capitalism Possible?" In *Is Capitalism Sustainable? Political Economy and the Politics of Ecology*, edited by Martin O'Connor, 152–75. New York: Guilford Press.

Pickering, Leslie James. 2007. *The Earth Liberation Front, 1997–2002*. Tempe, AZ: Arissa Media Group.

Potter, Will. 2011. *Green Is the New Red: An Insider's Account of a Social Movement under Siege*. San Francisco, CA: City Lights Books.

Schnaiberg, Allan, and Kenneth Alan Gould. 2000. *Environment and Society: The Endur-ing Conflict.* Caldwell, NJ: The Blackburn Press.

Schuster, Henry. 2005. "Domestic Terror: Who's Most Dangerous?" CNN.com. http://www.cnn.com/2005/US/08/24/schuster.column/index.html (accessed August 24, 2009).

Seager, Joni. 1994. *Earth Follies: Coming to Feminist Terms with the Global Environmental Crisis.* New York: Routledge.

SHAC.net. N.d. "Who Are HLS?" http://www.shac.net/HLS/who_are_hls.html (accessed August 15, 2011).

Sheen, Nicoal. 2012. "Total Liberation Provides Agency for All." North American Ani-mal Liberation Press Office, September 12.

Snow, David A., and Robert D. Benford. 1988. "Ideology, Frame Resonance, and Partic-ipant Mobilization." *International Social Movement Research* 1: 197–218.

Snow, David A., E. Burke Rochford Jr., Steven K. Worden, and Robert D. Benford. 1986. "Frame Alignment Processes, Micro-Mobilization, and Movement Participa-tion." *American Sociological Review* 51: 464–81.

Taylor, Dorceta. 1997. "Women of Color, Environmental Justice, and Ecofeminism." In *Ecofeminism: Women, Culture, Nature,* edited by Karen J. Warren, 38–81. Blooming-ton: Indiana University Press.

———. 2000. "The Rise of the Environmental Justice Paradigm: Injustice Framing and the Social Construction of Environmental Discourses." *American Behavioral Scientist* 43: 508–80.

Torres, Bob. 2007. *Making a Killing: The Political Economy of Animal Rights.* Oakland, CA: AK Press.

Ulsperger, Jason S. 2002. "Geezers, Greed, Grief, and Grammar: Frame Transformation in the Nursing Home Reform Movement." *Sociological Spectrum* 22 (4): 385–406.

United Nations. 2005. *Millennium Ecosystem Assessment. Ecosystems and Human Well-Being.* San Francisco, CA: Island Press and the United Nations.

Walker, Gordon. 2009. "Globalizing Environmental Justice: The Geography and Politics of Frame Contextualization and Evolution." *Global Social Policy* 9: 355–82.

Wang, Dan J., and Sarah Soule. 2012. "Social Movement Organizational Collaboration: Networks of Learning and the Diffusion of Protest Tactics, 1960–1995." *American Journal of Sociology* 117 (6): 1674–722.

Warren, Karen J. 1990. "The Power and the Promise of Ecological Feminism." *Environ-mental Ethics* 12: 125–46.

———. 1997. "Taking Empirical Data Seriously: An Ecofeminist Philosophical Perspec-tive." In *Ecofeminism: Women, Culture, Nature,* edited by Karen J. Warren, 20–38. Bloomington: Indiana University Press.

White, Rob. 2003. "Environmental Issues and the Criminological Imagination." *Theoret-ical Criminology* 7: 483–506.

Index

Note: Page references for figures are *italicized*.